Lecture Notes in Computer Science   8865

Commenced Publication in 1973
Founding and Former Series Editors:
Gerhard Goos, Juris Hartmanis, and Jan van Leeuwen

More information about this series at http://www.springer.com/series/7407

Kai-Uwe Schmidt · Arne Winterhof (Eds.)

Sequences and Their Applications – SETA 2014

8th International Conference
Melbourne, VIC, Australia, November 24–28, 2014
Proceedings

 Springer

Editors
Kai-Uwe Schmidt
Otto-von-Guericke University
Magdeburg
Germany

Arne Winterhof
Austrian Academy of Sciences
Linz
Austria

ISSN 0302-9743   ISSN 1611-3349 (electronic)
ISBN 978-3-319-12324-0   ISBN 978-3-319-12325-7 (eBook)
DOI 10.1007/978-3-319-12325-7

Library of Congress Control Number: 2014953219

LNCS Sublibrary: SL1 – Theoretical Computer Science and General Issues

Springer Cham Heidelberg New York Dordrecht London

Printed on acid-free paper

Springer is part of Springer Science+Business Media (www.springer.com)

Preface

This volume contains the refereed proceedings of the Eighth International Conference on Sequences and Their Applications (SETA 2014) held in Melbourne, Australia, November 24–28, 2014. The previous seven conferences were held in Singapore 1998, Bergen (Norway) 2001, Seoul (South Korea) 2004, Beijing (China) 2006, Lexington (USA) 2008, Paris (France) 2010, and Waterloo (Canada) 2012.

SETA 2014 invited submissions of previously unpublished work on technical aspects of sequences and their applications in communications, cryptography, coding, and combinatorics, including:

- Randomness of sequences
- Aperiodic and periodic correlation of sequences
- Combinatorial aspects of sequences, including difference sets
- Sequences with applications in coding theory and cryptography
- Sequences over finite fields/rings/function fields
- Linear and nonlinear feedback shift register sequences
- Sequences for radar, synchronization, and identification
- Sequences for wireless communications
- Linear and nonlinear complexity of sequences
- Pseudorandom sequence generators
- Correlation and transformation of Boolean functions
- Multidimensional sequences and their correlation properties

Invited talks were given by Josef Dick (University of New South Wales, Australia), Tor Helleseth (University of Bergen, Norway), Kathy Horadam (RMIT University, Australia), and Bernhard Schmidt (Nanyang Technological University, Singapore).

The Program Committee of SETA 2014 has received 36 qualified submissions and each was refereed by at least two experts. The Program Committee selected 24 of them for presentation at the conference and for the inclusion in these proceedings. In addition, these proceedings contain two refereed invited papers, which are based on the talks given by Josef Dick and Kathy Horadam.

Our sincere thanks go to the Program Committee for their dedication in the challenging task of refereeing the submissions. Special thanks go to the General Chair, Udaya Parampalli.

We gratefully acknowledge the School of Engineering of the University of Melbourne and the Australian Mathematical Sciences Institute for their generous financial support.

November 2014

Kai-Uwe Schmidt
Arne Winterhof

Organization

General Chair

Udaya Parampalli   University of Melbourne, Australia

Program Co-chairs

Kai-Uwe Schmidt   Otto-von-Guericke University, Germany
Arne Winterhof   Austrian Academy of Sciences, Austria

Program Committee

Christoph Aistleitner	Technische Universität Graz, Austria
Serdar Boztaş	RMIT University, Australia
Claude Carlet	University of Paris 8, France
Pascale Charpin	Inria-Rocquencourt, France
Jim Davis	University of Richmond, USA
Cunsheng Ding	Hong Kong University of Science and Technology, Hong Kong
Tuvi Etzion	Technion, Israel
Guang Gong	University of Waterloo, Canada
Philip Hawkes	Qualcomm, Australia
Honggang Hu	University of Science and Technology of China, China
Tom Høholdt	Technical University of Denmark, Denmark
Jonathan Jedwab	Simon Fraser University, Canada
Alexander Kholosha	University of Bergen, Norway
Andrew Klapper	University of Kentucky, USA
Gohar Kyureghyan	Otto-von-Guericke University, Germany
Wilfried Meidl	Sabancı University, Turkey
Sihem Mesnager	University of Paris 8, France
Wai Ho Mow	Hong Kong University of Science and Technology, Hong Kong
Jong-Seon No	Seoul National University, South Korea
Alina Ostafe	University of New South Wales, Australia
Matthew Parker	University of Bergen, Norway
Friedrich Pillichshammer	Johannes Kepler Universität Linz, Austria
Alexander Pott	Otto-von-Guericke University, Germany
Hong-Yeop Song	Yonsei University, South Korea
Xiaohu Tang	Southwest Jiaotong University, China

Alev Topuzoğlu   Sabancı University, Turkey
Steven Wang   Carleton University, Canada
Kyeongcheol Yang   Pohang University of Science and Technology,
                                  South Korea
Nam Yul Yu   Lakehead University, Canada
Zhengchun Zhou   Southwest Jiaotong University, China

Sponsoring Institutions

School of Engineering, University of Melbourne
Australian Mathematical Sciences Institute

Contents

Crosscorrelation of Sequences

Prime Numbers in Sequences

OFDM and CDMA

Frequency-Hopping Sequences

Invited Paper

Relationships Between CCZ and EA Equivalence Classes and Corresponding Code Invariants

Kathy J. Horadam[1(✉)] and Mercé Villanueva[2]

[1] RMIT University, Melbourne, VIC 3001, Australia
kathy.horadam@rmit.edu.au
[2] Universitat Autònoma de Barcelona, 08193 Bellaterra, Spain

Abstract. The purpose of this paper is to provide a brief survey of CCZ and EA equivalence for functions $f : G \to N$ where G and N are finite and N is abelian, and, for the case $f : \mathbb{Z}_p^m \rightarrow \mathbb{Z}_p^m$, to investigate two codes derived from f, inspired by these equivalences. In particular we show the dimension of the kernel of each code determines a new invariant of the corresponding equivalence class. We present computational results for $p = 2$ and small m.

Keywords: EA-equivalence class · CCZ equivalence class · Code invariant · APN function · Differential cryptanalysis

1 Introduction

The usefulness of any equivalence relation for functions between finite groups depends on the groups, the types of functions and the purpose of the classification. The resulting equivalence classes will have value when each class consists of functions sharing common properties or invariants. If a potentially new function satisfying desirable conditions is found, it is important to be able to show whether or not it is equivalent to a known function.

For functions between finite rings and fields, as functions between the underlying finite abelian groups, such classifications are needed for applications in finite geometry, coding and cryptography. The equivalence classes should preserve properties such as planarity or invariants such as differential uniformity or maximum nonlinearity.

Two quite separate approaches to defining equivalence for functions over $\mathbb{F}_{p^n}$, which preserve important algebraic or combinatorial properties across a wide range of interesting functions, have been used.

The first of these approaches involves pre- and post-composition of a given function $f : G \to G$, $G = (\mathbb{F}_{p^n}, +)$, with other functions having specified characteristics, to define an equivalent function. In 1964, Cavior [11] introduced *weak equivalence* between f and f' as

$$f' = \tau \circ f \circ \sigma \tag{1}$$

© Springer International Publishing Switzerland 2014
K.-U. Schmidt and A. Winterhof (Eds.): SETA 2014, LNCS 8865, pp. 3–17, 2014.
DOI: 10.1007/978-3-319-12325-7_1

for any elements τ, σ of the symmetric group $\mathrm{Sym}(G)$ of G. Mullen [22] restricted τ and σ to (possibly equal) subgroups of $\mathrm{Sym}(G)$, so defining a relative form of weak equivalence. *Linear equivalence* between f and f' is defined by

$$f' = \tau \circ f \circ \sigma + \chi, \qquad (2)$$

where τ, σ are *linear* permutations and χ is linear, so is a coarsening of weak equivalence relative to linear permutations, by addition of a linear function.

When χ in (2) is extended to include affine functions, it defines *extended affine (EA) equivalence*, introduced in [9] for $p = 2$, and now one of the main classifying equivalences for cryptographic functions.

The second approach involves defining equivalence between functions in terms of an equivalence between their graphs. This approach was originally proposed by Carlet, Charpin and Zinoviev [10, Proposition 3] for $p = 2$ (as cited in [9]), and is called *CCZ equivalence*. More generally, for a function $f : G \to N$ between finite abelian groups G and N, Pott [24] suggests using properties of its graph $\{(x, f(x)), \ x \in G\}$ as a means of measuring combinatorial and spectral properties of f.

Horadam [17] generalises these two types of equivalence to functions $f : G \to N$ between arbitrary finite groups G and N, and both types of equivalence are shown to have a common source in the equivalence relation for splitting semiregular relative difference sets. It is shown to be sufficient to restrict to those functions $f : G \to N$ for which $f(1) = 1$, which form a group $C^1(G, N)$ under the operation of pointwise multiplication of functions, and we will assume this is the case throughout.

We further assume throughout that N is abelian, and is written multiplicatively unless context dictates otherwise. For the non-abelian case see [17,18].[1]

The affineness in an EA or CCZ equivalence of f is captured by a *shift $f \cdot r$* of f for some $r \in G$, where

$$f \cdot r(x) = f(r)^{-1}f(rx), \ x \in G.$$

Definition 1. *Two functions $f, f' \in C^1(G, N)$ are EA equivalent if there exist $r \in G$, $\theta \in Aut(G)$, $\gamma \in Aut(N)$ and $\chi \in Hom(G, N)$ such that*

$$f' = (\gamma \circ (f \cdot r) \circ \theta) \chi. \qquad (3)$$

The graph of f is $\mathcal{G}_f = \{(x, f(x)) \ : \ x \in G\}$. Two functions $f, f' \in C^1(G, N)$ are CCZ equivalent if there exist $r \in G$ and $\alpha \in Aut(G \times N)$ such that

$$\alpha(\mathcal{G}_{f \cdot r}) = \mathcal{G}_{f'}. \qquad (4)$$

If $r = 1$, we say f and f' are EA isomorphic and CCZ isomorphic, respectively.

---

[1] In [17,18], EA equivalence is called bundle equivalence and CCZ equivalence is called graph equivalence.

In particular, suppose $G = N = (\mathbb{F}_{p^n}, +) \cong \mathbb{Z}_p^n$. Every $f \in C^1(G, G)$ is the evaluation map of some polynomial $f(x) \in \mathbb{F}_{p^n}[x]$ of degree $\leq p^n - 2$ with $f(0) = 0$. The homomorphisms $\mathrm{Hom}(G, G)$ are the linearised polynomials $\sum_{j=0}^{n-1} a_j x^{p^j}$, $a_j \in \mathbb{F}_{p^n}$, and $\mathrm{Aut}(G)$ consists of the linearised permutation polynomials. Weak equivalence (1) relative to $\mathrm{Aut}(G)$ is the case $r = 0$, $\chi \equiv \mathbf{0}$ of (3) and linear equivalence (2) is the case $r = 0$ of (3). In [9], CCZ equivalence uses translation by $e \in G \times G$ on the right, rather than on the left as in (4), but composition with the inner automorphism defined by e shows they give the same CCZ equivalence classes.

The equivalence defined by (3) is known implicitly to finite geometers, because planar functions equivalent by (3) will determine isomorphic planes [12]. Planarity of f is preserved by addition of a linearised polynomial of G or pre- or post-composition with a linearised permutation polynomial, or by linear transformation. For instance, if $r \in G$, then $f \cdot r$ is a linear transformation.

A very large number of cryptographically strong functions over $\mathbb{F}_{2^n}$ have been found in the past decade, and it is important to be able to tell if they are genuinely new. The choice of equivalence relation best suited to classify cryptographic functions has attracted considerable attention in this period. This has been prompted by the observation that if f is invertible, then its compositional inverse $\mathrm{inv}(f)$ has the same cryptographic robustness as f with respect to several measures of nonlinearity, so the inverse of a function is often regarded as being equivalent to it. However, $\mathrm{inv}(f)$ is not always EA equivalent to f.

CCZ equivalence is a coarser equivalence than EA equivalence and includes permutations and their inverses in the same equivalence class. It is currently very difficult to decide, either theoretically or computationally, whether two functions are CCZ equivalent, and if so, whether they are EA-inequivalent.

The paper is organised as follows. In Sect. 2 we survey briefly the main results known about CCZ and EA equivalence and their interrelationships. We will need the *coboundary* function $\partial f : G \times G \to N$ defined for $f : G \to N$ by

$$\partial f(x, y) = f(x)^{-1} f(y)^{-1} f(xy), \quad x, y \in G, \tag{5}$$

which measures how much f differs from a homomorphism. Section 3 discusses two codes inspired by these equivalences for functions over $\mathbb{Z}_p^m$: the graph code $\mathcal{G}_f$ and the coboundary code $\mathcal{D}_f = \mathrm{im} \, \partial f$. We survey known results and show that the dimension of the kernel of each code determines a new invariant of the corresponding class. In Sect. 4 new computational results about the codes and their invariants, and some open problems, are presented.

2 Equivalence of Functions Between Groups

Let G be a finite group and N a finite abelian group, written multiplicatively. If $\alpha \in \mathrm{Aut}(G \times N)$, it has a unique factorisation $\alpha = \eta \times \iota$, where its action on the first component $G \times \{1\}$ determines a monomorphism $\eta = (\eta_2, \eta_1) : G \rightarrowtail G \times N$

and its action on the second component $\{1\} \times N$ determines a monomorphism $\imath = (\imath_2, \imath_1) : N \rightarrowtail G \times N$ which commutes with (η_2, η_1), with

$$\alpha(x, a) = (\eta \times \imath)(x, a) = (\imath_2(a)\ \eta_2(x),\ \imath_1(a)\ \eta_1(x)). \qquad (6)$$

CCZ equivalence has the following functional form, which is a mix of weak equivalence (1) and EA equivalence (3).

Proposition 1. [17] *Two functions $f, f' \in C^1(G, N)$ are CCZ equivalent if and only if there exist $\alpha = \eta \times \imath \in Aut(G \times N)$ and $r \in G$ such that: the function $\rho := (\imath_2 \circ (f \cdot r))\ \eta_2$ that they define with f is a permutation of G; and*

$$f' = (\imath_1 \circ (f \cdot r) \circ \sigma)\ (\eta_1 \circ \sigma), \qquad (7)$$

where $\sigma = inv(\rho)$.

Corollary 1. [17] *For functions in $C^1(G, N)$, EA equivalence implies CCZ equivalence.*

Proof. If (3) holds, define α in Proposition 1 by setting $\imath = (1, \gamma)$ and $\eta = (inv(\theta), \chi \circ inv(\theta))$. □

If $\alpha = \eta \times \imath \in Aut(G \times N)$ in Proposition 1 fixes the subgroup $\{1\} \times N$ then $\imath_2 = 1$ so $\eta_2 \in Aut(G)$ and (3) holds. This correspondence, proved in [9] for $p = 2$, can be used as an alternative definition of EA equivalence.

Corollary 2. [17] *Two functions $f, f' \in C^1(G, N)$ are EA equivalent if and only if there exist $r \in G$ and $\alpha \in Aut(G \times N)$ such that*

1. *$\alpha(\mathcal{G}_{f \cdot r}) = \mathcal{G}_{f'}$ and*
2. *$\alpha(\{1\} \times N) = \{1\} \times N$.* □

In a few cases (as well as those in Lemma 1 below) it is known that the converse of Corollary 1 holds.

Corollary 3. *The CCZ class of $f \in C^1(G, N)$ is its EA class in the following cases:*

1. *if $f \in Hom(G, N)$;*
2. *if $\gcd(|G|, |N|) = 1$.*

Proof. Case 1 follows by definition. Case 2 follows from Corollary 2 because any automorphism of $G \times N$ must fix $\{1\} \times N$ (and $G \times \{1\}$ by symmetry). The argument is due to Pott and Zhou [25] for G abelian but holds in general, and in particular, includes the case $G \cong \mathbb{Z}_p^n$, $N \cong \mathbb{Z}_q^m$, p, q different primes. □

The restricted set of automorphisms used to redefine EA equivalence in Corollary 2 are not the only automorphisms preserving the graphs of EA equivalent functions. It is possible to say exactly when a CCZ equivalence in (7) can be rewritten as an EA equivalence in (3). Note that for any $r \in G$, f and $f \cdot r$ are trivially EA equivalent by (3), and thus CCZ equivalent by Corollary 1, so here we give the case for $r = 1$ and EA and CCZ isomorphism. The results extend straightforwardly to the general case.

Theorem 1. [18] *Set $r = 1$ in (7) and (3). The CCZ isomorphism between f and f' determined by α in (7) can be rewritten as an EA isomorphism (3) if and only if*

1. *$\rho \in Aut(G)$ and*
2. *there exists $\delta \in Aut(N)$ such that the permutation $\hat{\delta}$ of $G \times N$ defined by*

$$\hat{\delta} \circ \alpha((x, f(x)a)) = \alpha((x, f(x))) (1, \delta(a)), \quad x \in G, \ a \in N, \qquad (8)$$

is an automorphism of $G \times N$.

In this case, the rewriting as an EA isomorphism is

$$f' = (\delta \circ f \circ \sigma) (\chi_\delta \circ \sigma),$$

where $\chi_\delta := (\delta \circ f)^{-1}(f' \circ \rho)$. $\qquad\qquad\square$

2.1   The Case $N \cong \mathbb{Z}_p^m$

When N is elementary abelian, Condition 2 in Theorem 1 always holds. If we find an automorphism of $G \times \mathbb{Z}_p^m$ which proves two functions are CCZ equivalent, this gives us more flexibility than Corollary 2 does to determine if they are EA equivalent. A direct proof is given for convenience.

Theorem 2. *Let $N = \mathbb{Z}_p^m$. Suppose f and f' are CCZ isomorphic. For $\alpha \in Aut(G \times N)$ as in Proposition 1 (with $r = 1$), write $f' = f^\alpha$.*

   Then f and f' are EA isomorphic

1. *$\Leftrightarrow$ **there exists** α with $f' = f^\alpha$ for which $\alpha(\{1\} \times N) = \{1\} \times N$*
2. *$\Leftrightarrow$ **there exists** α with $f' = f^\alpha$ for which $\rho \in Aut(G)$.*

Proof. 1 $\Rightarrow$ 2. Suppose $\alpha(\{1\} \times N) = \{1\} \times N$. Then in (6), for all $x \in G$, $\iota_2(x) = 1$ so $\rho = \eta_2$ and is an automorphism of G.

2 $\Rightarrow$ 1. Suppose $\rho \in Aut(G)$. Let $\iota : N \to \{1\} \times N$ be given by $\iota(a) = (1, a)$, $a \in N$. Set $J = \alpha(\iota(N)) \cap \iota(N)$, $M = \mathrm{inv}(\alpha \circ \iota)(J) \leq N$ and $M' = \mathrm{inv}(\iota)(J) \leq N$, and let $\check{\alpha} : M \to M'$ be the isomorphism induced by α, ie.

$$\check{\alpha}(a) = \mathrm{inv}(\iota) \circ \alpha \circ \iota(a), \ a \in M.$$

Calculation using (5) shows $\mathrm{im}\,\partial f \subseteq M$ and $\check{\alpha}(\partial f) = \partial(f' \circ \rho)$. Then $\check{\alpha}$ can be extended, by extension of a minimal generating set for M to one for N, to at least one $\delta \in Aut(N)$. Thus $\partial(f' \circ \rho) = \check{\alpha}(\partial f) = \delta(\partial f) = \partial(\delta \circ f)$, so $\partial((\delta \circ f)^{-1}(f' \circ \rho)) = 1$. Consequently, $\chi_\delta = (\delta \circ f)^{-1}(f' \circ \rho) \in \mathrm{Hom}(G, N)$. Calculation using (8) shows $\hat{\delta} \circ \alpha((x, a)) = (\rho(x), \delta(a)\chi_\delta(x))$, so that $\hat{\delta} \circ \alpha((1, a)) = (1, \delta(a))$ and $f' = f^{\hat{\delta} \circ \alpha}$. $\qquad\square$

It is worth noting that two functions that are CCZ equivalent as in Proposition 1 may still be EA equivalent without the automorphism α satisfying $\rho \in Aut(G)$. The following example is due to Hou [20]. A particular instance is $f : \mathbb{Z}_5 \to \mathbb{Z}_5$ defined by $f(\pm 1) = \mp 1$ and $f(x) = x$ for all $x \in \mathbb{Z}_5 \setminus \{\pm 1\}$; that is, $f(x) = -x^3$.

Example 1. Let $f : \mathbb{Z}_p^m \to \mathbb{Z}_p^m$ be such that $f = \operatorname{inv}(f)$ but f is not linear. Let $\alpha \in Aut(\mathbb{Z}_p^m \times \mathbb{Z}_p^m)$ be defined by $\alpha(x, a) = (a, x)$ for all $(x, a) \in \mathbb{Z}_p^m \times \mathbb{Z}_p^m$. Then $\alpha(x, f(x)) = (f(x), x) \; \forall x \in \mathbb{Z}_p^m$, so $\alpha(\mathcal{G}_f) = \mathcal{G}_{\operatorname{inv}(f)}$. Here f is necessarily EA equivalent to itself $(= \operatorname{inv}(f))$, but ρ is not linear. □

2.2   The Case $G = \mathbb{Z}_p^n$ and $N = \mathbb{Z}_p^m$

From now on, we write G and N additively. It is known [7] that CCZ equivalence implies EA equivalence for functions $\mathbb{Z}_2^n \to \mathbb{Z}_2$. This is not always true for functions $\mathbb{Z}_2^m \to \mathbb{Z}_2^m$, however, as a permutation and its inverse under composition lie in the same CCZ class, but permutations over $\mathbb{Z}_2^m$ exist which are EA-inequivalent to their inverses.

Recall that if $n \geq m$, a function $f : \mathbb{Z}_p^n \to \mathbb{Z}_p^m$ is PN (perfect nonlinear) if for each $a \neq \mathbf{0} \in \mathbb{Z}_p^n$ the function $\partial(f)(a, x)$ takes each value of $\mathbb{Z}_p^m$ exactly p^{n-m} times. A function $f : \mathbb{Z}_2^m \to \mathbb{Z}_2^m$ is APN (almost perfect nonlinear) if for each $a \neq \mathbf{0}$, $b \in \mathbb{Z}_2^m$ the equation $\partial(f)(a, x) = b$ has no more than two solutions x in $\mathbb{Z}_2^m$. In some important instances of PN and APN functions, CCZ equivalence does imply EA equivalence.

Lemma 1. *Over $\mathbb{Z}_p^m$, CCZ equivalence implies EA equivalence in the following cases.*

1. *[19] If $p = 2$ and $m \leq 3$.*
2. *[21] If p is odd, two PN functions are CCZ equivalent if and only if they are EA equivalent.*
3. *[4,29] If $p = 2$ and $m \geq 2$, two quadratic APN functions are CCZ equivalent if and only if they are EA equivalent.* □

More generally, for $G = \mathbb{Z}_p^n$ with n large enough and $N = \mathbb{Z}_p^m$ with $m > 1$, CCZ equivalence does not imply EA equivalence.

Theorem 3 (Budaghyan, Carlet, Helleseth [7,8]). *Let p be an odd (even) prime, $n \geq 3$ ($n \geq 5$) and $k > 1$ the smallest divisor of n. Then for any $m \geq k$, CCZ equivalence of functions from $\mathbb{Z}_p^n$ to $\mathbb{Z}_p^m$ is strictly more general than EA equivalence.* □

Even though the two equivalences can be compared directly using either the functional or the graphical approach, it is more computationally difficult to check functions for CCZ equivalence than for EA equivalence, and more computationally difficult to generate CCZ equivalence classes than EA equivalence classes.

One advantage of determining either equivalence lies in the properties shared by equivalent functions, and the chance it provides of replacing a complex function by a simpler equivalent function to improve efficiency in applications.

A recent illustration of this appears in [27] for $G = \mathbb{Z}_2^m \times \mathbb{Z}_2^m$. It is shown, after mapping each element of $\mathbb{Z}_{2^m}$ to the coefficient vector of its binary representation, that addition modulo 2^m is CCZ equivalent to a very simple quadratic vectorial Boolean function. This is applied to simplify attacks on cryptosystems which employ addition modulo 2^m.

Conversely, finding more complex functions which are EA-inequivalent to known simple functions but which nonetheless possess similar desirable properties can improve cryptographic security or enlarge the known set of sequences with optimal correlation properties.

A recent illustration of this appears in [15] where it is shown that for $p \geq 5$ and m an integer that does not divide $p^m + 1$, then the function $f(x) = x^{p^m+2}$ over $\mathbb{Z}_p^m$ is an Alltop function (that is, its differential functions are PN) which is EA-inequivalent to the Alltop function $f'(x) = x^3$, even though ∂f and $\partial f'$ are EA equivalent PN functions.

2.3   The Case $G = N = \mathbb{Z}_2^m$

EA equivalence partitions the set of (non-affine) functions over $\mathbb{Z}_2^m$ into classes with the same nonlinearity, differential uniformity and algebraic degree [9]. CCZ equivalence partitions the set of functions over $\mathbb{Z}_2^m$ into classes with the same Walsh spectrum, differential uniformity and resistance to algebraic cryptanalysis [9,10] but not necessarily the same algebraic degree.

It remains very difficult to tell when CCZ equivalent functions are EA-inequivalent. Some results for APN functions in small orders are known. Computation has shown [5] that there is 1 CCZ class of APN functions over $\mathbb{Z}_2^4$, containing 2 EA classes; and 3 CCZ classes of APN functions over $\mathbb{Z}_2^5$, containing 1, 3 and 3 EA classes, respectively. There are at least 14 CCZ classes of APN functions over $\mathbb{Z}_2^6$ [5], at least 302 over $\mathbb{Z}_2^7$ and at least 33 over $\mathbb{Z}_2^8$ [28], and at least 11 over $\mathbb{Z}_2^9$ [14]. Edel [13] has computed the partition of many of them into EA classes. He shows, for example, that, for $n = 5, 6, 7, 8$ and 9 the CCZ class of the Gold quadratic APN function $f(x) = x^3$ contains $3, 3, 3, 2$ and 5 EA classes, respectively. Summaries appear in [6,21].

3   Code Invariants of EA and CCZ Classes of Functions Over $\mathbb{Z}_p^m$

For cryptographic applications, the focus is to find functions over $\mathbb{Z}_2^m$ which have simultaneously low differential uniformity (APN or 4-uniform), high nonlinearity and algebraic degree ≥ 4 and which are, preferably, permutations. This aim can be aided by working with specific codes they generate. The graph code for APN functions was introduced in [6] and the coboundary code was introduced in [19].

Definition 2. *Let $f : \mathbb{Z}_p^m \to \mathbb{Z}_p^m$ satisfy $f(0) = 0$.*
Define the graph code *of f to be the p-ary code*
   $\mathcal{G}_f = \{(x, f(x)) \; : \; x \in \mathbb{Z}_p^m\} \subseteq \mathbb{Z}_p^{2m}$.
Define the coboundary code *of f to be the p-ary code*
   $\mathcal{D}_f = \{\partial f(x,y) \; : \; x, \; y \in \mathbb{Z}_p^m\} \subseteq \mathbb{Z}_p^m$.
The linear codes they generate are denoted $\langle \mathcal{G}_f \rangle$ and $\langle \mathcal{D}_f \rangle$, respectively.

Let $n(f) = \mathrm{rank}_p\, \mathcal{D}_f = \dim_p \langle \mathcal{D}_f \rangle$ and $s(f) = \mathrm{rank}_p\, \mathcal{G}_f = \dim_p \langle \mathcal{G}_f \rangle$; that is $|\langle \mathcal{D}_f \rangle| = p^{n(f)}$ and $|\langle \mathcal{G}_f \rangle| = p^{s(f)}$.

For the remainder of this Section we will investigate the properties of, and relationships between, these codes. The following simple properties of their dimensions appear in [19, Theorem 4].

Theorem 4. *1. $0 \le n(f) \le m$ and $m \le s(f) \le 2m$;*
2. $n(f) = 0 \Leftrightarrow f$ is linear $\Leftrightarrow \mathcal{G}_f = \langle \mathcal{G}_f \rangle \Leftrightarrow s(f) = m$;
3. $\{0\} \times \langle \mathcal{D}_f \rangle < \langle \mathcal{G}_f \rangle$ and $n(f) < s(f)$;
4. if $n(f) = m$ then $s(f) = 2m$; i.e. if $\mathcal{D}_f$ generates $\mathbb{Z}_p^m$ then $\mathcal{G}_f$ generates
$\mathbb{Z}_p^{2m}$.   □

Both these dimensions are related to the differential uniformity $\Delta(f)$, which is defined to be the maximum over $a \ne \mathbf{0} \in \mathbb{Z}_p^m$ of the number of solutions of

$$- f(x) + f(x + a) = b;\ b \in \mathbb{Z}_p^m . \tag{9}$$

Lemma 2 [19]. *For each f, $n(f) \ge m - \lfloor \log_p \Delta(f) \rfloor$. In particular,*

if p is odd and $1 \le \Delta(f) < p$, $n(f) = m$;
if $p = 2$ and $\Delta(f) = 2$, $n(f) = m$ or $n(f) = m - 1$;
if $p = 2$ and $\Delta(f) = 4$, $n(f) = m$ or $n(f) = m - 1$ or $n(f) = m - 2$. □

A further parameter of each of the codes $\mathcal{D}_f$ and $\mathcal{G}_f$ is the dimension of its kernel. Recall that the *p-kernel* of a code C over $\mathbb{Z}_p$ of length n is defined [23] as

$$K(C) = \{x \in \mathbb{Z}_p^n\ :\ x + C = C\}.$$

If $\mathbf{0} \in C$, then $K(C)$ is a linear subspace of C and C can be written as the union of cosets of $K(C)$. If so, $K(C)$ is the largest such linear code for which this is true. For $p = 2$, the kernel was introduced in [2].

Definition 3. *Let $f : \mathbb{Z}_p^m \to \mathbb{Z}_p^m$ satisfy $f(0) = \mathbf{0}$, so $K(\mathcal{G}_f)$ is a linear subcode of $\mathcal{G}_f$ and $K(\mathcal{D}_f)$ is a linear subcode of $\mathcal{D}_f$. Set $K(f) = \dim_p K(\mathcal{G}_f)$.*
Set $k(f) = \dim_p K(\mathcal{D}_f)$ and let $M(f)$ be the multiset $\{\{k(f \cdot r),\ r \in \mathbb{Z}_p^m\}\}$,
denoted $M(f) = \{0^\wedge a_0, 1^\wedge a_1, \ldots, m^\wedge a_m\}$, for some $a_0, \ldots, a_m$ with $\sum_{i=0}^m a_i = p^m$.

It is known that differential uniformity $\Delta(f)$ is a combinatorial invariant of the EA equivalence class of f [16, Corollary 9.52.1]. In fact this is a consequence of it being a combinatorial invariant of the CCZ equivalence class of f.

Lemma 3. *If f and f' are CCZ equivalent, then $\Delta(f) = \Delta(f')$.*

Proof. Differential uniformity is a combinatorial invariant of CCZ isomorphism [18, Lemma 5], so if $\alpha(\mathcal{G}_{f \cdot r}) = \mathcal{G}_{f'}$ as in (4) then $\Delta(f \cdot r) = \Delta(f')$. It remains only to show that $\Delta(f \cdot r) = \Delta(f)$. Suppose $a \ne \mathbf{0} \in \mathbb{Z}_p^m$. Then for each $b \in \mathbb{Z}_p^m$, $\{x \in \mathbb{Z}_p^m\ :\ -(f \cdot r)(x) + (f \cdot r)(x + a) = b\} = \{x \in \mathbb{Z}_p^m\ :\ -f(r + x) + f(r + x + a) = b\} = \{y \in \mathbb{Z}_p^m\ :\ -f(y) + f(y + a) = b\}$ and the set sizes are identical.   □

We show that the dimensions $n(f)$ and $s(f)$ are algebraic invariants of the EA and CCZ equivalence classes of f, respectively.

Theorem 5. *If f and f' are EA equivalent, then $n(f) = n(f')$. If f and f' are CCZ equivalent, then $s(f) = s(f')$.*

Proof. The dimensions $n(f)$ and $s(f)$ are algebraic invariants of EA and CCZ isomorphism, respectively [19, Theorem 5], so that we only need to consider $f' = f \cdot r$, $r \neq 0$ and note $(f \cdot r) \cdot (-r) = f$. Then $\partial(f \cdot r)(x, y) = \partial f(r + x, y) - \partial f(r, y) \in \langle D_f \rangle$ so by symmetry $\langle D_{f \cdot r} \rangle = \langle D_f \rangle$. Also $G_{f \cdot r} = G_f - (r, f(r))$ so $\langle G_{f \cdot r} \rangle = \langle G_f \rangle$. □

Now we show that $M(f)$ and $K(f)$ are algebraic invariants of the EA and CCZ equivalence classes of f, respectively.

Theorem 6. *If f and f' are EA equivalent, then $M(f) = M(f')$. If f and f' are CCZ equivalent, then $K(f) = K(f')$.*

Proof. If f and f' are EA equivalent, suppose θ, $\gamma \in \mathrm{Aut}(\mathbb{Z}_p^m)$, $\chi \in \mathrm{Hom}(\mathbb{Z}_p^m, \mathbb{Z}_p^m)$ and $r \in \mathbb{Z}_p^m$ are such that $f' = \gamma \circ (f \cdot r) \circ \theta + \chi$, so that $\partial f'(\vartheta(x), \vartheta(y)) = \gamma(\partial(f \cdot r)(x, y))$ for all x, $y \in \mathbb{Z}_p^m$, where $\vartheta = \mathrm{inv}(\theta)$. Suppose $c \in K(D_{f \cdot r})$, so that $c = \partial(f \cdot r)(a, b)$ for some a, $b \in \mathbb{Z}_p^m$ and $c + \partial(f \cdot r)(x, y) = \partial(f \cdot r)(x', y')$. Then $\gamma(c) + \gamma(\partial(f \cdot r)(x, y)) = \gamma(\partial(f \cdot r)(x', y'))$ so $\gamma(c) \in K(D_{f'})$. Thus γ is an isomorphism between $K(D_{f \cdot r})$ and $K(D_{f'})$, so that $k(f') = k(f \cdot r) \in M(f)$. By symmetry, $k(f) \in M(f')$ and $M(f) = M(f')$.

If f and f' are CCZ isomorphic, $\alpha \in \mathrm{Aut}(\mathbb{Z}_p^{2m})$ and $\alpha(G_f) = G_{f'}$, suppose $c \in K(G_f)$. Then $c = (a, f(a))$ for some $a \in \mathbb{Z}_p^m$ and if $c + (x, f(x)) = (x', f(x'))$ then $\alpha(c) + \alpha((x, f(x))) = \alpha((x', f(x')))$ and $\alpha(c) \in K(G_{f'})$. Thus α is an isomorphism between $K(G_f)$ and $K(G_{f'})$. Finally, $K(G_f) = K(G_{f \cdot r})$ for all r. □

When $p = 2$, we are interested in additional properties of the codes G_f and D_f.

Definition 4. *Let $f : \mathbb{Z}_2^m \to \mathbb{Z}_2^m$ satisfy $f(0) = 0$. Let H be an $m \times (2^m - 1)$ parity check matrix of the Hamming code $\mathcal{H}^m$, that is, its columns are the transposes $x^\top$ of the non-zero row vectors $x \in \mathbb{Z}_2^m$. Define*

$$H_f = \begin{pmatrix} H \\ H^{(f)} \end{pmatrix} = \begin{pmatrix} \cdots & x^\top & \cdots \\ \cdots & f(x)^\top & \cdots \end{pmatrix}.$$

Let $\mathcal{C}_f$ be the linear code of length $2^m - 1$ admitting H_f as a parity check matrix.

Note that $\mathcal{C}_f$ is a subcode of $\mathcal{H}^m$. Since $G_f = H_f^\top \cup \{(0, 0)\}$, $\langle H_f^\top \rangle = \langle G_f \rangle$, and $\langle H_f \rangle$ is the dual of $\mathcal{C}_f$. The dimension of $\mathcal{C}_f$, or equivalently the dimension of the extended code $\mathcal{C}_f^*$, is $2^m - 1 - s(f)$. Therefore, the rank of G_f can also be computed using the dimension of C_f^*.

Proposition 2 [6]. *Let f and f' be maps from $\mathbb{Z}_2^m$ to $\mathbb{Z}_2^m$ with $\dim_2\langle H_f \rangle = \dim_2\langle H_{f'} \rangle = 2m$. Then, f and f' are CCZ equivalent if and only if their extended codes C_f^* and $C_{f'}^*$ are equivalent.* □

If f is APN, the dimension $s(f)$ is already known to be maximal, ie. $s(f) = 2m$, by [6]. It has also been proved that $s(f) = 2m$ for another class of functions, the AF permutations [26], but the AF property itself is not an invariant of CCZ equivalent permutations.

Consideration of Theorem 4 raises the possibility that $s(f)$ and $n(f)$ are not independent invariants, which we formulate as a conjecture in the next Section. However we demonstrate computationally that $K(f)$ and $M(f)$ are independent.

4   Examples for $p = 2$ with Low Dimensions

In this Section, we concentrate on computations for $p = 2$ and functions which are either monomial power functions or have differential uniformity 4. Let S_n be the symmetric group of permutations of length n, where $n = 2^m - 1$.

4.1   Monomial Power Functions

Table 1 shows the classification of all monomial power functions into CCZ equivalence classes for all $3 \le m \le 7$. Additional properties have been computed, and included in the table. These are: whether they are APN; the pair (rank, kernel dimension) $= (s(f), K(f))$ for the binary code $\mathcal{G}_f$; and the possibilities for the number of solutions of (9) for $p = 2$.

For $m = 5$, it is known that the three CCZ classes of APN functions in Table 1 contain 3, 3 and 1 EA classes respectively [5]. Two of the EA classes in the CCZ equivalence class of x^3 contain the monomials x^3 and x^{11}, respectively, and two of the EA classes in the CCZ equivalence class of x^5 contain the monomials x^5 and x^7, respectively. Non-monomial representatives of each other EA class are given in [5].

For $m = 7$, it is only necessary to check the cases x^7 and x^{21} computationally, since the other CCZ classes of non-APN monomials can be distinguished by the number of solutions of (9).

For $m = 8$, a classification of monomial power functions by cyclotomic coset, differential uniformity and nonlinearity is given in [1, Table 3]. After combining cyclotomic cosets containing f with those containing $\text{inv}(f)$ (recall that f and $\text{inv}(f)$ are CCZ equivalent [10]) and comparing the number of solutions of (9) for representative power functions, the only power functions which still need distinguishing are x^{15} and x^{45}. The graph codes corresponding to these two functions have $s(f) = 2m = 16$, and as the two extended codes are inequivalent, the functions are CCZ inequivalent by Proposition 2. The classification in [1, Table 3] reduces to a list of 28 CCZ classes of monomial power functions. These are given in Table 2, together with their differential uniformity $\Delta(f)$ and the values $(s(f), K(f))$.

We have computed the invariant multiset $M(f)$ for every $f(x) = x^i$ in Tables 1 and 2. The results appear in Table 3. In these cases we have very simple and uniform results in terms of the cyclotomic coset C_i of $i \bmod 2^m - 1$. For instance, for $m = 4$, $M(x^5) = \{2^{\wedge}16\}$ and for $m = 6$, $M(x^9) = \{3^{\wedge}64\}$.

Table 1. Classification of all monomial power functions $f(x) = x^i$ for $3 \leq m \leq 7$ into CCZ equivalence classes, and some properties of these classes.

m	f	APN	j, for all x^j CCZ equivalent	$(s(f),$ $K(f))$	Number of solutions of (9)
3	x^1	no	1,2,4	(3,3)	$\{0^\wedge 49, 8^\wedge 7\}$
3	x^3	yes	3,5,6	(6,0)	$\{0^\wedge 28, 2^\wedge 28\}$
4	x^1	no	1,2,4,8	(4,4)	$\{0^\wedge 225, 16^\wedge 15\}$
4	x^3	yes	3,6,9,12	(8,0)	$\{0^\wedge 120, 2^\wedge 120\}$
4	x^5	no	5,10	(6,0)	$\{0^\wedge 180, 4^\wedge 60\}$
4	x^7	no	7,11,13,14	(8,0)	$\{0^\wedge 135, 2^\wedge 90, 4^\wedge 15\}$
5	x^1	no	1,2,4,8,16	(5,5)	$\{0^\wedge 961, 32^\wedge 31\}$
5	x^3	yes	3,6,11,12,13,17,21,22,24,26	(10,0)	$\{0^\wedge 496, 2^\wedge 496\}$
5	x^5	yes	5,7,9,10,14,18,19,20,25,28	(10,0)	$\{0^\wedge 496, 2^\wedge 496\}$
5	x^{15}	yes	15,23,27,29,30	(10,0)	$\{0^\wedge 496, 2^\wedge 496\}$
6	x^1	no	1,2,4,8,16,32	(6,6)	$\{0^\wedge 3969, 64^\wedge 63\}$
6	x^3	yes	3,6,12,24,33,48	(12,0)	$\{0^\wedge 2016, 2^\wedge 2016\}$
6	x^5	no	5,10,13,17,19,20,26,34,38,40,41,52	(12,0)	$\{0^\wedge 3024, 4^\wedge 1008\}$
6	x^7	no	7,14,28,35,49,56	(12,0)	$\{0^\wedge 2205, 2^\wedge 1701, 4^\wedge 63, 6^\wedge 63\}$
6	x^9	no	9,18,36	(9,0)	$\{0^\wedge 3528, 8^\wedge 504\}$
6	x^{11}	no	11,22,23,25,29,37,43,44,46,50,53,58	(12,0)	$\{0^\wedge 2520, 2^\wedge 1323, 6^\wedge 126, 10^\wedge 63\}$
6	x^{15}	no	15,30,39,51,57,60	(12,0)	$\{0^\wedge 2205, 2^\wedge 1764, 8^\wedge 63\}$
6	x^{21}	no	21,42	(8,0)	$\{0^\wedge 3780, 12^\wedge 126, 20^\wedge 126\}$
6	x^{27}	no	27,45,54	(9,0)	$\{0^\wedge 3528, 2^\wedge 63, 6^\wedge 189, 8^\wedge 63, 12^\wedge 189\}$
6	x^{31}	no	31,47,55,59,61,62	(12,0)	$\{0^\wedge 2079, 2^\wedge 1890, 4^\wedge 63\}$
7	x^1	no	1,2,4,8,16,32,64	(7,7)	$\{0^\wedge 10129, 128^\wedge 127\}$
7	x^3	yes	3,6,12,24,43,45,48,53,65,85,86,90,96,106	(14,0)	$\{0^\wedge 8128, 2^\wedge 8128\}$
7	x^5	yes	5,10,20,27,33,40,51,54,66,77,80,89,102,108	(14,0)	$\{0^\wedge 8128, 2^\wedge 8128\}$
7	x^7	no	7,14,28,55,56,59,67,91,93,97,109,110,112,118	(14,0)	$\{0^\wedge 9906, 2^\wedge 5461, 6^\wedge 889\}$
7	x^9	yes	9,15,17,18,30,34,36,60,68,71,72,99,113,120	(14,0)	$\{0^\wedge 8128, 2^\wedge 8128\}$
7	x^{11}	yes	11,13,22,26,35,44,49,52,69,70,81,88,98,104	(14,0)	$\{0^\wedge 8128, 2^\wedge 8128\}$
7	x^{19}	no	19,25,38,47,50,61,73,76,87,94,100,107,117,122	(14,0)	$\{0^\wedge 10795, 2^\wedge 2794, 4^\wedge 2667\}$
7	x^{21}	no	21,31,37,41,42,62,74,79,82,84,103,115,121,124	(14,0)	$\{0^\wedge 9906, 2^\wedge 5461, 6^\wedge 889\}$
7	x^{23}	yes	23,29,39,46,57,58,75,78,83,92,101,105,114,116	(14,0)	$\{0^\wedge 8128, 2^\wedge 8128\}$
7	x^{63}	yes	63,95,111,119,123,125,126	(14,0)	$\{0^\wedge 8128, 2^\wedge 8128\}$

Thus $M(f)$ can distinguish between some, but not all, representatives of distinct CCZ classes for these special cases. Furthermore, for each CCZ class in these Tables which consists of APN functions but contains more than one EA class, we computed $M(f)$ for a representative function from each EA class, and obtained exactly the same $M(f)$ for each EA class. In other words, in all these cases, $k(f)$ itself is an invariant of EA class. It is determined by the size of a corresponding cyclotomic coset, and does not distinguish between different EA classes in the same CCZ class of APN functions.

However, this does not hold in general, as we shall see in the following Subsection.

4.2 Differentially 4-uniform Permutations

For $m = 4$, in general (not only considering monomial power permutations), it is well known that there are no APN permutations.

According to [19] there are 5 EA equivalent classes of differentially 4-uniform permutations, and as they all have different extended Walsh spectra, they each

Table 2. Classification of representative functions $f(x) = x^i$ for $m = 8$ into CCZ equivalence classes, and some invariants of these classes. Classes with $\Delta(f) = 2$ are the APN functions.

i	$\Delta(f)$	$(s(f), K(f))$	i	$\Delta(f)$	$(s(f), K(f))$	i	$\Delta(f)$	$(s(f), K(f))$	i	$\Delta(f)$	$(s(f), K(f))$
1	256	(8,8)	15	14	(16,0)	31	16	(16,0)	63	6	(16,0)
3	2	(16,0)	17	16	(12,0)	39	2	(16,0)	85	84	(10,0)
5	4	(16,0)	19	16	(16,0)	43	30	(16,0)	87	30	(16,0)
7	6	(16,0)	21	4	(16,0)	45	14	(16,0)	95	4	(16,0)
9	2	(16,0)	23	16	(16,0)	51	50	(12,0)	111	4	(16,0)
11	10	(16,0)	25	6	(16,0)	53	16	(16,0)	119	22	(12,0)
13	12	(16,0)	27	26	(16,0)	55	12	(16,0)	127	4	(16,0)

Table 3. Invariant multiset $M(f)$ for the monomial power functions $f(x) = x^i$ for all $3 \le m \le 8$ in Tables 1 and 2, where C_i is the cyclotomic coset of $i \bmod 2^m - 1$.

i	$M(f(x) = x^i)$
$i \in C_1$	$\{0^{\wedge}2^m\}$
$i \notin C_1$	$\{|C_i|^{\wedge}2^m\}$

form a single CCZ equivalence class. On the other hand, using MAGMA [3] and checking all differentially 4-uniform permutations in S_{15}, there are exactly 10 CCZ equivalence classes, given by the following permutations:

$$\sigma_1 = (5,6,7,8)(10,12,11,15,13,14) \quad (= f_3 \text{ in } [19]),$$
$$\sigma_2 = (5,6,7,8)(10,12,14,13)(11,15) \quad (= f_4 \text{ in } [19]),$$
$$\sigma_3 = (5,6,8)(7,10,12)(9,11,15,14,13) \quad (= f_5 \text{ in } [19]),$$
$$\sigma_4 = (5,6,8)(7,10,12)(9,11,15,14) \quad (= f_6 \text{ in } [19]),$$
$$\sigma_5 = (5,6,8)(7,11,14,10,12,13),$$
$$\sigma_6 = (5,6,8)(7,11,14)(10,12,13) \quad (= f_7 \text{ in } [19]),$$
$$\sigma_7 = (5,6,8)(7,11,13,15)(9,12,10),$$
$$\sigma_8 = (5,6,8)(7,11,13)(9,12,14,10),$$
$$\sigma_9 = (5,6,8)(7,11,13,10,9,12,14),$$
$$\sigma_{10} = (5,6,8)(7,11)(9,12,10,13,15,14).$$

Table 4 corrects [19, Table 2], where the CCZ classes of $\sigma_1, \sigma_2, \sigma_3, \sigma_4, \sigma_6$ were claimed to exhaust the differentially 4-uniform classes of permutations fixing $\mathbf{0}$ over $\mathbb{Z}_2^4$. For every f in Table 4, the dimension $K(f)$ of the kernel of the binary code $\mathcal{G}_f$ is the minimum value 0 and the rank $s(f)$ is the maximum value $2m = 8$, so Proposition 2 applies. Computation of $n(f)$ confirms that $n(f) = 4 = m$ in all cases. Table 4 lists invariants of the 10 CCZ equivalence classes: the order of the automorphism group of C_f^*; the minimum distance and covering radius of C_f^* as a pair (d, ρ); the weight distribution of the dual of C_f^*; and the possibilities for the number of solutions of (9).

All these functions have $n(f) = 4 = m$ so that Theorem 4 applies, and we observe that $n(f) = s(f) - 4$. A computational check of the 7 CCZ classes of functions over $\mathbb{Z}_2^3$ ([19, Table 1]) shows that even though only f_2 and f_4 have $n(f) = 3$, it remains true that $n(f) = s(f) - 3$. We conjecture that this holds in general.

Table 4. Classification of all differentially 4-uniform permutations of order 15 into CCZ equivalence classes, and some invariants of these classes.

f	$|Aut(C_f^*)|$	(d, ρ)	Weight distribution of the dual of C_f^*	Number of solutions of (9)
σ_1	4	$(4, 5)$	$1 + x^2 + 28x^4 + 119x^6 + 214x^8 + 119x^{10} + 28x^{12} + x^{14} + x^{16}$	$\{0^{\wedge}141, 2^{\wedge}78, 4^{\wedge}21\}$
σ_2	96	$(4, 5)$	$1 + x^2 + 30x^4 + 111x^6 + 226x^8 + 111x^{10} + 30x^{12} + x^{14} + x^{16}$	$\{0^{\wedge}144, 2^{\wedge}72, 4^{\wedge}24\}$
σ_3	1152	$(4, 4)$	$1 + 36x^4 + 96x^6 + 246x^8 + 96x^{10} + 36x^{12} + x^{16}$	$\{0^{\wedge}144, 2^{\wedge}72, 4^{\wedge}24\}$
σ_4	16	$(4, 5)$	$1 + 32x^4 + 112x^6 + 222x^8 + 112x^{10} + 32x^{12} + x^{16}$	$\{0^{\wedge}138, 2^{\wedge}84, 4^{\wedge}18\}$
σ_5	12	$(4, 5)$	$1 + 32x^4 + 112x^6 + 222x^8 + 112x^{10} + 32x^{12} + x^{16}$	$\{0^{\wedge}138, 2^{\wedge}84, 4^{\wedge}18\}$
σ_6	4	$(4, 5)$	$1 + 30x^4 + 120x^6 + 210x^8 + 120x^{10} + 30x^{12} + x^{16}$	$\{0^{\wedge}135, 2^{\wedge}90, 4^{\wedge}15\}$
σ_7	28	$(4, 5)$	$1 + x^2 + 28x^4 + 119x^6 + 214x^8 + 119x^{10} + 28x^{12} + x^{14} + x^{16}$	$\{0^{\wedge}141, 2^{\wedge}78, 4^{\wedge}21\}$
σ_8	20	$(4, 5)$	$1 + 30x^4 + 120x^6 + 210x^8 + 120x^{10} + 30x^{12} + x^{16}$	$\{0^{\wedge}135, 2^{\wedge}90, 4^{\wedge}15\}$
σ_9	16	$(4, 5)$	$1 + 30x^4 + 120x^6 + 210x^8 + 120x^{10} + 30x^{12} + x^{16}$	$\{0^{\wedge}135, 2^{\wedge}90, 4^{\wedge}15\}$
σ_{10}	720	$(4, 4)$	$1 + 30x^4 + 120x^6 + 210x^8 + 120x^{10} + 30x^{12} + x^{16}$	$\{0^{\wedge}135, 2^{\wedge}90, 4^{\wedge}15\}$

Conjecture 1. Let $f : \mathbb{Z}_p^m \to \mathbb{Z}_p^m$ satisfy $f(\mathbf{0}) = \mathbf{0}$. Then $n(f) = s(f) - m$.

However, it is not the case that all parameters of the codes $\mathcal{D}_f$ and $\mathcal{G}_f$ must be related. For instance $K(f) = 0$ for every f in Table 4, but $M(f)$ varies.
We have calculated
$M(\sigma_1) = \{0^{\wedge}8, 1^{\wedge}4, 4^{\wedge}4\}$,
$M(\sigma_2) = \{1^{\wedge}6, 4^{\wedge}10\}$,
$M(\sigma_3) = \{4^{\wedge}16\}$,
$M(\sigma_4) = \{0^{\wedge}4, 4^{\wedge}12\}$,
$M(\sigma_5) = \{0^{\wedge}6, 4^{\wedge}10\}$,
$M(\sigma_6) = \{0^{\wedge}4, 4^{\wedge}12\}$,
$M(\sigma_7) = \{0^{\wedge}15, 4\}$,
$M(\sigma_8) = \{0^{\wedge}10, 4^{\wedge}6\}$,
$M(\sigma_9) = \{0^{\wedge}8, 4^{\wedge}8\}$,
$M(\sigma_{10}) = \{4^{\wedge}16\}$.

So the two invariants $K(f)$ and $M(f)$ are independent in general. Furthermore, in these examples, the dimension $k(f \cdot r)$ of the kernel of the code $\mathcal{D}_{f \cdot r}$ does vary with the affine term r within an EA equivalence class.

4.3   Open Questions

It seems to us that $\mathcal{D}_f$ provides a new code-based technique for investigating EA equivalence classes, while $\mathcal{G}_f$ can be used for investigating CCZ classes and, in some cases, EA classes [4,6]. For future work, we expect that further study of the relationship between the invariants $n(f)$ and $s(f)$, and $M(f)$ and $K(f)$, will clarify how CCZ classes partition into EA classes, particularly for functions with low differential uniformity. Does $K(f)$ take any other values than the two, 0 and m, observed so far? Can $n(f)$ or $M(f)$ distinguish between two CCZ equivalent functions which are EA-inequivalent, especially for APN functions? The cases $n(f) = m$, $m - 1$ and $m - 2$ (for both odd and even p) are the most interesting. Do APN functions f exist for which $n(f) = m - 1$? Of course, if the answer to Conjecture 1 is "yes" then the answer to this question is "no". We can ask if, for the EA equivalence class of a power function f, the constant value of $k(f \cdot r)$, $r \in \mathbb{Z}_2^m$ and its dependence on a cyclotomic coset that we have observed in low dimensions, can be proved to hold in general. We can also ask if there is a relationship between $n(f)$ or $M(f)$ and the algebraic degree of f, since they are all invariants of EA equivalence class.

Acknowledgements. We thank the anonymous referee for helpful comments which improved the content and clarity of the paper. The second author has been partially supported by the Spanish MICINN under Grants TIN2010-17358 and TIN2013-40524-P, and by the Catalan AGAUR under Grant 2014SGR-691. She thanks RMIT School of Mathematical and Geospatial Sciences for hosting her while this research was undertaken.

References

1. Aslan, B., Sakalli, M.T., Bulus, E.: Classifying 8-bit to 8-bit S-boxes based on power mappings from the point of DDT and LAT distributions. In: von zur Gathen, J., Imaña, J.L., Koç, Ç.K. (eds.) WAIFI 2008. LNCS, vol. 5130, pp. 123–133. Springer, Heidelberg (2008)
2. Bauer, H., Ganter, B., Hergert, F.: Algebraic techniques for nonlinear codes. Combinatorica **3**, 21–33 (1983)
3. Bosma, W., Cannon, J., Playoust, C.: The MAGMA algebra system I: the user language. J. Symb. Comput. **24**, 235–265 (1997)
4. Bracken, C., Byrne, E., McGuire, G., Nebe, G.: On the equivalence of quadratic APN functions. Des. Codes Cryptogr. **61**, 261–272 (2011)
5. Brinkmann, M., Leander, G.: On the classification of APN functions up to dimension 5. Des. Codes Cryptogr. **49**, 273–288 (2008)
6. Browning, K.A., Dillon, J.F., Kibler, R.E., McQuistan, M.T.: APN polynomials and related codes. J. Comb. Inf. Syst. Sci. **34**, 135–159 (2009). Special issue honoring the 75th birthday of Prof. D.K. Ray-Chaudhuri

7. Budaghyan, L., Carlet, C.: CCZ-equivalence of single and multi-output Boolean functions, Post-proceedings of the 9th International Conference on Finite Fields and Their Applications Fq'09, Contemporary Mathematics, vol. 518, pp. 43–54 (2010)
8. Budaghyan, L., Helleseth, T.: Planar functions and commutative semifields. Tatra. Mt. Math. Publ. **45**, 15–45 (2010)
9. Budaghyan, L., Carlet, C., Pott, A.: New classes of almost bent and almost perfect nonlinear polynomials. IEEE Trans. Inf. Theory **52**, 1141–1152 (2006)
10. Carlet, C., Charpin, P., Zinoviev, V.: Codes, bent functions and permutations suitable for DES-like cryptosystems. Des. Codes Cryptogr. **15**, 125–156 (1998)
11. Cavior, S.R.: Equivalence classes of functions over a finite field. Acta Arith. **10**, 119–136 (1964)
12. Coulter, R.S., Matthews, R.W.: Planar functions and planes of Lenz-Barlotti Class II. Des. Codes Cryptogr. **10**, 167–184 (1997)
13. Edel, Y.: Personal correspondence, March 2010
14. Edel, Y., Pott, A.: A new almost perfect nonlinear function which is not quadratic. Adv. Math. Comput. **3**, 59–81 (2009)
15. Hall, J., Rao, A., Gagola III, S.M.: A family of Alltop functions that are EA-inequivalent to the cubic function. IEEE Trans. Commun. **61**(11), 4722–4727 (2013)
16. Horadam, K.J.: Hadamard Matrices and Their Applications. Princeton University Press, Princeton (2007)
17. Horadam, K.J.: Relative difference sets, graphs and inequivalence of functions between groups. J. Comb. Des. **18**, 260–273 (2010)
18. Horadam, K.J.: Equivalence classes of functions between finite groups. J. Algebraic Comb. **35**, 477–496 (2012)
19. Horadam, K.J., East, R.: Partitioning CCZ classes into EA classes. Adv. Math. Commun. **6**, 95–106 (2012)
20. Hou, X.-D.: Personal correspondence, May 2013
21. Kyureghyan, G.M., Pott, A.: Some theorems on planar mappings. In: von zur Gathen, J., Imaña, J.L., Koç, Ç.K. (eds.) WAIFI 2008. LNCS, vol. 5130, pp. 117–122. Springer, Heidelberg (2008)
22. Mullen, G.L.: Weak equivalence of functions over a finite field. Acta Arith. **35**, 259–272 (1979)
23. Phelps, K.T., Rifà, J., Villanueva, M.: Kernels and p-kernels of p^r-ary 1-perfect codes. Des. Codes Cryptogr. **37**, 243–261 (2001)
24. Pott, A.: Nonlinear functions in abelian groups and relative difference sets. Discr. Appl. Math. **138**, 177–193 (2004)
25. Pott, A., Zhou, Y.: CCZ and EA equivalence between mappings over finite Abelian groups. Des. Codes Cryptogr. **66**(1–3), 99–109 (2013)
26. Rifà, J., Solov'eva, F.I., Villanueva, M.: Intersection of Hamming codes avoiding Hamming subcodes. Des. Codes Cryptogr. **62**, 209–223 (2012)
27. Schulte-Geers, E.: On CCZ-equivalence of addition mod 2^n. Des. Codes Cryptogr. **66**(1–3), 111–127 (2013)
28. Weng, G., Tan, Y., Gong, G.: On quadratic almost perfect nonlinear functions and their related algebraic object, CACR Technical report 18 (2013). http://cacr.uwaterloo.ca/techreports/2013/cacr2013-18.pdf
29. Yoshiara, S.: Equivalences of quadratic APN functions. J. Algebraic Comb. **35**, 461–475 (2012)

Boolean Functions

Results on Constructions of Rotation Symmetric Bent and Semi-bent Functions

Claude Carlet[1], Guangpu Gao[2,3]([✉]), and Wenfen Liu[2,3]

[1] LAGA (UMR 7539), University of Paris 8 and University of Paris 13, CNRS,
2 Rue de la Liberté, 93526 Saint-Denis, Cedex, France
claude.carlet@univ-paris8.fr
[2] State Key Laboratory of Mathematical Engineering and Advanced Computing,
Zhengzhou, China
guangpu.gao@gmail.com
[3] State Key Laboratory of Networking and Switching Technology,
Beijing University of Posts and Telecommunications, Beijing, China

Abstract. In this paper, we introduce a class of cubic rotation symmetric (RotS) functions and prove that it can yield bent and semi-bent functions. To the best of our knowledge, this is the second primary construction of an infinite class of nonquadratic RotS bent functions which could be found and the first class of nonquadratic RotS semi-bent functions. We also study a class of idempotents (giving RotS functions through the choice of a normal basis of $GF(2^n)$ over $GF(2)$). We derive a characterization of the bent functions among these idempotents and we relate their precise determination to a problem studied in the framework of APN functions. Incidentally, the proofs of bentness given here are useful for a paper studying a construction of idempotents from RotS functions, entitled "A secondary construction and a transformation on rotation symmetric functions, and their action on bent and semi-bent functions" by the same authors, to appear in the journal JCT series A.

Keywords: Rotation symmetric Boolean function · Bent · Semi-bent · Maiorana-McFarland class · Idempotent · Permutation

1 Introduction

Boolean functions play a critical role in cryptography as well as in the design of circuits and chips for digital computers. They can be defined over the finite field $GF(2^n)$ and represented as univariate polynomials, or over the vector space $GF(2)^n$ and represented as $f(x_0, x_1, \ldots, x_{n-1})$, the latter representation being deduced from the former (and vice versa) through the choice of a basis of the

The work of G. Gao, and W. Liu is supported in part by 973 Program under Grant No. 2012CB315905 and Open Foundation of State key Laboratory of Networking and Switching Technology (Beijing University of Posts and Telecommunications)(SKLNST-2013-1-06).

© Springer International Publishing Switzerland 2014
K.-U. Schmidt and A. Winterhof (Eds.): SETA 2014, LNCS 8865, pp. 21–33, 2014.
DOI: 10.1007/978-3-319-12325-7_2

$GF(2)$-vector space $GF(2^n)$. Idempotents, introduced by Filiol and Fontaine in [12,13] are polynomials over $GF(2^n)$ such that $f(z) = f(z^2)$, for all $z \in GF(2^n)$. Rotation symmetric (RotS) Boolean functions, introduced by Pieprzyk and Qu [24], are invariant under circular translation of indices. They can be obtained from idempotents (and vice versa) through the choice of a normal basis of $GF(2^n)$. Such class of Boolean functions is of interest because of its smaller search space ($\approx 2^{\frac{2^n}{n}}$) comparably to the whole space ($= 2^{2^n}$), which allows investigating functions for a number of variables larger (by a factor of 2), and also because of the more compact representation of RotS functions. It has been experimentally demonstrated that the class of RotS Boolean functions is extremely rich in terms of cryptographically significant Boolean functions. For example, Kavut *et al.* have found Boolean functions on 9 variables with nonlinearity 241 [17], which solved an almost three-decade old open problem. Motivated by this study, important cryptographic properties such as nonlinearity, balancedness, correlation immunity, algebraic degree and algebraic immunity of these functions have been investigated at the same time and encouraging results have been obtained [10,14,27,28]. Note that RotS functions are also interesting for the design of Substitution Boxes in block ciphers (see [16,25]).

Plateaued functions [29] represent much interest for the study of Boolean functions in cryptography, as they can possess desirable cryptographic properties such as high onlinearity, resiliency, propagation criteria, low additive autocorrelation and high algebraic degree. Their class is larger than that of "partially bent functions" introduced in [3]. Two important classes of plateaued functions are those of bent functions and of semi-bent functions, due to their algebraic and combinatorial properties. An n-variable (n even) bent function is a Boolean function with the maximum possible nonlinearity $2^{n-1} - 2^{n/2-1}$. Such functions provide the best resistance against attacks by affine approximations, such as the fast correlation cryptanalysis (but are weak against other attacks like the Siegenthaler correlation attack and the fast algebraic attack). They have been extensively investigated in cryptography (Rothaus who introduced them in [26] worked in this framework), spread spectrum, coding theory (the Kerdock codes are made of affine functions and bent functions) and combinatorial design (in relation with difference sets). A lot of research has been devoted to designing constructions of bent functions. The two best known constructions produce the so-called Maiorana-McFarland class, denoted by $\mathcal{M}$ [11,21] and the $\mathcal{PS}$ class [11]. A survey on bent functions can be found in [2].

It is well known that the Walsh transform of a bent function only takes on the values $\pm 2^{\frac{n}{2}}$. Hence, bent functions are unbalanced and exist only for even number of variables. For even n, a semi-bent function has Walsh transform taking values 0 and $\pm 2^{\frac{n}{2}+1}$ only; it can also be called 3-valued almost optimal. Semi-bent functions can provide protection against fast correlation attack and more general cryptanalysis by affine approximation [22], and unlike bent functions can also be balanced and resilient. A number of constructions of semi-bent functions have been developed. For detailed discussion please see [5,9,23] and the references therein.

In [15], the authors presented a class of cubic RotS bent functions. But such examples of bent RotS functions are very few. Further research is needed to find other classes of cryptographically important RotS functions. In [6], the authors studied the following transformation of RotS functions into idempotents: given, $f(x_0, x_1, \ldots, x_{n-1})$ a RotS function over $GF(2)^n$, the function f' is defined over $GF(2^n)$ as: $f'(z) = f(z, z^2, \ldots, z^{2^{n-1}})$. If the ANF of f is $f(x_0, x_1, \ldots, x_{n-1}) = \sum_{u \in GF(2)^n} a_u x^u$, where $x_0, x_1, \ldots, x_{n-1}$ and a_u belong to $GF(2)$, we have: $f'(z) = \sum_{u \in GF(2)^n} a_u \prod_{i=0}^{n-1} (z^{2^i})^{u_i} = \sum_{u \in GF(2)^n} a_u z^{\sum_{i=0}^{n-1} u_i 2^i}$. The transformation $f \mapsto f'$ maps any RotS Boolean function f to a Boolean idempotent f' over $GF(2^n)$. The algebraic degree is preserved. All Boolean idempotents are obtained this way, with uniqueness. This transformation, contrary to the decomposition of an idempotent over a normal basis, allows obtaining infinite classes from infinite classes. The question whether such infinite classes exist for all situations "f bent / not bent" and "f' bent / not bent" is studied in [6]. The proofs given in the present paper allow to reply positively.

We organize this paper as follows. Section 2 is an introductory part providing some preliminary definitions and results. In Sect. 3, we characterize the Walsh transform of a class of cubic RotS functions f_t. Necessary and sufficient conditions for f_t to be bent or semi-bent functions are obtained. Section 4 presents a class of idempotent bent functions.

2   Preliminaries

We first recall some general definitions about Boolean functions. Denote by $GF(2)^n$ the n-dimensional vector space over the finite field $GF(2)$ and by $+$ the addition operation over $GF(2)$. Let $\mathbf{0}$ and $\mathbf{1}$ be the all-zero vector and the all-one vector of $GF(2)^n$ respectively. An n-variable Boolean function $f(x)$, where $x = (x_0, x_1, \ldots, x_{n-1}) \in GF(2)^n$, is a mapping from $GF(2)^n$ to $GF(2)$, which can be represented uniquely as a polynomial, called its algebraic normal form (ANF), of the form:

$$f(x_0, x_1, \ldots, x_{n-1}) = \sum_{u \in GF(2)^n} \lambda_u (\prod_{i=0}^{n-1} x_i^{u_i}), \qquad \lambda_u \in GF(2).$$

The number of variables in the highest order product term with nonzero coefficient is called its *algebraic degree*. A Boolean function is said to be *affine* if its degree does not exceed 1. The set of all n-variable affine functions is denoted by $A_n(x)$. We call a function nonlinear if it is not in $A_n(x)$. The *Hamming weight* $w_H(x)$ of a binary vector $x \in GF(2)^n$ is the number of its nonzero coordinates, and the Hamming weight $w_H(f)$ of a Boolean function f is the size of its support $\{x \in GF(2)^n | f(x) = 1\}$. If $w_H(f) = 2^{n-1}$, we call $f(x)$ *balanced*. We say two n-variable Boolean functions $f(x)$ and $g(x)$ are *affinely equivalent* if $g(x) = f(Ax + b)$ where b is an element of $GF(2)^n$ and A is an $n \times n$ nonsingular binary matrix. It is easy to see that if $f(x)$ and $g(x)$ are affinely equivalent then

$w_H(f) = w_H(g)$. Let $x = (x_0, x_1, \ldots, x_{n-1})$ and $w = (w_0, w_1, \ldots, w_{n-1})$ both belong to $GF(2)^n$ and $w \cdot x$ be an inner product in $GF(2)^n$, for instance the usual inner product $w_0 x_0 + w_1 x_1 + \cdots + w_{n-1} x_{n-1}$. Then the *Walsh transform* of $f(x)$ is the real valued function over $GF(2)^n$ defined as: $W_f(w) = \sum\limits_{x \in GF(2)^n} (-1)^{f(x) + w \cdot x}$.

Definition 1. *Let n be even. A Boolean function $f(x)$ on $GF(2)^n$ is called bent if its Walsh transform satisfies $W_f(w) = \pm 2^{\frac{n}{2}}$, for all $w \in GF(2)^n$.*

Definition 2. *Let n be any positive integer. A Boolean function $f(x)$ on $GF(2)^n$ is called semi-bent if its Walsh transform satisfies $W_f(w) = 0, \pm 2^{\lceil \frac{n+1}{2} \rceil}$, for all $w \in GF(2)^n$.*

Maiorana and McFarland [21] introduced independently a class of bent functions by concatenating affine functions. We call the Maiorana-McFarland class $\mathcal{M}$ the set of all the Boolean functions on $GF(2)^{2m} = \{(x, y) \,|\, x, y \in GF(2)^m\}$, of the form:

$$f(x, y) = \pi(x) \cdot y + h(x), \tag{1}$$

where π is any mapping from $GF(2)^m$ to $GF(2)^m$ and $h(x)$ is any Boolean function on $GF(2)^m$. Then f is bent if and only if π is bijective.

Let $x_i \in GF(2)$ for $0 \le i \le n - 1$. For $0 \le k \le n - 1$, we define the *left k-cyclic shift operator* ρ_n^k as $\rho_n^k(x_i) = x_{(i+k) \bmod n}$ (this is an abuse of notation since $x_{(i+k) \bmod n}$ does not depend on x_i but on another coordinate of x; but this notation will simplify the presentation below). Let $(x_0, x_1, \ldots, x_{n-1}) \in GF(2)^n$, we can extend the definition of ρ_n^k on tuples as follows: $\rho_n^k(x_0, x_1, \ldots, x_{n-1}) = (\rho_n^k(x_0), \rho_n^k(x_1), \ldots, \rho_n^k(x_{n-1}))$, and on monomials as follows: $\rho_n^k(x_{i_0} x_{i_1} \ldots x_{i_l}) = \rho_n^k(x_{i_0}) \rho_n^k(x_{i_1}) \ldots \rho_n^k(x_{i_l})$ with $0 \le i_0 < i_1 < \cdots < i_l \le n - 1$.

Definition 3. *A Boolean function f on $GF(2)^n$ is called rotation symmetric if for each input $(x_0, x_1, \ldots, x_{n-1}) \in GF(2)^n$, we have:*

$$f(\rho_n^k(x_0, x_1, \ldots, x_{n-1})) = f(x_0, x_1, \ldots, x_{n-1}), \quad \text{for } 0 \le k \le n - 1.$$

Let us denote by $G_n(x_{i_0} x_{i_1} \ldots x_{i_l}) = \{\rho_n^k(x_{i_0} x_{i_1} \ldots x_{i_l}), \text{ for } 0 \le k \le n - 1\}$ the *orbit* of the monomial $x_{i_0} x_{i_1} \ldots x_{i_l}$. We select the representative element of $G_n(x_{i_0} x_{i_1} \ldots x_{i_l})$ as the lexicographically first element. For instance, the representative element of the orbit $\{x_0 x_1 x_2, x_1 x_2 x_3, x_2 x_3 x_0, x_3 x_0 x_1\}$ is $x_0 x_1 x_2$. For a RotS function f, the existence of a representative term $x_0 x_{i_1} \ldots x_{i_l}$ implies the existence of all the terms from $G_n(x_0 x_{i_1} \ldots x_{i_l})$ in the ANF of f.

3   Constructions of Rotation Symmetric Bent and Semi-bent Functions

The lemma below is straightforward and well-known.

Lemma 1. *Assume that a Boolean function $f : GF(2)^{2m} \to GF(2)$ can be expressed in the form (1). Then the following conditions hold.*

1. *If π is a 2-to-1 mapping, then f is a semi-bent function.*
2. *If, for every $b \in GF(2)^m$, the set $S_b = \{x \in GF(2)^m | \pi(x) = b\}$ is either empty or an s-dimensional affine subspace of $GF(2)^m$, then f is semi-bent if and only if $s = 1$, or $s = 2$ and the restriction of h to S_b, viewed as a 2-variable function, has algebraic degree 2 (i.e. has odd Hamming weight).*

Now, we are able to prove our main theorem.

Theorem 1. *Let $f_t(x)$ be the n-variable RotS Boolean function of the form:*

$$f_t(x) = \sum_{i=0}^{n-1} \rho_n^i (x_0 x_r x_{2r}) + \sum_{i=0}^{2r-1} \rho_n^i (x_0 x_{2r} x_{4r}) + \sum_{i=0}^{\nu(t)-1} \rho_n^i (x_0 x_t) \qquad (2)$$

where ρ_n^i is the left i-cyclic shift operator, and $n = 2m = 6r$ with $r \geq 1$, $t \leq m$, $\nu(t) = n$ if $0 < t < m$; $\nu(t) = m$ if $t = m$. Then we have

1. *If $0 < t < m$, then $f_t(x)$ is semi-bent if and only if $\gcd(2t, m) = 1$ or if $\gcd(2t, m) = 2$ and $\gcd(t, m) = 1$.*
2. *If $t = m$, then $f_t(x)$ is a bent function.*

Proof. We first note that

$$\begin{aligned}
f_t(x) = \ & (x_0 + x_{3r})(x_r + x_{4r})(x_{2r} + x_{5r}) \\
& + (x_1 + x_{3r+1})(x_{r+1} + x_{4r+1})(x_{2r+1} + x_{5r+1}) \\
& \ \ \vdots \\
& + (x_{r-1} + x_{4r-1})(x_{2r-1} + x_{5r-1})(x_{3r-1} + x_{6r-1}) + \sum_{i=0}^{\nu(t)-1} \rho_n^i (x_0 x_t).
\end{aligned}$$

Let

$$E = \{x \in GF(2)^n | x_i + x_{m+i} = 0, \forall\, i = 0, \ldots, m-1\}$$

and

$$W = \{x \in GF(2)^n | x_{m+i} = 0, \forall\, i = 0, \ldots, m-1\},$$

then E and W are two supplementary m-dimensional vector subspaces of $GF(2)^n$, that is, any vector $x \in GF(2)^n$ can then be uniquely represented as $x = a + y$ with $a \in W$ and $y \in E$. By replacing x by $a + y$ above, we deduce that:

1. If $0 < t < m$, then

$$\begin{aligned}
f_t(x) = f_t(a+y) &= a_0 a_r a_{2r} + a_1 a_{r+1} a_{2r+1} + \cdots + a_{r-1} a_{2r-1} a_{3r-1} \\
& \quad + \sum_{i=0}^{n-1} \rho_n^i (a_0 + y_0)(a_t + y_t) \\
&= \sum_{i=0}^{r-1} \rho_m^i (a_0 a_r a_{2r}) + \sum_{i=0}^{n-1} \rho_n^i (a_0 a_t + a_0 y_t + a_t y_0 + y_0 y_t).
\end{aligned}$$

Using $a_{m+i} = 0$ and $y_i = y_{m+i}$ for $0 \leq i \leq m-1$, we have:

$$\sum_{i=0}^{n-1} \rho_n^i(a_0 a_t) = \sum_{i=0}^{m-t-1} \rho_n^i(a_0 a_t)$$

$$= \sum_{i=0}^{m-t-1} \rho_m^i(a_0 a_t) \text{ (this is an abuse of notation)},$$

$$\sum_{i=0}^{n-1} \rho_n^i(a_0 y_t) = a_0 y_t + \cdots + a_{m-t-1} y_{m-1} + a_{m-t} y_0 + \cdots + a_{m-1} y_{t-1}$$

$$= \sum_{i=0}^{m-1} \rho_m^i(a_0 y_t) = \sum_{i=0}^{m-1} \rho_m^i(a_{m-t} y_0),$$

$$\sum_{i=0}^{n-1} \rho_n^i(a_t y_0) = \sum_{i=0}^{n-1} \rho_n^i(a_0 y_{n-t}) = \sum_{i=0}^{m-1} \rho_m^i(a_0 y_{m-t}) = \sum_{i=0}^{m-1} \rho_m^i(a_t y_0).$$

Therefore, since $\sum_{i=0}^{n-1} \rho_n^i(y_0 y_t) = 2 \sum_{i=0}^{m-1} \rho_n^i(y_0 y_t) \pmod 2 = 0$:

$$f_t(x) = f_t(a+y)$$

$$= \sum_{i=0}^{r-1} \rho_m^i(a_0 a_r a_{2r}) + \sum_{i=0}^{m-t-1} \rho_m^i(a_0 a_t) + \sum_{i=0}^{m-1} \rho_m^i((a_t + a_{m-t}) y_0)$$

$$= \pi(a) \cdot y + h(a),$$

where

$$\pi(a) = (a_t + a_{m-t}, a_{t+1} + a_{m-t+1}, \ldots, a_{t-1} + a_{m-t-1}),$$

and

$$h(a) = \sum_{i=0}^{r-1} \rho_m^i(a_0 a_r a_{2r}) + \sum_{i=0}^{m-t-1} \rho_m^i(a_0 a_t).$$

If $t = m/2$, then $\pi = 0$ and the function is neither semi-bent nor bent. For $t \neq m/2$, according to the expression obtained for $\pi(a)$, we can assume without loss of generality that $0 < t < m/2$. Let $s = \gcd(2t, m)$. It follows from Theorem 1 of [20, p. 190] that π is a 2^s-to-1 mapping since $\gcd(x^t + x^{m-t}, x^m + 1) = x^s + 1$. This is equivalent to saying that S_w is either an empty set or an s-dimensional affine subspace of $GF(2)^m$. By Case 2 of Lemma 1, we deduce that f_t can be semi-bent only if $s = 1$, or $s = 2$.

– If $s = 1$, then π is a 2-to-1 mapping, which implies f_t is semi-bent by Case 1 of Lemma 1.
– If $s = 2$, denote by G the kernel of π, then

$$G = \{\mathbf{0}, \mathbf{1}, (1, 0, 1, 0, \ldots, 1, 0), (0, 1, 0, 1, \ldots, 0, 1)\} \subset GF(2)^m.$$

Suppose that S_w is nonempty. Then, for any $a \in S_w$, there exists some vector $b \in GF(2)^m$ such that $\{b + e | e \in G\}$ (b can be unique if we require for instance that $b_0 = b_1 = 0$). Then the restriction g of h to S_w is:

$$g = \sum_{i=0}^{r-1} \rho_m^i ((b_0 + e_0)(b_r + e_r)(b_{2r} + e_{2r}) + \sum_{i=0}^{m-t-1} \rho_m^i ((b_0 + e_0)(b_t + e_t))$$

$$= \sum_{i=0}^{r-1} \rho_m^i (b_0 b_r b_{2r} + b_0 b_r e_{2r} + b_0 b_{2r} e_r + b_r b_{2r} e_0$$

$$+ b_0 e_r e_{2r} + b_r e_0 e_{2r} + b_{2r} e_0 e_r + e_0 e_r e_{2r})$$

$$+ \sum_{i=0}^{m-t-1} \rho_m^i (b_0 b_t + b_0 e_t + b_t e_0 + e_0 e_t).$$

Since $\gcd(2t, m) = 2$, then $\gcd(t, m) = 1, 2$ and r is even. Using $e_i = e_j$ if $i \equiv j \pmod 2$, we shall calculate the non-linearized part B of g relative to e for the cases $\gcd(t, m) = 1$ and $\gcd(t, m) = 2$ respectively.
- If $\gcd(t, m) = 2$, then t is even. We have

$$B = \sum_{i=0}^{r-1} \rho_m^i (e_0 e_r e_{2r} + b_0 e_r e_{2r} + b_r e_0 e_{2r} + b_{2r} e_0 e_r) + \sum_{i=0}^{m-t-1} \rho_m^i (e_0 e_t)$$

$$= \sum_{i=0}^{r-1} \rho_m^i (e_0 e_0 e_0 + b_0 e_0 e_0 + b_r e_0 e_0 + b_{2r} e_0 e_0) + \sum_{i=0}^{m-t-1} \rho_m^i (e_0 e_0)$$

$$= \sum_{i=0}^{r/2-1} ((1 + b_{2i} + b_{r+2i} + b_{2r+2i}) e_0$$

$$+ (1 + b_{2i+1} + b_{r+2i+1} + b_{2r+2i+1}) e_1)$$

$$+ (\frac{m-t}{2} \bmod 2)(e_0 + e_1).$$

It shows that g is an affine function on $b + G$. According to Case 2 of Lemma 1, f_t can not be semi-bent if $\gcd(t, m) = 2$.
To complete our proof, it will suffice to check that g is quadratic when $\gcd(t, m) = 1$. In this case, t is odd and so is $m - t$.
- If $\gcd(t, m) = 1$, then

$$B = \sum_{i=0}^{r-1} \rho_m^i (e_0 e_r e_{2r} + b_0 e_r e_{2r} + b_0 e_r e_{2r} + b_r e_0 e_{2r} + b_{2r} e_0 e_r)$$

$$+ \sum_{i=0}^{m-t-1} \rho_m^i (e_0 e_t)$$

$$= \sum_{i=0}^{r/2-1} ((1 + b_{2i} + b_{r+2i} + b_{2r+2i}) e_0$$

$$+(1 + b_{2i+1} + b_{r+2i+1} + b_{2r+2i+1})e_1)$$
$$+(m - t \bmod 2)(e_0 e_1)$$
$$= e_0 e_1 + \sum_{i=0}^{r/2-1} ((1 + b_{2i} + b_{r+2i} + b_{2r+2i})e_0$$
$$+(1 + b_{2i+1} + b_{r+2i+1} + b_{2r+2i+1})e_1).$$

Hence g has algebraic degree 2. We conclude that $f_t(x)$ is semi-bent if $\gcd(2t, m) = 2$ and $\gcd(t, m) = 1$, completing the proof of Case 1 of Theorem 1.

2. If $t = m$, by a straightforward computation, we have

$$f_m(x) = f_m(a + y)$$
$$= \sum_{i=0}^{r-1} \rho_m^i(a_0 a_r a_{2r}) + \sum_{i=0}^{m-1} \rho_m^i((a_0 + y_0)(a_m + y_m))$$
$$= \sum_{i=0}^{r-1} \rho_m^i(a_0 a_r a_{2r}) + \sum_{i=0}^{m-1} \rho_m^i((a_0 + y_0)a_m + (a_0 + y_0)y_m)$$
$$= \sum_{i=0}^{r-1} \rho_m^i(a_0 a_r a_{2r}) + \sum_{i=0}^{m-1} \rho_m^i((a_0 + y_0)y_m)$$
$$= \sum_{i=0}^{r-1} \rho_m^i(a_0 a_r a_{2r}) + \sum_{i=0}^{m-1} \rho_m^i((a_0 + y_0)y_0)$$
$$= \sum_{i=0}^{r-1} \rho_m^i(a_0 a_r a_{2r}) + \sum_{i=0}^{m-1} \rho_m^i((a_0 + 1)y_0)$$

Obviously, $f_m(x)$ is a bent function from the class $\mathcal{M}$, completing the proof.

Remark 1. From the proof of Theorem 2, one can claim that the homogenous RotS function $\sum_{i=0}^{n-1} \rho_n^i(x_0 x_r x_{2r}) + \sum_{i=0}^{2r-1} \rho_n^i(x_0 x_{2r} x_{4r})$ can not be bent. It is conjectured that there are no homogenous RotS bent functions [27].

4 Rotation Symmetric Functions Obtained as Idempotents over $GF(2^n)$

In this section we identify the vector space $GF(2)^n$ with the finite field $GF(2^n)$. For any positive integer k dividing n, we denote the trace function from $GF(2^n)$ to $GF(2^k)$ by $Tr_k^n(z) = z + z^{2^k} + \cdots + z^{2^{n-k}}$. Note that for every integer k dividing n, the trace function Tr_k^n satisfies the transitivity property $Tr_1^n = Tr_1^k \circ Tr_k^n$. Every nonzero Boolean function f defined over $GF(2^n)$ has a unique representation of the form: $f(z) = \sum_{i=0}^{2^n-1} u_i z^i$ where $u_i \in GF(2^n)$. Thanks to the fact that

f is Boolean, that is, satisfies $(f(z))^2 = f(z)$ [mod $z^{2^n} + z$], it can be written in the form (called its univariate polynomial form or trace form):

$$f(z) = \sum_{j \in \Gamma_n} Tr_1^{o(j)}(a_j z^j) + \varepsilon(1 + z^{2^n - 1}), \qquad (3)$$

where Γ_n is the set of integers obtained by choosing one element in each cyclotomic coset of 2 modulo $2^n - 1$ (the most usual choice for j is the smallest element in its cyclotomic class, called the coset leader of the class), $o(j)$ is the size of the corresponding cyclotomic coset containing j, $a_j \in GF(2^{o(j)})$ and $\varepsilon \in GF(2)$. The algebraic degree of f equals the maximum 2-weight of those j such that $a_j \neq 0$, where the 2-weight of j is the Hamming weight of its binary expansion (see e.g. [2]). Let us denote by $\varphi_u(z) = Tr_1^n(uz), u \in GF(2^n)$, the general linear Boolean function on $GF(2^n)$. The Walsh transform of f is defined as

$$W_f(u) = \sum_{z \in GF(2^n)} (-1)^{f(z) + Tr_1^n(uz)}, \quad u \in GF(2^n).$$

Thanks to the identification between the vectors pace $GF(2)^n$ and the field $GF(2^n)$, the Maiorana-McFarland class $\mathcal{M}$ of Boolean functions over $GF(2^{2m})$ can be expressed in the form: $f(x, y) = Tr_1^m(\pi(x)y + h(x))$, where π and h are mappings from $GF(2^m)$ to $GF(2^m)$. A function $f(z)$ given by (3) is an idempotent if and only if every coefficient a_j in every term $Tr^{o(j)}(a_j z^j)$ belongs to $GF(2)$.

4.1   The Bentness of Some Cubic Idempotents

It is known that the monomial function $Tr_1^{2m}(\lambda x^d)$, when cubic, can yield bent functions in $\mathcal{M}$ only if $m = 3r, d = 1 + 2^r + 2^{2r}$ [1], or $d = 1 + 2^j + 2^m$ with $1 \leq j < m$ [8] respectively. But [1, Theorem 3] and [8, Theorem 5.1] imply that such cubic bent monomial functions can not be idempotent (i.e. such that $\lambda = 1$). In this subsection, we characterize the bentness of the idempotent functions of the form:

$$f_k^{(c)}(z) = Tr_1^n(z^{1 + 2^k + 2^m}) + \sum_{i=1}^{m-1} c_i Tr_1^n(z^{1 + 2^i}) + c_m Tr_1^m(z^{1 + 2^m}), \qquad (4)$$

where $n = 2m, 0 < k < m$, and $c = (c_1, \ldots, c_m) \in GF(2)^m$.

The next theorem will show that function $f_k^{(c)}(z)$ is from the class $\mathcal{M}$, and then the bentness of $f_k^{(c)}(z)$ can be related to the bijectivity of some quadratic polynomial of the form $z^{1 + 2^k} + L(z)$, where $L(z)$ is a linearized polynomial over $GF(2^m)$. Such polynomials have received attention for their importance in constructing quadratic APN permutations [19].

Theorem 2. *Let $f_k^{(c)}(z)$ be defined over $GF(2^n)$ by relation (4) and let $L(z) = z^{2^{k-1}} + c_m z + \sum_{i=1}^{\lfloor \frac{m-1}{2} \rfloor} (c_i + c_{m-i})(z^{2^i} + z^{2^{m-i}})$. Then $f_k^{(c)}(z)$ is bent if and only if $z^{1+2^k} + L(z)$ is a permutation polynomial of $GF(2^m)$.*

Proof. Let $V = GF(2^m)$ and denote by U a subspace supplementary to V in the vector space $GF(2^n)$. We have $GF(2^n) = \bigcup_{u \in U} (u + V)$. Then, for any $u \in U$ and $y \in V$, we have

$$
\begin{aligned}
f_k^{(c)}(z) &= f_k^{(c)}(u + y) \\
&= Tr_1^n((u + y)^{1+2^k+2^m}) + \sum_{i=1}^{m-1} c_i Tr_1^n((u + y)^{1+2^i}) \tag{5} \\
&\quad + c_m Tr_1^m((u + y)^{1+2^m}) \\
&= Tr_1^n(u^{1+2^k+2^m}) + Tr_1^n(u^{2^m} y^{1+2^k} + uy^{2^k+2^m} + u^{2^k} y^{1+2^m}) \\
&\quad + Tr_1^n(u^{1+2^m} y^{2^k} + u^{2^k+2^m} y + u^{1+2^k} y^{2^m}) + Tr_1^n(y^{1+2^k+2^m}) \\
&\quad + \sum_{i=1}^{m-1} c_i Tr_1^n(u^{1+2^i} + uy^{2^i} + u^{2^i} y + y^{1+2^i}) \\
&\quad + c_m Tr_1^m(u^{2^m+1} + u^{2^m} y + uy^{2^m} + y^{2^m+1}). \tag{6}
\end{aligned}
$$

Since $u^{1+2^m}, u + u^{2^m}, y \in GF(2^m)$, we have:

$$
Tr_1^n(u^{2^m} y^{1+2^k} + uy^{2^k+2^m}) = Tr_1^n((u + u^{2^m})y^{1+2^k}) = 0,
$$

and

$$
Tr_1^n(y^{1+2^i}) = Tr_1^n(y^{1+2^k+2^m}) = Tr_1^n(u^{1+2^m} y^{2^k}) = 0.
$$

By using the transitivity of the trace function, the part depending on y is

$$
\begin{aligned}
A &= Tr_1^n(u^{2^k} y^{1+2^m} + u^{2^k+2^m} y + u^{1+2^k} y^{2^m}) + \sum_{i=1}^{m-1} c_i Tr_1^n(uy^{2^i} + u^{2^i} y) \\
&\quad + c_m Tr_1^m(u^{2^m} y + uy^{2^m} + y^{2^m+1}) \\
&= Tr_1^n(u^{2^{k-1}} y + u^{2^k}(u + u^{2^m})y) + \sum_{i=1}^{m-1} c_i Tr_1^n((u^{2^{n-i}} + u^{2^i})y) \\
&\quad + c_m Tr_1^m((u^{2^m} + u + 1)y) \\
&= Tr_1^m(((u + u^{2^m})^{2^{k-1}} + (u + u^{2^m})^{2^k+1})y) \\
&\quad + \sum_{i=1}^{m-1} c_i Tr_1^m(((u + u^{2^m})^{2^i} + (u + u^{2^m})^{2^{m-i}})y) + c_m Tr_1^m((u + u^{2^m} + 1)y) \\
&= Tr_1^m(\pi(u)y),
\end{aligned}
$$

where

$$
\begin{aligned}
\pi(u) &= (u + u^{2^m})^{2^{k-1}} + (u + u^{2^m})^{2^k+1} + \sum_{i=1}^{m-1} c_i((u + u^{2^m})^{2^i} + (u + u^{2^m})^{2^{m-i}}) \\
&\quad + c_m(u + u^{2^m} + 1).
\end{aligned}
$$

Let

$$h(u) = Tr_1^m(u^{2^m+1}(u + u^{2^m})^{2^k}) + \sum_{i=1}^{m-1} c_i Tr_1^n(u^{2^i+1}) + c_m Tr_1^m(u^{2^m+1}).$$

Then the sum in Relation (5) is simplified as follows:

$$f_k^{(c)}(u+y) = Try_1^m(\pi(u)y) + h(u).$$

Denoting $u + u^{2^m}$ by ξ, we have:

$$\pi(u) = \xi^{2^k+1} + \xi^{2^{k-1}} + c_m\xi + \sum_{i=0}^{m-1} c_i(\xi^{2^i} + \xi^{2^{m-i}}) + c_m$$

$$= \xi^{2^k+1} + \xi^{2^{k-1}} + c_m\xi + \sum_{i=1}^{\lfloor \frac{m-1}{2} \rfloor} (c_i + c_{m-i})(\xi^{2^i} + \xi^{2^{m-i}}) + c_m$$

$$= \xi^{2^k+1} + L(\xi) + c_m. \qquad (7)$$

This completes the proof.

Reference [19] addresses the problem of the bijectivity of functions of the form $z^{2^k+1} + L(z)$. But it does not address completely the case where k is not co-prime with m:

Lemma 2. *[19] Let* $\gcd(d, 2^m - 1) > 1$ *and* $L(z)$ *be a linearized polynomial on* $GF(2^m)$. *Then if* $L(z)$ *is not a permutation on* $GF(2^m)$, *then* $z^d + L(z)$ *is not a permutation. If* $d = 1 + 2^k$ *with* $\gcd(k, m) = 1$, *then* $z^{1+2^k} + L(z)$ *is a permutation polynomial if and only if* m *is odd and* $L(z) = \alpha^{2^i} z + \alpha z^{2^i}$ *for some* $\alpha \in GF(2^m)^*$.

Proposition 1. *Let* $\pi(z)$ *be given by (7). Then the following statements hold:*

1. $\pi(z)$ *is a permutation only if* $c_m = 1$ *and* $m/\gcd(m,k)$ *is odd.*
2. *If* $k = 1$, *then* π *is a permutation only if* $c_i + c_{m-i} = 0$ *for all* $i = 1 \ldots \lfloor \frac{m-1}{2} \rfloor$.

Proof. 1. If $c_m = 0$, then $\pi(z)$ can not be a permutation for $\pi(0) = \pi(1)$. Now we can assume that $c_m = 1$. Then $L(z)$ can not be a permutation on $GF(2^m)$ since $L(0) = L(1)$. And, if $m/\gcd(m,k)$ is even, then $\gcd(2^k+1, 2^m-1) > 1$. Hence $\pi(z)$ is not a permutation by Lemma 2.

2. From the conclusions above, we can suppose that $c_m = 1$. If $k = 1$, then
$$\pi(z) = z^3 + \sum_{i=1}^{\lfloor \frac{m-1}{2} \rfloor} (c_i + c_{m-i})(z^{2^i} + z^{2^{m-i}}) + 1.$$
By Lemma 2, π can not be bijective if there exists some $1 \leq i \leq \lfloor \frac{m-1}{2} \rfloor$ such that $c_i + c_{m-i} \neq 0$. This closed the proof.

References

1. Canteaut, A., Charpin, P., Kyureghyan, G.: A new class of monomial bent functions. Finite Fields Appl. **14**(1), 221–241 (2008)
2. Carlet, C.: Boolean functions for cryptography and error correcting codes. In: Crama, Y., Hammer, P. (eds.) Boolean Models and Methods in Mathematics, Computer Science, and Engineering, pp. 257–397. Cambridge University Press, Cambridge (2010)
3. Carlet, C.: Partially-bent functions. Des. Codes Cryptogr. **3**, 135–145 (1993)
4. Carlet, C., Mesnager, S.: On Dillon's class $\mathcal{H}$ of bent functions Niho bent functions and o-polynomials. J. Combin. Theory Ser. A **118**(8), 2392–2410 (2011)
5. Carlet, C., Mesnager, S.: On semi-bent Boolean functions. IEEE Trans. Inform. Theory **58**, 3287–3292 (2012)
6. Carlet, C., Gao, G., Liu, W.: A secondary construction and a transformation on rotation symmetric functions, and their action on bent and semi-bent functions. J. Combin. Theory Ser. A **127**, 161–175 (2014)
7. Charpin, P., Gong, G.: Hyperbent functions, Kloosterman sums and Dickson polynomials. IEEE Trans. Inform. Theory **54**(9), 4230–4238 (2008)
8. Charpin, P., Kyureghyan, G.: On cubic monomial bent functions in the class $\mathcal{M}$. SIAM J. Discrete Math. **22**(2), 650–665 (2008)
9. Charpin, P., Pasalic, E., Tavernier, C.: On bent and semi-bent quadratic Boolean functions. IEEE Trans. Inf. Theory **51**, 4286–4298 (2005)
10. Dalai, D.K., Maitra, S., Sarkar, S.: Results on rotation symmetric bent functions. Discrete Math. **309**, 2398–2409 (2009)
11. Dillon, J.: Elementary Hadamard difference sets. Ph.D. Dissertation, University of Maryland (1974)
12. Filiol, É., Fontaine, C.: Highly nonlinear balanced boolean functions with a good correlation-immunity. In: Nyberg, K. (ed.) EUROCRYPT 1998. LNCS, vol. 1403, pp. 475–488. Springer, Heidelberg (1998)
13. Fontaine, C.: On some cosets of the first-order Reed-Muller code with high minimum weight. IEEE Trans. Inform. Theory **45**, 1237–1243 (1999)
14. Fu, S., Qu, L., Li, C., Sun, B.: Blanced $2p$-variable rotation symmetric Boolean functions with maximum algebraic immunity. Appl. Math. Lett. **24**, 2093–2096 (2011)
15. Gao, G., Zhang, X., Liu, W., Carlet, C.: Constructions of quadratic and cubic rotation symmetric bent functions. IEEE Trans. Inform. Theory **58**, 4908–4913 (2012)
16. Gao, G., Cusick, T.W., Liu, W.: Families of rotation symmetric functions with useful cryptographic properties, to appear in IET Information Security
17. Kavut, S., Maitra, S., Yücel, M.D.: Search for Boolean functions with excellent profiles in the rotation symmetric class. IEEE Trans. Inform. Theory **53**, 1743–1751 (2007)
18. Khoo, K., Gong, G., Stinson, D.: A new characterization of semi-bent and bent functions on finite fields. Des. Codes Cryptogr. **38**, 279–295 (2006)
19. Li, Y., Wang, M.: On EA-equivalence of certain permutations to power mappings. Des. Codes Cryptogr. **58**, 259–269 (2011)
20. MacWilliams, F.J., Sloane, J.: The Theory of Error-Correcting Codes. North Holland, Amsterdam (1977)
21. McFarland, R.L.: A family of noncyclic difference sets. J. Combin. Theory Ser. A **15**, 1–10 (1973)

22. Meier, W., Staffelbach, O.: Fast correlation attacks on stream ciphers. In: Günther, C.G. (ed.) EUROCRYPT 1988. LNCS, vol. 330, pp. 301–314. Springer, Heidelberg (1988)
23. Mesnager, S.: Semi-bent functions from Dillon and Niho exponents, Kloosterman sums, and Dickson polynomials. IEEE Trans. Inform. Theory 57, 7443–7458 (2011)
24. Pieprzyk, J., Qu, C.: Fast Hashing and rotation symmetric functions. J. Univers. Comput. Sci. 5, 20–31 (1999)
25. Rijmen, V., Barreto, P., Gazzoni, D.: Filho, Rotation symmetry in algebraically generated cryptographic substitution tables. Inf. Process. Lett. 106, 246–250 (2008)
26. Rothaus, O.S.: On bent functions. J. Combin. Theory Ser. A 20, 300–305 (1976)
27. Stănica, P., Maitra, S.: Rotation symmetric Boolean functions-count and cryptographic properties. Discrete Appl. Math. 156, 1567–1580 (2008)
28. Stănică, P., Maitra, S., Clark, J.A.: Results on rotation symmetric bent and correlation immune boolean functions. In: Roy, B., Meier, W. (eds.) FSE 2004. LNCS, vol. 3017, pp. 161–177. Springer, Heidelberg (2004)
29. Zheng, Y., Zhang, X.-M.: Plateaued functions. In: Varadharajan, V., Mu, Y. (eds.) ICICS 1999. LNCS, vol. 1726, pp. 284–300. Springer, Heidelberg (1999)

Properties of a Family of Cryptographic Boolean Functions

Qichun Wang[✉] and Chik How Tan

Temasek Laboratories, National University of Singapore,
Singapore 117411, Singapore
{tslwq,tsltch}@nus.edu.sg

Abstract. In 2008, Carlet and Feng studied a class of functions with good cryptographic properties. Based on that function, [18] proposed a family of cryptographically significant Boolean functions which contains the functions proposed by [28,30]. However, their study is not in-depth. In this paper, we investigate the properties of those functions further, and find that they can be divided into some affine equivalent classes. The bent functions proposed by [18] are in fact in the same class with the function proposed by [30]. We then prove that those functions have optimum algebraic immunity if and only if a combinatorial conjecture is correct, which gives a new direction to prove the conjecture. Furthermore, we improve upon the lower bound on the nonlinearity, and our bound is higher than all other similar bounds. Finally, we extend the construction to a balanced function, and give an example of a 12-variable function which has the best cryptographic properties among all currently known functions.

Keywords: Boolean function · Algebraic immunity · Nonlinearity

1 Introduction

To resist the main known attacks, Boolean functions used in stream ciphers should be balanced, with high algebraic degree, with high algebraic immunity, with high nonlinearity and with good immunity to fast algebraic attacks. It is hard to construct Boolean functions satisfying all these criteria.

Many classes of Boolean functions with optimum algebraic immunity have been introduced [1,5,12,13,20,21,24,25]. However, the nonlinearity of these functions are not good, and we do not know whether they can behave well against fast algebraic attacks. In 2008, Carlet and Feng studied a class of functions which had been introduced by [16], and they found that these functions seem to satisfy all the cryptographic criteria [6]. This is a breakthrough in the field of Boolean functions. Based on the Carlet-Feng function, some researchers proposed several classes of cryptographically significant Boolean functions [4,26–31,34–36].

Functions constructed by [30] have the optimum algebraic immunity if a combinatorial conjecture is correct, and the nonlinearity of them are very high.

© Springer International Publishing Switzerland 2014
K.-U. Schmidt and A. Winterhof (Eds.): SETA 2014, LNCS 8865, pp. 34–46, 2014.
DOI: 10.1007/978-3-319-12325-7_3

However, they are weak against fast algebraic attacks [3,32]. Based on a similar conjecture which has been proved to be correct by [7], in [28], the authors constructed another class of functions which seems to satisfy all the criteria. In [18], the authors proposed a family of Boolean functions which contains the functions proposed by [28,30]. However, their study is not in-depth, and the proof they gave is quite similar to [6,30]. In this paper, we investigate the properties of those functions further, and get some new results.

The paper is organized as follows. In Sect. 2, the necessary background is established. We discuss affine equivalent classes and prove that a family of Boolean functions has optimum algebraic immunity if and only if a combinatorial conjecture is correct in Sect. 3. In Sect. 4, we give a new bound on the nonlinearity. We then extend the construction to a balanced function in Sect. 5. We end in Sect. 6 with conclusions.

2 Preliminaries

Let $\mathbb{F}_2^n$ be the n-dimensional vector space over the finite field $\mathbb{F}_2$. We denote by B_n the set of all n-variable Boolean functions, from $\mathbb{F}_2^n$ into $\mathbb{F}_2$.

Any Boolean function $f \in B_n$ can be uniquely represented as a multivariate polynomial in $\mathbb{F}_2[x_1, \cdots, x_n]$,

$$f(x_1, \ldots, x_n) = \sum_{K \subseteq \{1,2,\ldots,n\}} a_K \prod_{k \in K} x_k,$$

which is called algebraic normal form (ANF). The algebraic degree of f, denoted by $\deg(f)$, is the number of variables in the highest order term with nonzero coefficient.

A Boolean function is *affine* if there exists no term of degree strictly greater than 1 in the ANF and the set of all affine functions is denoted by A_n.

Let

$$1_f = \{x \in \mathbb{F}_2^n | f(x) = 1\}, \ 0_f = \{x \in \mathbb{F}_2^n | f(x) = 0\}.$$

The cardinality of 1_f is called the *Hamming weight* of f, and will be denoted by $wt(f)$. The *Hamming distance* between two functions f and g is the Hamming weight of $f+g$, and will be denoted by $d(f,g)$. We say that an n-variable Boolean function f is *balanced* if $wt(f) = 2^{n-1}$.

Let $f \in B_n$. The *nonlinearity* of f is its minimum distance from the set of all n-variable affine functions, i.e.,

$$nl(f) = \min_{g \in A_n} d(f,g).$$

The nonlinearity of an n-variable Boolean function is bounded above by $2^{n-1} - 2^{n/2-1}$, and a function is said to be *bent* if it achieves this bound. Clearly, bent functions exist only for even n and it is known that the algebraic degree of a bent function is bounded above by $\frac{n}{2}$ [2,11].

For any $f \in B_n$, a nonzero function $g \in B_n$ is called an *annihilator* of f if $fg = 0$, and the *algebraic immunity* of f, denoted by $\mathcal{AI}(f)$, is the minimum value of d such that f or $f+1$ admits an annihilator of degree d [23]. It is known that the algebraic immunity of an n-variable Boolean function is bounded above by $\lceil \frac{n}{2} \rceil$ [9].

If we can find g of low degree and h of algebraic degree not much larger than $n/2$ such that $fg = h$, then f is considered to be weak against fast algebraic attacks [8,17]. Let

$$\mathcal{FAI}(f) = \min_{\substack{fg=h \\ 1 \leq \deg(g) < n/2}} \{\deg(g) + \deg(h)\}.$$

It is known that $\mathcal{FAI}(f) \leq n$ and the equality can be achieved only when n is one more than a power of two [22].

The *Walsh transform* of a given function $f \in B_n$ is the integer-valued function over $\mathbb{F}_{2^n}$ defined by

$$W_f(\omega) = \sum_{x \in \mathbb{F}_{2^n}} (-1)^{f(x) + tr(\omega x)},$$

where $\omega \in \mathbb{F}_{2^n}$ and $tr(x)$ denotes the trace function from $\mathbb{F}_{2^n}$ to $\mathbb{F}_2$. The nonlinearity of f can then be determined by

$$nl(f) = 2^{n-1} - \frac{1}{2} \max_{\omega \in \mathbb{F}_{2^n}} |W_f(\omega)|.$$

3   Affine Equivalent Classes and Algebraic Immunity of a Family of Boolean Functions

In [29], the authors proposed the following conjecture:

Conjecture 1. Let $k \geq 2$, $1 \leq u < 2^k - 1$ and $(u, 2^k - 1) = 1$. For any $0 < t < 2^k - 1$, let

$$C_t = \{(a,b) | 0 \leq a, b < 2^k - 1, \ a - ub = t \ (mod \ 2^k - 1),$$
$$wt(a) + wt(b) \leq k - 1\}$$

Then $|C_t| \leq 2^{k-1}$.

They verified it experimentally for $k \leq 15$ (it is noticed that in the conjecture of [29] $a - ub$ is replaced by $ua \pm b$, and that conjecture is equivalent to Conjecture 1). Taking $u = 1$ induces a special case of Conjecture 1, which has been proved by [7]. Based on this fact, [28] constructed two classes of Boolean functions with optimum algebraic immunity. Taking $u = 2^k - 2$ induces the conjecture proposed by [30], which has been investigated by [10,14,15] and it is still unsolved.

Base on Conjecture 1, [18] proposed the following function with optimum algebraic immunity:

Construction 1: Let $n = 2k \geq 4$ and α be a primitive element of $\mathbb{F}_{2^k}$. Let $\Delta_s = \{\alpha^s, \cdots, \alpha^{2^{k-1}+s-1}\}$, where $0 \leq s < 2^k - 1$. Then a function $f_1 \in B_n$ is constructed as follows:

$$f_1(x, y) = g(x^u y),$$

where the support of g is Δ_s and $1 \leq u < 2^k - 1$.

Taking $u = 2^k - 2$, we can get the first class constructed by [30], which is a bent function. Taking $u = 1$, we can get the first class constructed by [28], which is a function with optimum algebraic immunity and high algebraic degree and nonlinearity.

In [18], the authors have discussed the cryptographic properties of the functions given by Construction 1. However, the study is not in-depth. We now investigate this construction further.

Proposition 1. $f_1(x, y)$ is affine equivalent to $f_1(x^2, y)$.

Proof. Let β be a primitive element of $\mathbb{F}_{2^k}$. Taking $(1, \beta, \cdots, \beta^{k-1})$ as a basis, we can identify $x = \sum_{i=1}^{k} x_i \beta^{i-1}$ with the k-tuple of its coordinates $(x_1, \cdots, x_k) \in \mathbb{F}_2^k$. Similarly, $y = \sum_{i=1}^{k} y_i \beta^{i-1}$ can be identified with $(y_1, \cdots, y_k) \in \mathbb{F}_2^k$. $f_1(x, y)$ can then be represented as an n-variable polynomial $f_1(x_1, \cdots, x_k, y_1, \cdots, y_k)$ over $\mathbb{F}_2$. Clearly,

$$\begin{aligned}
x^2 &= (x_1 + x_2\beta + \cdots + x_k\beta^{k-1})^2 \\
&= x_1 + x_2\beta^2 + \cdots + x_k\beta^{2(k-1)} \\
&= x_1' + x_2'\beta + \cdots + x_k'\beta^{k-1},
\end{aligned}$$

where $(x_1', \cdots, x_k') = (x_1, \cdots, x_k)B$ and B is a $k \times k$ invertible matrix over $\mathbb{F}_2$. Therefore,

$$f_1(x^2, y) = f_1(xB, y) = f_1((x, y)A),$$

where

$$A = \begin{pmatrix} B & 0 \\ 0 & 1 \end{pmatrix},$$

and the result follows.

Remark 1. In Sect. 4.4 of [18], the authors put forward a class of bent functions with optimum algebraic immunity. By Proposition 1, those functions are in fact affine equivalent to the function proposed by [30].

Remark 2. In [28], the authors found that $g(xy)$ has good cryptographic properties. By Proposition 1, for any $0 \leq i < k$, $g(x^{2^i} y)$ shares the same cryptographic properties with $g(xy)$.

Remark 3. By Proposition 1, each affine equivalent class contained in Construction 1 is corresponding to a monic nonlinear irreducible polynomial over $\mathbb{F}_2$ whose roots are in $\mathbb{F}_{2^k}$ and it is a one-to-one correspondence. That is, it is corresponding to a monic irreducible polynomial of degree $d > 1$, where $d|k$. By [33], the number of monic irreducible polynomials of degree $\leq k$ over $\mathbb{F}_2$ is

$$\frac{2^{k+1}}{k} + \frac{2^{k+1}}{k^2} + O(\frac{2^{k+1}}{k^3}), \quad k \to \infty.$$

Table 1. Cryptographic properties of $f_1 \in B_{12}$

u	deg	$\mathcal{AI}$	nl	$\mathcal{FAI}$
1	10	6	1988	10
3	9	6	1976	11
5	9	6	1992	11
7	10	6	1964	10
9	9	6	1976	11
11	8	6	2000	10
13	9	6	1984	11
15	9	6	1992	10
21	10	6	1880	10
23	8	6	1984	10
27	9	6	1976	10
31	6	6	2016	8

Therefore, the number of affine equivalent classes is less than that value. If k is a prime, then the number of affine equivalent classes is equal to the number of k-variable irreducible polynomials of degree k, i.e.,

$$\frac{1}{k} \sum_{d|k} \mu(k/d) 2^d = \frac{2^k - 2}{k},$$

where d runs over the set of all positive divisors of k including 1 and k, and μ is the Möbius function.

Using Proposition 1, we can divide functions contained in Construction 1 into some affine equivalent classes. A natural question to ask is whether these classes are not affine equivalent to each other. We give an example for $k = 6$.

Example 1: Let $k = 6$. By Proposition 1, Construction 1 contains 12 classes of 12-variable functions, and functions in the same class are affine equivalent to each other. In Table 1, one can find the cryptographic properties of these classes. The function constructed by [28] is in the class $u = 1$ and the function constructed by [30] is in the class $u = 31$. Since algebraic degree, algebraic immunity, nonlinearity and $\mathcal{FAI}$ are all affine invariants, these classes are not affine equivalent to each other excluding the classes $u = 3$ and $u = 9$. We do not know whether the class $u = 3$ is affine equivalent to the class $u = 9$, and we leave it as an open problem.

In [18, 28, 30], the authors proved that those functions have optimum algebraic immunity if the corresponding combinatorial conjectures are correct. But they did not discuss the inverse propositions which are in fact also true.

Theorem 1. *The function f_1 has optimum algebraic immunity if and only if Conjecture 1 is correct.*

Proof. Similar to the proof in [18, 28, 30], it can be proved that f_1 has optimum algebraic immunity if Conjecture 1 is correct. Now, we prove the inverse proposition. Suppose f_1 has optimum algebraic immunity and Conjecture 1 is not correct. That is, there is a $0 < t_1 < 2^k - 1$ such that $|C_{t_1}| > 2^{k-1}$. We will construct an h of degree less than k such that $f_1 * h = 0$.

Let $h(x, y) = \sum_{i=0}^{2^k-1} \sum_{j=0}^{2^k-1} h_{i,j} x^i y^j$ be a function satisfying

(1) $h(x, \gamma x^{2^k-1-u}) = 0$ for $\forall x \in \mathbb{F}_{2^k}^*, \gamma \in \Delta_s$;
(2) $h_{i,j} = 0$ if $w_2(i) + w_2(j) \geq k$.

Then we have

$$h(x, \gamma x^{2^k-1-u}) = \sum_{i=0}^{2^k-2} \sum_{j=0}^{2^k-2} h_{i,j} \gamma^j x^{i-uj}$$

$$= \sum_{t=0}^{2^k-2} h_t(\gamma) x^t,$$

where

$$h_t(\gamma) = \sum_{\substack{0 \leq i,j \leq 2^k-2 \\ i-uj \equiv t \ (\mathrm{mod} \ 2^k-1)}} h_{i,j} \gamma^j.$$

Therefore, the condition (1) holds if and only if $h_t(\gamma) = 0$, $0 \leq t \leq 2^k - 2$. Consider the case $t = t_1$. To satisfy the condition (2), $h_{i,j}$ can be nonzero only when $w_2(i) + w_2(j) \leq k - 1$. Since $|C_{t_1}| > 2^{k-1}$, there are more than 2^{k-1} such $h_{i,j}$. Then $h_{t_1}(\gamma) = 0$ yields a system of homogeneous linear equations on $h_{i,j}$, which has 2^{k-1} number of equations and more than 2^{k-1} number of variables. Hence there exists at least one nonzero solution. Taking these $h_{i,j}$ to be one such solution and all other $h_{i,j} = 0$, we get a function h which is an annihilator of f_1 with $\deg(h) < k$. This is contradictory to the assumption that f_1 has optimum algebraic immunity, and the result follows.

Corollary 1. *Given $1 \leq u < 2^k - 1$, Conjecture 1 is correct for this u if and only if it is correct for $u * 2^i$, where $0 \leq i \leq k - 1$.*

Proof. By Theorem 1, $f_1(x^{2^i}, y)$ has the optimum algebraic immunity if and only if Conjecture 1 is correct for $u * 2^i$. Then by Proposition 1, $f_1(x^{2^i}, y)$ is affine equivalent to $f_1(x, y)$ which has the optimum algebraic immunity if and only if Conjecture 1 is correct for u, and the result follows.

Remark 4. It has been proved that Conjecture 1 is correct for $u = 1$. From Corollary 1, we know that it is also correct for $u = 2^i$. To prove Conjecture 1, we only need to prove it for $\frac{2^k-2}{k}$ number of u when k is a prime.

Remark 5. Theorem 1 gives a new direction to prove Conjecture 1, i.e., proving those functions have the optimum algebraic immunity directly.

4   New Bound on the Nonlinearity

Let $\chi(x)$ be a group homomorphism of $\mathbb{F}_{2^k}^*$ into the unit circle, extended to $x = 0$ where it takes the value 0. Then χ is a mapping of $\mathbb{F}_{2^k}$ into C, which is called a character. There are $2^k - 1$ characters and the set of characters forms a cyclic group. Let χ be the primitive character defined by $\chi(\alpha^j) = \zeta^j$ $(0 \leq j \leq 2^k - 2)$ and $\chi(0) = 0$, where $\zeta = e^{\frac{2\pi\sqrt{-1}}{2^k-1}}$. Then

$$G(\chi^\mu) = \sum_{x \in \mathbb{F}_{2^k}} \chi^\mu(x)(-1)^{tr(x)}$$

is a Gauss sum, where $0 \leq \mu \leq 2^k - 2$. We have $G(\chi^0) = -1$ and $\mid G(\chi^\mu) \mid = 2^{\frac{k}{2}}$ for $1 \leq \mu \leq 2^k - 2$ [19].

Lemma 1. *For $0 < x < \frac{\pi}{4}$, we have $y = x^2 \sin x + 5 \sin x - 5x > 0$. That is,* $\frac{1}{\sin x} < \frac{1}{x} + \frac{x}{5}$.

Proof. Clearly, $y' = x^2 \cos x + 2x \sin x + 5 \cos x - 5$ and $y'' = -x^2 \sin x + 4x \cos x - 3 \sin x$. Therefore,

$$\frac{y''}{4x \sin x} = \cot x - \frac{x^2 + 3}{4x}.$$

For $0 < x < \frac{\pi}{4}$, $y_1 = \cot x$ is decreasing and convex, and $y_2 = \frac{x^2+3}{4x}$ is decreasing and concave. Clearly, $\cot x > \frac{x^2+3}{4x}$ when $x \to 0$ and $\cot x < \frac{x^2+3}{4x}$ when $x = \pi/4$. Therefore, $y'' > 0$ on the interval $(0, x_0)$ and $y'' < 0$ on the interval $(x_0, \pi/4)$, where $0 < x_0 < \frac{\pi}{4}$ and $\cot x_0 = \frac{x_0^2+3}{4x_0}$. Hence, y' is increasing on $(0, x_0)$, and is decreasing on $(x_0, \pi/4)$. Since $y'(0) = 0$ and $y'(\pi/4) > 0$, we have $y' > 0$ on the interval $(0, \pi/4)$. Therefore, y is increasing on $(0, \pi/4)$, and the result follows.

Lemma 2. *Let*

$$\Gamma_u = \sum_{\gamma \in \Delta_s} \sum_{x \in \mathbb{F}_{2^k}^*} (-1)^{tr(\frac{1}{x} + \gamma x^u)},$$

where $1 \leq u < 2^k - 1$ and $k \geq 5$. Then $|\Gamma_u| < (\frac{\ln 2}{\pi} k + 0.267)2^k + \frac{16}{31}$.

Proof. Since

$$(-1)^{tr(\alpha^j)} = \frac{1}{2^k - 1} \sum_{\mu=0}^{2^k-2} G(\chi^\mu)\overline{\chi}^\mu(\alpha^j),$$

where $0 \leq j \leq 2^k - 2$, then

$$\Gamma_u = \sum_{\gamma \in \Delta_s} \sum_{j=0}^{2^k-2} (-1)^{tr(\alpha^{-j})}(-1)^{tr(\gamma\alpha^{uj})}$$

$$= \sum_{i=s}^{2^{k-1}+s-1} \sum_{j,\mu,\nu=0}^{2^k-2} \frac{G(\chi^\mu)G(\chi^\nu)}{(2^k - 1)^2} \zeta^{\mu j - \nu(i+uj)}$$

$$= \sum_{\mu,\nu=0}^{2^k-2} \frac{G(\chi^\mu)G(\chi^\nu)}{(2^k - 1)^2} \sum_{i=s}^{2^{k-1}+s-1} \zeta^{-\nu i} \sum_{j=0}^{2^k-2} \zeta^{(\mu-\nu u)j}.$$

Clearly,

$$\sum_{i=s}^{2^{k-1}+s-1} \zeta^{-\nu i} = \begin{cases} 2^{k-1} & \text{if } \nu = 0, \\ \zeta^{-\nu s}\frac{\zeta^{-\nu 2^{k-1}}-1}{\zeta^{-\nu}-1} & \text{otherwise} \end{cases}$$

and

$$\sum_{j=0}^{2^k-2} \zeta^{(\mu-\nu u)j} = \begin{cases} 2^k - 1 & \text{if } \mu = \nu u, \\ 0 & \text{otherwise.} \end{cases}$$

Therefore,

$$|\Gamma_u| = |\frac{2^{k-1}}{2^k-1} + \sum_{\nu=1}^{2^k-2} \frac{G(\chi^\nu)G(\chi^{\nu u})}{2^k-1}\zeta^{-\nu s}\frac{\zeta^{-\nu 2^{k-1}}-1}{\zeta^{-\nu}-1}|$$

$$\leq \frac{2^{k-1}}{2^k-1} + \frac{2^k}{2^k-1}\sum_{\nu=1}^{2^k-2} |\frac{\zeta^{-\nu 2^{k-1}}-1}{\zeta^{-\nu}-1}|$$

$$= \frac{2^{k-1}}{2^k-1} + \frac{2^k}{2^k-1}\sum_{\nu=1}^{2^k-2} |\frac{\zeta^{-\nu 2^{k-2}}-\zeta^{\nu 2^{k-2}}}{\zeta^{-\frac{\nu}{2}}-\zeta^{\frac{\nu}{2}}}|$$

$$\leq \frac{16}{31} + \frac{2^k}{2^k-1}\sum_{\nu=1}^{2^k-2} |\frac{\sin\frac{\nu\pi 2^{k-1}}{2^k-1}}{\sin\frac{\nu\pi}{2^k-1}}| = \frac{16}{31} +$$

$$\frac{2^{k+1}}{2^k-1}(\sum_{\nu=1}^{2^{k-2}} \frac{\sin\frac{2^{k-1}-\nu}{2^k-1}\pi}{\sin\frac{2\nu-1}{2^k-1}\pi} + \sum_{\nu=1}^{2^{k-2}-1} \frac{\sin\frac{\nu\pi}{2^k-1}}{\sin\frac{2\nu\pi}{2^k-1}})$$

$$= \frac{16}{31} + \frac{2^{k+1}}{2^k-1}(\sum_{\nu=1}^{2^{k-2}} \frac{\cos\frac{(2\nu-1)\pi}{2(2^k-1)}}{\sin\frac{(2\nu-1)\pi}{2^k-1}} + \sum_{\nu=1}^{2^{k-2}-1} \frac{1}{2\cos\frac{\nu\pi}{2^k-1}})$$

$$= \frac{16}{31} + \frac{2^k}{2^k-1}(\sum_{\nu=1}^{2^{k-2}} \frac{1}{\sin\frac{(2\nu-1)\pi}{2(2^k-1)}} + \sum_{\nu=1}^{2^{k-2}-1} \frac{1}{\cos\frac{\nu\pi}{2^k-1}}).$$

By Lemma 1 and $1 + \frac{1}{3} + \cdots + \frac{1}{2^{k-1}-1} < \frac{k-1}{2}\ln 2 + 0.6353$ $(k \geq 5)$, we have

$$\sum_{\nu=1}^{2^{k-2}} \frac{1}{\sin\frac{(2\nu-1)\pi}{2(2^k-1)}} < \sum_{\nu=1}^{2^{k-2}} (\frac{2(2^k-1)}{(2\nu-1)\pi} + \frac{(2\nu-1)\pi}{10(2^k-1)}) <$$

$$\frac{2(2^k-1)}{\pi}(\frac{k-2}{2}\ln 2 + 0.6353) + \frac{\pi 2^{2k}}{160(2^k-1)},$$

and

$$\sum_{\nu=1}^{2^{k-2}-1} \frac{1}{\cos \frac{\nu\pi}{2^k-1}} \leq \sum_{\nu=1}^{2^{k-2}-1} \left(\frac{2(2^k-1)}{(2^k-1)\pi - 2\nu\pi} + \right.$$

$$\left.\frac{(2^k-1-2\nu)\pi}{10(2^k-1)}\right) < \frac{2(2^k-1)}{\pi}0.35 + \frac{\pi}{10}(2^{k-2}-1)$$

$$-\frac{\pi 2^{k-2}}{10(2^k-1)}(2^{k-2}-1) < \frac{71}{250}2^k.$$

Therefore,

$$|\Gamma_u| < \left(\frac{(k-2)\ln 2 + 1.2706}{\pi} + \frac{\pi}{160} + \frac{71}{250}\right)2^k + 1$$

$$< \left(\frac{\ln 2}{\pi}k + 0.267\right)2^k + \frac{16}{31}.$$

Theorem 2. *The nonlinearity of the function $f_1(x,y)$ satisfies*

$$nl(f_1) > 2^{n-1} - \left(\frac{\ln 2}{\pi}k + 0.267\right)2^k - \frac{16}{31}.$$

Proof. For any $(a,b) \in \mathbb{F}_{2^k} \times \mathbb{F}_{2^k} - \{(0,0)\}$, we have

$$W_{f_1}(a,b) = \sum_{x,y \in \mathbb{F}_{2^k}} (-1)^{f_1(x,y)+tr(ax+by)}$$

$$= -2 \sum_{x,y \in 1_{f_1}} (-1)^{tr(ax+by)}$$

$$= -2 \sum_{x \in \mathbb{F}_{2^k}^*} (-1)^{tr(ax)} \sum_{\gamma \in \Delta_s} (-1)^{tr(b\gamma/x^u)}$$

$$= \begin{cases} 2^k & \text{if } a = 0, b \in \mathbb{F}_{2^k}^* \\ 2^k & \text{if } b = 0, a \in \mathbb{F}_{2^k}^* \\ -2 \sum_{\gamma \in \Delta_s} \sum_x (-1)^{tr(\frac{1}{x}+ab\gamma x^u)} & \text{if } a,b \in \mathbb{F}_{2^k}^*. \end{cases}$$

Then by Lemma 3,

$$|W_{f_1}(a,b)| < 2\left(\frac{\ln 2}{\pi}k + 0.267\right)2^k + \frac{32}{31},$$

and the result follows.

Remark 6. The bound on the nonlinearity given by [18] is $2^{n-1} - \frac{2\ln 2}{\pi}k2^k$. Clearly, the new bound improves it largely.

Table 2. Comparison of the bounds on nonlinearity

n	Bound in [6]	Bound in [28]	Our bound
8	70	102	106
10	366	458	464
12	1700	1929	1940
14	7382	7931	7952
16	30922	32195	32236
18	126927	129823	129903
20	515094	521577	521735
22	2076956	2091288	2091603
24	8344600	8376003	8376632
26	33459185	33527429	33528684
28	134012775	134160165	134162673

5   Extending to a Balanced Function

Construction 2: Let m be odd and $n = 2k = 2^t m \geq 4$. We construct $f_2 \in B_n$ as follows

$$f_2(x, y) = \begin{cases} f_1(x, y) & \text{if } x \neq 0 \\ w(y) & \text{otherwise,} \end{cases}$$

where w is a k-variable balanced function satisfying $w(0) = 0$, $\deg(w) = k - 1$ and

$$|W_w(a)| \leq \begin{cases} 2^{\frac{m+1}{2}} & \text{if } t = 1 \\ \sum_{i=0}^{t-2} 2^{2^i m} + 2^{\frac{m+1}{2}} & \text{otherwise.} \end{cases}$$

Similar to the above proof, it is easy to verify that f_2 has the following properties.

Theorem 3. f_2 is a balanced function, $\deg(f_2) = n - 1$, $\mathcal{AI}(f_2) = k$ and

$$nl(f_2) > \begin{cases} 2^{n-1} - (\frac{\ln 2}{\pi} k + 0.267)2^k - 2^{\frac{k-1}{2}} - \frac{16}{31}, & \text{if } t = 1 \\ 2^{n-1} - (\frac{\ln 2}{\pi} k + 0.267)2^k - \sum_{i=0}^{t-2} 2^{2^i m - 1} \\ \qquad -2^{\frac{m-1}{2}} - \frac{16}{31}, & \text{otherwise.} \end{cases}$$

The lower bound on nonlinearity in Theorem 3 improves upon those bounds deduced by others. In Table 2, one can find the comparison of our bound with others.

There are many Boolean functions with very good cryptographic properties in this family. In Table 3, one can find the cryptographic properties of the functions $f_2 \in B_{12}$, where $w = 566C27782E175359$. In the following, we give an example of a 12-variable function which has the best cryptographic properties among all currently known functions.

Table 3. Cryptographic properties of $f_2 \in B_{12}$

u	deg	$\mathcal{AI}$	nl	$\mathcal{FAI}$
1	11	6	1982	11
3	11	6	1970	11
5	11	6	1986	11
7	11	6	1982	11
9	11	6	1978	11
11	11	6	1994	10
13	11	6	1978	11
15	11	6	1986	11
21	11	6	1912	11
23	11	6	1978	10
27	11	6	1986	11
31	11	6	2010	8

Example 2: Let $k = 6$, $f_1 = g(x^5 y) \in B_{12}$ and $w = 566C27782E175359$. Then f_2 is balanced, $\deg(f_2) = 11$, $\mathcal{AI}(f_2) = 6$, $nl(f_2) = 1986$ and $\mathcal{FAI}(f_2) = 11$. As a comparison, the nonlinearity of the Carlet-Feng function is 1974 and the function constructed by [28] has the nonlinearity 1982. The function f_2 is balanced and with the optimum algebraic degree, optimum algebraic immunity and optimum $\mathcal{FAI}$. It has the highest nonlinearity among all those known functions with the above properties.

6 Conclusion

This paper studies a family of Boolean functions with optimum algebraic immunity. We find that they can be divided into some affine equivalent classes, and give a new direction to prove a combinatorial conjecture. The bound on the nonlinearity we deduced is higher than other similar bounds. Moreover, we extend the construction to a balanced function, and give an example of a 12-variable function which has the best cryptographic properties among all currently known functions.

We divide functions contained in Construction 1 into some affine equivalent classes. But among these classes, we do not know whether they are affine inequivalent to each other. We leave this as an open problem.

Acknowledgment. The first author would like to thank the financial support from the National Natural Science Foundation of China (Grant No. 61202463).

References

1. Braeken, A., Preneel, B.: On the algebraic immunity of symmetric Boolean functions. In: Maitra, S., Veni Madhavan, C.E., Venkatesan, R. (eds.) INDOCRYPT 2005. LNCS, vol. 3797, pp. 35–48. Springer, Heidelberg (2005)

2. Carlet, C.: Boolean functions for cryptography and error correcting codes. In: Chapter of the monography Boolean Models and Methods in Mathematics, Computer Science, and Engineering, pp. 257–397. Cambridge University Press (2010). http://www-roc.inria.fr/secret/Claude.Carlet/pubs.html
3. Carlet, C.: On a weakness of the Tu-Deng function and its repair. Cryptology ePrint Archive, 2009/606 [Online]. Available: eprint.iacr.org/2009/606
4. Carlet, C.: Comments on 'Constructions of cryptographically significant Boolean functions using primitive polynomials. IEEE Trans. Inf. Theory **57**, 7 (2011)
5. Carlet, C., Dalai, D.K., Gupta, K.C., Maitra, S.: Algebraic immunity for cryptographically significant Boolean functions: analysis and construction. IEEE Trans. Inf. Theory **52**(7), 3105–3121 (2006)
6. Carlet, C., Feng, K.: An infinite class of balanced functions with optimal algebraic immunity, good immunity to fast algebraic attacks and good nonlinearity. In: Pieprzyk, J. (ed.) ASIACRYPT 2008. LNCS, vol. 5350, pp. 425–440. Springer, Heidelberg (2008)
7. Cohen, G., Flori, J.: On a generalized combinatorial conjecture involving addition mod $2^k - 1$. Cryptology ePrint Archive, 2011/400 [Online]. Available: eprint.iacr.org/2011/400
8. Courtois, N.T.: Fast algebraic attacks on stream ciphers with linear feedback. In: Boneh, D. (ed.) CRYPTO 2003. LNCS, vol. 2729, pp. 176–194. Springer, Heidelberg (2003)
9. Courtois, N.T., Meier, W.: Algebraic attacks on stream ciphers with linear feedback. In: Biham, E. (ed.) EUROCRYPT 2003. LNCS, vol. 2656, pp. 345–359. Springer, Heidelberg (2003)
10. Cusick, T.W., Li, Y., Stanica, P.: On a combinatorial conjecture. Integers **11**(2), 185–203 (2011)
11. Cusick, T.W., Stănică, P.: Cryptographic Boolean Functions and Applications. Elsevier-Academic Press, Amsterdam (2009)
12. Dalai, D.K., Gupta, K.C., Maitra, S.: Cryptographically significant Boolean functions: construction and analysis in terms of algebraic immunity. In: Gilbert, H., Handschuh, H. (eds.) FSE 2005. LNCS, vol. 3557, pp. 98–111. Springer, Heidelberg (2005)
13. Dalai, D.K., Maitra, S., Sarkar, S.: Baisc theory in construction of Boolean functions with maximum possible annihilator immunity. Des. Codes Crypt. **40**(1), 41–58 (2006)
14. Flori, J.P., Randriam, H.: On the Number of Carries Occuring in an Addition mod $2^k - 1$. Cryptology ePrint Archive, 2010/170 [Online]. Available: eprint.iacr.org/2010/170
15. Flori, J.-P., Randriam, H., Cohen, G., Mesnager, S.: On a conjecture about binary strings distribution. In: Carlet, C., Pott, A. (eds.) SETA 2010. LNCS, vol. 6338, pp. 346–358. Springer, Heidelberg (2010)
16. Feng, K., Liao, Q., Yang, J.: Maximum values of generalized algebraic immunity. Des. Codes Crypt. **50**(2), 243–252 (2009)
17. Hawkes, P., Rose, G.G.: Rewriting variables: the complexity of fast algebraic attacks on stream ciphers. In: Franklin, M. (ed.) CRYPTO 2004. LNCS, vol. 3152, pp. 390–406. Springer, Heidelberg (2004)
18. Jin, Q., Liu, Z., Wu, B., Zhang, X.: A general conjecture similar to T-D conjecture and its applications in constructing Boolean functions with optimal algebraic immunity. Cryptology ePrint Archive, 2011/515 [Online]. Available: eprint.iacr.org/2011/515

19. Lidl, R., Niederreiter, H.: Introduction to Finite Fields and Their Applications. Cambridge University Press, Cambridge (1986)
20. Li, N., Qi, W.-F.: Construction and analysis of Boolean functions of $2t+1$ variables with maximum algebraic immunity. In: Lai, X., Chen, K. (eds.) ASIACRYPT 2006. LNCS, vol. 4284, pp. 84–98. Springer, Heidelberg (2006)
21. Li, N., Qu, L., Qi, W., Feng, G., Li, C., Xie, D.: On the construction of Boolean functions with optimal algebraic immunity. IEEE Trans. Inf. Theory **54**(3), 1330–1334 (2008)
22. Liu, M., Zhang, Y., Lin, D.: Perfect algebraic immune functions. In: Wang, X., Sako, K. (eds.) ASIACRYPT 2012. LNCS, vol. 7658, pp. 172–189. Springer, Heidelberg (2012)
23. Meier, W., Pasalic, E., Carlet, C.: Algebraic attacks and decomposition of Boolean functions. In: Cachin, C., Camenisch, J.L. (eds.) EUROCRYPT 2004. LNCS, vol. 3027, pp. 474–491. Springer, Heidelberg (2004)
24. Pasalic, E.: Almost fully optimized infinite classes of Boolean functions resistant to (fast) algebraic cryptanalysis. In: Lee, P.J., Cheon, J.H. (eds.) ICISC 2008. LNCS, vol. 5461, pp. 399–414. Springer, Heidelberg (2009)
25. Qu, L., Feng, K., Liu, F., Wang, L.: Constructing symmetric Boolean functions with maximum algebraic immunity. IEEE Trans. Inf. Theory **55**(5), 2406–2412 (2009)
26. Rizomiliotis, P.: On the resistance of Boolean functions against algebraic attacks using univariate polynomial representation. IEEE Trans. Inf. Theory **56**(8), 4014–4024 (2010)
27. Tan, C., Goh, S.: Several classes of even-variable balanced Boolean functions with optimal algebraic immunity. IEICE Trans. **E94**(A:1), 165–171 (2011)
28. Tang, D., Carlet, C., Tang, X.: Highly nonlinear Boolean functions with optimal algebraic immunity and good behavior against fast algebraic attacks. IEEE Trans. Inf. Theory **59**(1), 653–664 (2013)
29. Tang, D., Carlet, C., Tang, X.: Highly Nonlinear Boolean Functions with Optimal Algebraic Immunity and Good Behavior Against Fast Algebraic Attacks. Cryptology ePrint Archive, 2011/366 [Online]. Available: eprint.iacr.org/2011/366
30. Tu, Z., Deng, Y.: A conjecture about binary strings and its applications on constructing Boolean functions with optimal algebraic immunity. Des. Codes Crypt. **60**(1), 1–14 (2011)
31. Wang, Q., Peng, J., Kan, H., Xue, X.: Constructions of cryptographically significant Boolean functions using primitive polynomials. IEEE Trans. Inf. Theory **56**(6), 3048–3053 (2010)
32. Wang, Q., Johansson, T.: A note on fast algebraic attacks and higher order nonlinearities. In: Lai, X., Yung, M., Lin, D. (eds.) Inscrypt 2010. LNCS, vol. 6584, pp. 404–414. Springer, Heidelberg (2011)
33. Wang, Q., Kan, H.: Counting irreducible polynomials over finite fields. Czech. Math. J. **60**(135), 881–886 (2010)
34. Wang, Q., Tan, C.H.: Balanced Boolean functions with optimum algebraic degree, optimum algebraic immunity and very high nonlinearity. Discrete Appl. Math. **1673**, 25–32 (2014)
35. Wang, Q., Tan, C.H.: A new method to construct Boolean functions with good cryptographic properties. Inform. Process. Lett. **113**(14), 567–571 (2013)
36. Zeng, X., Carlet, C., Shan, J., Hu, L.: More balanced Boolean functions with optimal algebraic immunity, and good nonlinearity and resistance to fast algebraic attacks. IEEE Trans. Inf. Theory **57**(9), 6310–6320 (2011)

A New Transform Related to Distance from a Boolean Function (Extended Abstract)

Andrew Klapper[✉]

Department of Computer Science, University of Kentucky, Lexington, USA
klapper@cs.uky.edu
http://www.cs.uky.edu/~klapper/

Abstract. We introduce a new transform on Boolean functions generalizing the Walsh-Hadamard transform. For Boolean functions q and f, the *q-transform of f* measures the proximity of f to the set of functions obtained from q by change of basis. This has implications for security against certain algebraic attacks. In this paper we derive the expected value and second moment (Parseval's equation) of the q-transform, leading to a notion of q-bentness. We also develop a Poisson Summation Formula, which leads to a proof that the q-transform is invertible.

Keywords: Boolean function · Walsh-Hadamard transform · Poisson summation formula

1 Introduction

In much of cryptography using linear operations leads to vulnerability to attack. Moreover, in many cases a nonlinear function that can be approximated by a linear function also is vulnerable. This occurs, for example, in various correlation attacks such as linear cryptanalysis.

In some cases there is a suitable measure of nonlinearity that quantifies resistance to the attack. Examples include algebraic degree, algebraic immunity, resilience, and nonlinearity. The latter measure is closely related to the *Walsh-Hadamard transform* (WHT). A Walsh-Hadamard (WH) coefficient of a Boolean function f is the correlation between f and a linear function (described in more detail in the next section). A function has large nonlinearity if and only if all its WH coefficients are small. Thus the WHT is a measure of linearity.

We also mention *algebraic cryptanalysis* [2]. Suppose we know a Boolean function f being used as a filter function for a linear sequence generator. The initial state of the generator is unknown, but we know a sequence of outputs

Thanks to Mark Goresky for useful conversations. This material is based upon work supported by the National Science Foundation under Grant No. CNS-1420227. Any opinions, findings, and conclusions or recommendations expressed in this material are those of the author and do not necessarily reflect the views of the National Science Foundation.

© Springer International Publishing Switzerland 2014
K.-U. Schmidt and A. Winterhof (Eds.): SETA 2014, LNCS 8865, pp. 47–59, 2014.
DOI: 10.1007/978-3-319-12325-7_4

from f. In order to find the initial state, we can think of f and the known output values as giving a nonlinear system of equations in the bits of the initial state. Unfortunately, such a system is in general hard to solve, so we linearize the system by treating it as a system of linear equations in the monomials of f. Even this is not good enough since in general the number of monomials to consider is exponential in the number of variables. To improve things, we instead try to find a low degree multiple g of f and work with g rather than f. If the degree of g is small enough, then the linear system we must solve becomes tractable. Many variations on this approach have been studied.

The attack can be further improved by finding a low degree approximation h to the filter function f. We then use h in place of f (or g) — treat the monomials of h as variables in a linear system to solve for the initial state. Since h is just an approximation to f some of these equations will be wrong, but if enough are right the system may be solvable.

For this to work, the approximation should be chosen from a set of functions that is preserved by composition on the right with (linear) state change functions. The smallest such set of functions is of the form $S_q = \{q_A : A$ is a nonsingular matrix$\}$ for some fixed Boolean function q, where $q_A(b) = q(bA)$ for $b \in \{0,1\}^n$. This paper concerns a generalization of the WHT, called the q-transform, that measures proximity to the set S_q just as WHT measures proximity to linear functions. We study basic properties of the q-transform, obtain expressions for the mean and second moment (which leads to a notion of q-bent functions), and develop Poisson summation formulas. Many proofs are omitted for lack of space.

2   Basics on Transforms

See Carlet's book chapter [1] or Cusick and Stănică's book [3] for background on Boolean functions and WHTs. Let $\mathbb{F}_2 = \{0,1\}$. Let n be a positive integer, let $V_n = \mathbb{F}_2^n$, let B_n denote the set of Boolean functions on V_n, and let R_n denote the set of real valued functions on V_n. We denote the vector of all 0s by 0. If $f \in B_n$, then we let $\Phi_f(a) = (-1)^{f(a)}$. Thus $\Phi_f \in R_n$. If $G \in R_n$, then we let

$$Z(G) = \sum_{a \in V_n} G(a).$$

If $f \in B_n$, then the *imbalance* of f is $I_f = Z(\Phi_f)$. If $F, G \in R_n$, then the *correlation* between F and G is

$$C_{F,G} = Z(F \cdot G) = \sum_{a \in V_n} F(a)G(a).$$

Let $t_a \in B_n$ denote the linear function $t_a(b) = a \cdot b$ (inner product). Let $T_a = \Phi_{t_a}$. The WHT of $F \in R_n$ is the list of real numbers $W(F)(a) = C_{F,T_a}$. Each $W(F)(a)$ is called a *Walsh-Hadamard coefficient*. Our goal is to explore analogous transforms defined by sets of functions other than the set of linear functions. At the least such a set of functions must be rich enough so that the mapping from functions to their transforms is invertible.

Let $q \in B_n$, $Q = \Phi_q$, and $G \in R_n$. Let $GL_n = GL_n(\mathbb{F}_2)$ denote the set of invertible n by n matrices over $\mathbb{F}_2$. Recall that the cardinality of GL_n is $N = (2^n - 1)(2^n - 2)(2^n - 4) \cdots (2^n - 2^{n-1})$. For any invertible $A \in GL_n$ and $a \in V_n$, let $q_A(a) = q(aA)$ and let $G_A(a) = G(aA)$.

Definition 1. *Let $q \in B_n$ and $F \in R_n$. The q-transform of F is the list of real numbers*

$$W_q(F)(A) = C_{F,q_A} = \sum_{a \in V_n} F(a)(-1)^{q(aA)},$$

where A ranges over GL_n, together with

$$W_q(f)(0) = Z(F) = \sum_{a \in V_n} F(a).$$

We let $G = GL_n \cup \{0\}$. We refer to $W_q(F)(A)$ as the q-transform coefficient of F associated with A. If $f \in B_n$, then we let $\overline{W}_q(f)(A) = W_q(\Phi_f)(A)$.

The inclusion of $W_q(F)(0)$ is necessary for the q-transform to be an invertible map. This becomes apparent in the proof of Corollary 2.

Theorem 1. *Let $A \in GL_n$. Let $f \in B_n$, $G \in R_n$, and $B \in GL_n$. Then*

1. *$\overline{W}_q(f)(A) = \overline{W}_f(q)(A^{-1})$;*
2. *$W_{q_B}(G)(A) = W_q(G)(AB)$, so the q_B-transform of G is a permutation of the q-transform of G; and*
3. *$W_q(G_B)(A) = W_q(G)(B^{-1}A)$, so the q-transform of G_B is a permutation of the q-transform of G.*

Thus the q-transform is, up to a permutation, unchanged by a linear change of basis. Apparently there are $N + 1 \sim 2^{n^2}$ q-transform coefficients for a given $f \in B_n$. However, by part (2) of the theorem, if $H = \{B : q(b) = q(bB)\}$ is the stabilizer subgroup of q for the action of GL_n on B_n, then the effective number of q-transform coefficients is $N/|H| + 1$. For example, for $q(a) = a_1 a_2$ the effective number of coefficients is only $(2^n - 1)(2^{n-1} - 1) + 1$.

3   Relation to the Walsh-Hadamard Transform

In this section we assume that q is a non-zero linear function. If q' is another non-zero linear function, then there is an element $B \in GL_n$ so that $q(a) = q'(aB)$. Thus the q-transform of a function F is a permutation of the q'-transform of F. Hence for many purposes we may assume $q(c) = c_1$, where $c = (c_1, \cdots, c_n)$. Let us assume this now. If $A = [a_{i,j}]$, then $q(cA) = \sum_j a_{j,1} c_j$. Thus the q-transform of f associated with the matrix A is just the WHT of f associated with the vector $a = (a_{1,1}, \cdots, a_{n,1})$. Note that this depends only on the first column of A, so for each $a \neq 0^n$ there is a set of $N/(2^n - 1)$ distinct matrices $A \in GL_n$ such that for all F, $W_q(F)(A) = W(F)(a)$. Also, $W_q(F)(0) = W(F)(0)$, the zeroth WH coefficient of F. It follows that the information in the q-transform of F is the same as the information in the WHT, although each WH coefficient $W(F)(a)$, $a \neq 0$, occurs $N/(2^n - 1)$ times as a $W_q(F)(A)$, while $W(F)(0)$ occurs only once.

4   Statistics

In this section we compute the mean and second moment of the q-transform. In the classical setting, one takes expectations using a uniform distribution over the set of WH coefficients. The expected WHT of F is $F(0)$ and the second moment is 2^n by a straightforward calculation. However, in our more general setting we must be careful about the distribution used in computing the mean. Recall that when q is linear, each coefficient $W(F)(A)$ occurs $N/(2^n - 1)$ times, while $W(F)(0) = Z(F)$ occurs only once. It seems that this bias should be accounted for in the expectation. Therefore we compute the expectations in two ways: first we use a uniform distribution on the set of $W(F)(A)$, excluding $W(F)(0)$, then we include $W(F)(0)$, but weight it with $N/(2^n - 1)$. That is, we define a probability distribution ω on $G = GL_n \cup \{0\}$ by letting

$$\omega(A) = \frac{1}{N + N/(2^n - 1)} = \frac{2^n - 1}{2^n N}$$

for $A \in GL_n$ and

$$\omega(0) = \frac{N/(2^n - 1)}{N + N/(2^n - 1)} = \frac{1}{2^n}.$$

If Z is a random variable on G, then the mean of Z with respect to ω is called the *full mean* and is denoted by $E[Z]$. The mean of the restriction of Z to GL_n with the uniform distribution is called the *partial mean* and is denoted by $E'[Z]$. The full and partial means are related by

$$E[Z] = \frac{2^n - 1}{2^n} E'[Z] + \frac{1}{2^n} Z(0). \tag{1}$$

Let $f \in B_n$. Then the partial mean of the q-transform of F is

$$E'[W_q(F)] = \frac{1}{2^n - 1}(Z(F) - F(0))(I_q - Q(0)) + F(0)Q(0).$$

Thus the full mean is

$$E[W_q(F)] = \frac{I_q Z(F) + (1 - Q(0))Z(F) - F(0)I_q + 2^n F(0)Q(0)}{2^n}.$$

To describe the second moment of the q-transform for any $a, c \in V_n$, let

$$K_{a,c} = \sum_{B \in GL_n} (-1)^{q(aB) + q(cB)}.$$

Let

$$X = \frac{Q(0)I_q - 1}{2^n - 1} \quad \text{and} \quad Y = \frac{I_q^2 - 2^n - 2Q(0)I_q + 2}{(2^n - 1)(2^n - 2)}.$$

Lemma 1. *We have $K_{a,c} = N$ if $a = c$, $K_{a,c} = NX$ if $a \neq c = 0$ or $c \neq a = 0$, and $K_{a,c} = NY$ otherwise.*

The partial second moment of the q-transform of f is

$$E'[W_q(F)^2] = YZ(F)^2 + 2(X - Y)F(0)(Z(F) - F(0)) + (1 - Y)2^n.$$

The full second moment of the q-transform of f is $E[W_q(F)^2] = \frac{2^n - 1}{2^n}(YZ(F)^2 + 2(X - Y)F(0)(Z(F) - F(0)) + (1 - Y)2^n) + \frac{1}{2^n}Z(F)^2$.

When q is balanced we have $E'[W_q(F)] = -Q(0)Z(F)/(2^n - 1) + (2^n - 1 + Q(0)F(0))/(2^n - 1)$ and $E'[W_q(F)^2] = (2^{2n} - Z(F)^2)/(2^n - 1)$. Thus $E[W_q(F)] = (1 - Q(0))Z(F)/2^n + (2^n - 1 + Q(0))F(0)/2^n$ and $E[W_q(F)^2] = 2^n$.

Definition 2. *If q is balanced, then $f \in B_n$ is q-bent if for all $A \in GL_2$, $|\overline{W}_q(f)| = 2^{n/2}$.*

Theorem 2. *If $f \in B_n$ is balanced, then*

$$E'[\overline{W}_q(f)] = -\frac{1}{2^n - 1}(I_q - Q(0)) + F(0) < 2 \qquad and$$

$$E'[\overline{W}_q(f)^2] = \frac{2^{2n} - I_q^2}{2^n - 1} < 2^n + 2.$$

Definition 3. *A balanced $f \in B_n$ is q-nearly bent if for all $A \in GL_2$,*

$$|\overline{W}_q(f)| \leq \left\lceil \left(\frac{2^{2n} - I_q^2}{2^n - 1} \right)^{1/2} \right\rceil.$$

Theorem 3. *If q and f are balanced, then f is q-nearly bent iff q is f-nearly bent.*

5   Poisson Summation Formulas

For any subset $S \subseteq V_n$, let $S^\perp$ denote the set of vectors that are orthogonal to every vector in S, the *parity checks* for S. Recall that if S is a linear subspace of V_n, then $\dim(S^\perp) = n - \dim(S)$.

In classical theory of Boolean functions, the Poisson summation formula (PSF) says that if $F \in R_n$ and $S \subseteq V_n$ is a linear subspace and $d \in V_n$, then

$$\sum_{a \in S}(-1)^{d \cdot a}W(F)(a) = |S| \sum_{b \in d + S^\perp} F(b),$$

where $W(F)(a)$ is the Walsh-Hadamard coefficient of F at a. The PSF has been used in the analysis of many important properties of Boolean functions. Examples include bounds on the algebraic degree of a Boolean function from divisibility properties of its WHT, characterization of cryptographic properties such as correlation immunity, resilience, and propagation criteria in terms of the WHT, and relations between bent functions and their duals [1].

Here we consider summation formulas for the q-transform $W_q(F)(A)$ where $q \in B_n$ and $A \in GL_n$. Let $S \subset GL_n$ be a non-empty subset. Let $\chi : GL_n \to \mathbb{C}$. Then we define

$$\Gamma_S^\chi(F) = \sum_{A \in S} \chi(A) W_q(F)(A).$$

We want to compute $\Gamma_S^\chi(F)$ for various sets S and various functions χ. In general we have

$$\Gamma_S^\chi(F) = \sum_{A \in S} \chi(A) \sum_{b \in V_n} F(b)(-1)^{q(bA)} = \sum_{b \in V_n} F(b) \sum_{A \in S} \chi(A)(-1)^{q(bA)}.$$

Thus for fixed q we want to compute

$$\sigma_{S,\chi}(b) = \sum_{A \in S} \chi(A)(-1)^{q(bA)}.$$

For any subset $U \subseteq V_n$, let $Z_U(F) = \sum_{b \in U} F(u)$. If $f \in B_n$, let $I_{f,U} = Z_U(\Phi_f)$, the imbalance of f over U. Thus $Z(F) = Z_{V_n}(F)$ and $I_f = I_{f,V_n}$.

5.1   When q is Linear

When q is linear we are essentially in the case of the classical WHT. Suppose $q(b) = t_c(b) = b \cdot c$ for some nonzero $c \in V_n$. Then $q(bA) = bAc^t$. Let $\chi(A) = (-1)^{dAc^t}$ for some $d \in V_n$. Let L be a nontrivial linear subspace of V_n and let $S = \{A \in GL_n : Ac^t \in L^t\}$. Then

$$\chi(A)(-1)^{q(bA)} = (-1)^{dAc^t + bAc^t}.$$

We have $dAc^t + bAc^t \subseteq (d+b)L^t$. First suppose $b \in d+L^\perp$. Then $\chi(A)(-1)^{q(bA)} = 1$ for all $A \in S$, so

$$\sigma_{S,\chi}(b) = \sum_{A \in S} = |S| = (|L| - 1)N/(2^n - 1).$$

Now suppose $b \notin d+L^\perp$. As A varies in S, Ac^t takes each value in $L^* N/(2^n - 1)$ times, so $\sigma_{S,\chi}(b) = -N/(2^n - 1)$.

Theorem 4. *Suppose $q = t_c$ for some nonzero $c \in V_n$, $\chi(A) = (-1)^{dAc^t}$ for some $d \in V_n$, L is a nontrivial linear subspace of V_n, and $S = \{A \in GL_n : Ac^t \in L\}$. Then*

$$\Gamma_S^\chi(F) = \frac{N}{2^n - 1}((|L| - 1)Z_{d+L^\perp}(F) - Z_{V_n \setminus (d+L^\perp)}(F)).$$

Keeping in mind the term $A = 0$, this recovers the classical Poisson summation formula. In effect $W_q(F)(0)$ should again be counted with weight $N/(2^n - 1)$.

5.2 $\chi(A) = (-1)^{q(dA)}$, S Acting 2-uniformly

This time we take $\chi = \chi_d$, $d \in V_n$, defined by $\chi_d(A) = (-1)^{q(dA)}$. Then

$$\sigma_{S,\chi}(b) = \sum_{A \in S} (-1)^{q(dA)+q(bA)}.$$

Definition 4. *Let G be a group with a group action on a set R. Let $S \subseteq G$. Then S acts* uniformly *on R if for all $x, u \in R$ the cardinality of the set of $A \in S$ such that $A(x) = u$ is $|S|/|R|$. Moreover S acts* 2-uniformly *on R if for all $x \neq y, u \neq v \in R$ the cardinality of the set of $A \in S$ such that $A(x) = u$ and $A(y) = v$ is $|S|/(|R|(|R| - 1))$.*

It can be shown that if S acts 2-uniformly on R, then S acts uniformly on R.

We apply this definition with $G = GL_n$, $R = V_n^* = V_n \setminus \{0^n\}$ and $A(x) = xA$ for $A \subset GL_n$ and $x \in V_n^*$. For example, GL_n acts 2-uniformly on V_n^*. We let $S \subseteq GL_n$ act 2-uniformly on V_n^*. Thus for all $x \neq y \in V_n^*$ and $u \neq v \in V_n^*$, $|\{A \in S : xA = u \wedge yA = v\}| = |S|/((2^n - 1)(2^n - 2))$.

Lemma 2. *For any $b, d \subset V_n$ we have*

$$\sigma_{S,\chi}(b) = \begin{cases} |S| & \text{if } b = d \\[2mm] \dfrac{|S|Q(0^n)(I_q - Q(0^n))}{2^n - 1} & \text{if } b \neq d = 0^n \text{ or } d \neq b = 0^n \\[3mm] \dfrac{|S| \left(I_q^2 - 2Q(0^n)I_q - 2^n + 2\right)}{(2^n - 1)(2^n - 2)} & \text{otherwise.} \end{cases}$$

Corollary 1. *If q is balanced, then $\sigma_{S,\chi}(d) = |S|$ and $\sigma_{S,\chi}(b) = -|S|/(2^n - 1)$ if $b \neq d$.*

Theorem 5. *Let $F \in R_n$ and $\chi = \chi_d$. Then*

$$\Gamma_S^\chi(f) = \begin{cases} |S|F(0^n) + \dfrac{|S|}{2^n - 1} Q(0^n)(I_q - Q(0^n))(Z(F) - F(0^n)) \text{ if } d = 0^n \\[3mm] |S|F(d) + \dfrac{|S|}{2^n - 1} Q(0^n)(I_q - Q(0^n))F(0^n) \\[3mm] \quad + \dfrac{|S|}{(2^n - 1)(2^n - 2)} \left(I_q^2 - 2Q(0^n)I_q - 2^n + 2\right) \\[3mm] \quad \cdot (Z(F) - F(0^n) - F(d)) \qquad\qquad\qquad\qquad \text{otherwise.} \end{cases}$$

If q is balanced, then

$$\Gamma_S^\chi(F) = \begin{cases} \dfrac{|S|}{2^n - 1}((2^n - 2)F(0^n) - Z(F)) \text{ if } d = 0^n \\[3mm] \dfrac{|S|}{2^n - 1}(2^n F(d) - Z(F)) \qquad \text{otherwise.} \end{cases}$$

Corollary 2. *If $F \in R_n$, then F is uniquely determined by its q-transform. That is, the q-transform is invertible.*

Proof. Take $S = GL_n$. Recall that $W_q(F)(0) = Z(F)$. Thus the case $d = 0^n$ allows us to solve for $F(0^n)$. Then the second equation allows us to solve for $F(d)$ for any nonzero d.   □

5.3   Basing S on a Subspace

Suppose that q has rank $r \leq n$. By change of basis we may assume that $q(a)$ depends only on $a_1, \cdots, a_r$. We denote the restriction of q to its support by q'. That is, $q' \in B_r$ is defined by $q'(a_1, \cdots, a_r) = q(a_1, \cdots, a_n)$ for any $a_{r+1}, \cdots, a_n \in \mathbb{F}_2$. Thus $I_q = 2^{n-r} I_{q'}$. Let L be a subspace of V_n of dimension $k \geq r$. In this section we let $S = \{A \in GL_n : A_1, \cdots, A_r \in L\}$, where A_i is the ith column of A, and $\chi(A) = (-1)^{q(dA)}$ for some $d \in V_n$. We have $\sigma_{S,\chi}(b) = \sum_{A \in S} (-1)^{q(dA)+q(bA)}$.

Lemma 3. *Let $L \subseteq V_n$ be a subspace. Suppose that $b \in V_n \setminus L^{\perp}$. Then as $A_1, \cdots, A_r$ vary in L, $\psi_b(A_1, \cdots, A_r) = (b \cdot A_1, \cdots, b \cdot A_r) \in V_r$ takes on each value in V_r $|L|^r / 2^r$ times.*

Lemma 4. *Let $L \subseteq V_n$ be a subspace. Suppose that $d, b, d+b \in V_n \setminus L^{\perp}$. Then as $A_1, \cdots, A_r$ vary in L, $\tau_{d,b}(A_1, \cdots, A_r) = (d \cdot A_1, \cdots, d \cdot A_r, b \cdot A_1, \cdots, b \cdot A_r) \in V_{2r}$ takes on each value in V_{2r} $|L|^r / 2^{2r}$ times.*

The hypotheses of Lemma 4 imply that $\dim(L) \geq 2$, so that $|L|^r / 2^{2r} \in \mathbb{Z}$. If $d + b \in L^{\perp}$, then $q(dA) = q(bA)$ for all $A \in S$, so

$$\sigma_{S,\chi}(b) = |S| = \frac{N}{\prod_{i=0}^{r-1}(2^n - 2^i)} \prod_{i=0}^{r-1}(|L| - 2^i).$$

If $d \in L^{\perp}$ and $b \notin L^{\perp}$, then

$$\sigma_{S,\chi}(b) = \sum_{A \in S}(-1)^{q(0^n)+q(bA)} = \frac{NQ(0^n)}{\prod_{i=0}^{r-1}(2^n - 2^i)} \sum_{\substack{A_1, \cdots, A_r \in L \\ \text{linearly independent}}} (-1)^{q'(b \cdot A_1, \cdots, b \cdot A_r)}.$$

A similar equation (with b replaced by d on the right hand side) holds if $d \in L^{\perp}$ and $b \notin L^{\perp}$. If $d, b, d+b \notin L^{\perp}$, then

$$\sigma_{S,\chi}(b) = \sum_{A \in S}(-1)^{q(dA)+q(bA)}$$

$$= \frac{N}{\prod_{i=0}^{r-1}(2^n - 2^i)} \sum_{\substack{A_1, \cdots, A_r \in L \\ \text{linearly independent}}} (-1)^{q'(d \cdot A_1, \cdots, d \cdot A_r)+q'(b \cdot A_1, \cdots, b \cdot A_r)}. \quad (2)$$

We can evaluate these sums by first summing over all $A_1, \cdots, A_r$ (not just the linearly independent ones), then correcting by subtracting the terms that have been over counted. This leads to an inclusion/exclusion formula over subspaces of V_r. We need some notation.

If $v = (v_1, \cdots, v_r) \in V_r$ and $M = (A_1, \cdots, A_r) \in V_n^r$, let $vM = \sum_{i=1}^r v_i A_i \in V_n$. That is, we think of the A_i as rows of a matrix and do matrix multiplication. If T is a linear subspace of V_r, let $U_T = \{M \in L^r : \forall v \in T : vM = 0\}$. Let $\emptyset$ denote the zero dimensional subspace of any vector space. Then $U_\emptyset = L^r$.

For $b \in V_n$ and $M = (A_1, \cdots, A_r) \in V_n^r$, let $b \cdot M = (b \cdot A_1, \cdots, b \cdot A_r)$. That is, think of the A_i as columns of a matrix and do matrix multiplication. Let $b, d \in V_n$. Let

$$W_{T,b} = \sum_{M \in U_T} (-1)^{q'(b \cdot M)} \text{ and } X_{T,d,b} = \sum_{M \in U_T} (-1)^{q'(d \cdot M) + q'(b \cdot M)}.$$

If $b \in L^\perp$, then $W_{T,b} = |U_T| Q(0^n) = |L|^{r - \dim(T)} Q(0^n)$. If $b, d \in L^\perp$, then $X_{T,d,b} = |U_T| = |L|^{r - \dim(T)}$. If $T = \emptyset$, then $U_T = L^r$. By Lemma 3, if $b \notin L^\perp$, then $W_{\emptyset,b} = I_{q'} |L|^r / 2^r$. By Lemma 4, if $b, d, b+d \notin L^\perp$, then $X_{\emptyset,d,b} = I_{q'}^2 |L|^r / 2^{2r}$. Also, $W_{V_r,b} = Q(0^n)$ and $X_{V_r,d,b} = 1$.

For simplicity, let us write $\sigma_{S,\chi}(b) = \dfrac{N}{\prod_{i=0}^{r-1}(2^n - 2^i)} \sigma'_{S,\chi}(b)$.

If $d \in L^\perp$ and $b \notin L^\perp$, then $\sigma'_{S,\chi}(b) = Q(0^n) \sum_{j=0}^{r} (-1)^j c_j \sum_{\dim(T)=j} W_{T,b}.$

If $d \notin L^\perp$ and $b \in L^\perp$, then $\sigma'_{S,\chi}(b) = Q(0^n) \sum_{j=0}^{r} (-1)^j c_j \sum_{\dim(T)=j} W_{T,d}.$

If $d, b, d+b \notin L^\perp$, then $\sigma'_{S,\chi}(b) = \sum_{j=0}^{r} (-1)^j c_j \sum_{\dim(T)=j} X_{T,d,b}.$

The c_i are constants, independent of q, L, d, and b. In the first case we want to sum $(-1)^{q(b \cdot M)}$ over all M whose components are linearly independent. We take $c_0 = 1$, so the $j = 0$ term sums $(-1)^{q(b \cdot M)}$ over all M. Taking $c_1 = 1$, the $j = 1$ term subtracts $(-1)^{q(b \cdot M)}$ for all M satisfying a linear relation. But if M satisfies a 2 dimensional space of linear relations, then we have subtracted its corresponding term three times, hence we have counted it twice too much. Thus $c_2 = 2$. Now if M satisfies a 3 dimensional space of linear relations then we have counted its corresponding term once when $j = 0$, -7 times when $j = 1$ (a 3 dimensional space has 7 one dimensional subspaces), and 14 times when $j = 2$ (a 3 dimensional space has 7 two dimensional subspaces). Thus $c_3 = 8$. Now let

$$\tau(k, \ell) = \frac{(2^\ell - 1)(2^{\ell-1} - 1) \cdots (2^{\ell-k+1} - 1)}{(2^k - 1)(2^{k-1} - 1) \cdots (2 - 1)},$$

the number of k dimensional subspaces of an ℓ dimensional space. By similar reasoning, it follows that for all $0 < \ell \le r$ the c_j satisfy $0 = c_0 - c_1 \tau(1, \ell) + c_2 \tau(2, \ell) - + \cdots + (-1)^\ell c_\ell \tau(\ell, \ell)$. We claim that

$$c_j = 2^{\binom{j}{2}}. \tag{3}$$

To see this, we use the Möbius inversion formula for the lattice of subspaces of a finite dimensional vector space over a finite field [5, p. 301]. We recall this formula in the case when the finite field is $\mathbb{F}_2$. We write $J \leq K$ to indicate that J is a subspace of the vector space K. Suppose that f and g are functions on a vector space V with values in a ring R, satisfying

$$f(K) = \sum_{K \leq J \leq V} g(J) \tag{4}$$

for all subspaces $K \leq V$. Then also

$$g(K) = \sum_{K \leq J \leq V} \mu_V(K, J) f(J), \tag{5}$$

where μ_V is the Möbius μ function for the lattice of subspaces of V, defined by

$$\mu_V(K, J) = (-1)^{\dim(J) - \dim(K)} 2^{\binom{\dim(J) - \dim(K)}{2}}.$$

In our case we can apply this with $\dim(V) = \ell$, $f(K) = 1$ for all $K \leq V_r$, $g(V) = 1$, and $g(K) = 0$ if $K \neq V$. Then Eq. (4) holds, and so Eq. (5) with $K = \emptyset$ then implies Eq. (3). This proves the following theorem.

Theorem 6. If $d + b \in L^\perp$, then $\sigma_{S,\chi}(b) = |S| = \dfrac{N}{\prod_{i=0}^{r-1}(2^n - 2^i)} \displaystyle\prod_{i=0}^{r-1}(|L| - 2^i)$.

If $d \in L^\perp$, $b \notin L^\perp$, then $\sigma_{S,\chi}(b) = \dfrac{NQ(0^n)}{\prod_{i=0}^{r-1}(2^n - 2^i)} \displaystyle\sum_{j=0}^{r}(-1)^j 2^{\binom{j}{2}} \sum_{\dim(T)=j} W_{T,b}$.

If $d \notin L^\perp$, $b \in L^\perp$, then $\sigma_{S,\chi}(b) = \dfrac{NQ(0^n)}{\prod_{i=0}^{r-1}(2^n - 2^i)} \displaystyle\sum_{j=0}^{r}(-1)^j 2^{\binom{j}{2}} \sum_{\dim(T)=j} W_{T,d}$.

If $d, b, d + b \notin L^\perp$, then $\sigma_{S,\chi}(b) = \dfrac{N}{\prod_{i=0}^{r-1}(2^n - 2^i)} \displaystyle\sum_{j=0}^{r}(-1)^j 2^{\binom{j}{2}} \sum_{\dim(T)=j} X_{T,d,b}$.

Corollary 3. Let $F \in R_n$. If $d \in L^\perp$, then

$$\Gamma_S^\chi(F) = \frac{N}{\prod_{i=0}^{r-1}(2^n - 2^i)} \left(Z_{L^\perp}(F) \prod_{i=0}^{r-1}(|L| - 2^i) \right.$$

$$\left. + \sum_{b \in V_n \setminus L^\perp} F(b) Q(0^n) \sum_{j=0}^{r}(-1)^j 2^{\binom{j}{2}} \sum_{\dim(T)=j} W_{T,b} \right).$$

If $d \notin L^\perp$, then

$$\Gamma_S^\chi(F) = \frac{N}{\prod_{i=0}^{r-1}(2^n - 2^i)} \left(Z_{L^\perp}(F) Q(0^n) \sum_{j=0}^{r}(-1)^j 2^{\binom{j}{2}} \sum_{\dim(T)=j} W_{T,d} \right.$$

$$\left. + Z_{d+L^\perp}(F) \prod_{i=0}^{r-1}(|L| - 2^i) + \sum_{b \in V_n \setminus (L^\perp \cup d + L^\perp)} F(b) \sum_{j=0}^{r}(-1)^j 2^{\binom{j}{2}} \sum_{\dim(T)=j} X_{T,d,b} \right).$$

5.4   Rank 2 and 3 Quadratic Forms

First let $q \in B_n$ be quadratic of rank $r = 2$ [4]. By change of basis we may assume $q(a) = q'(a_1, a_2) = a_1 a_2$ or $a_1 a_2 + a_1 + a_2$. Let $\epsilon = 1$ if q in the first case and $\epsilon = -1$ in the second. Then $I_{q'} = 2\epsilon$. It follows that $W_{\emptyset,b} = \epsilon |L|^2/2$ and $W_{V_2,b} = 1$ if $b \notin L^\perp$, and that $X_{\emptyset,d,b} = |L|^2/4$ and $X_{V_r,d,b} = 1$ if $d, b, d+b \notin L^\perp$.

Let $\dim(T) = 1$. Suppose $(a_1, a_2) \in \{(0,1),(1,0)\}$. If $b \notin L^\perp$, then $W_{T,b} = |L|(1+\epsilon)/2$. If also $d, b+d \notin L^\perp$, then $X_{T,d,b} = |L|(1+\epsilon)/2$. Suppose $(a_1, a_2) = (1,1)$. If $b \notin L^\perp$, then $W_{T,b} = 0$. If also $d, b+d \notin L^\perp$, then $X_{T,d,b} = 0$.

The following theorem follows from Corollary 3.

Theorem 7. *Let $q(a) \in B_n$ be quadratic of rank 2. Let $\chi(A) = (-1)^{q(dA)}$. Then*

$$
\Gamma^\chi_S(f) = \begin{cases}
\frac{N}{(2^n-1)(2^n-2)} \left((|L|-1)(|L|-2)Z_{L^\perp}(F) \right. \\
\qquad\qquad \left. + \left(\epsilon \frac{|L|^2}{2} - (1+\epsilon)|L| + 2 \right) Z_{V_n \setminus L^\perp}(F) \right) & \text{if } d \in L^\perp \\[1em]
\frac{N}{(2^n-1)(2^n-2)} \left(\left(\epsilon \frac{|L|^2}{2} - (1+\epsilon)|L| + 2 \right) Z_{L^\perp}(F) \right. \\
\qquad\qquad (|L|-1)(|L|-2)Z_{d+L^\perp}(F) \\
\qquad\qquad \left. \left(\frac{|L|^2}{4} - (1+\epsilon)|L| + 2 \right) Z_{V_n \setminus (L^\perp \cup d+L^\perp)}(F) \right) & \text{if } d \notin L^\perp.
\end{cases}
$$

Let q be a rank 3 quadratic form. We may assume that $q(a) = q'(a_1, a_2, a_3) - a_1 a_2 + a_3$. Now $I_q = I_{q'} = 0$, so $W_{\emptyset,b} = 0$, $W_{V_r,b} = 1$ if $b \notin L^\perp$, $X_{\emptyset,d,b} = 0$, and $X_{V_r,d,b} = 1$ if $d, b, d+b \notin L^\perp$. There are 7 one dimensional subspaces $T \subseteq V_3$, of the form $T = \{0^n, a\}$. There are also 7 two dimensional subspaces $T \subseteq V_3$, the duals of the one dimensional subspaces. The values of $W_{T,b}$ (when $b \notin L^\perp$) and of $X_{T,d,b}$ (when $d, b, d+b \notin L^\perp$) are given in Tables 1 and 2.

Table 1. Values of $W_{T,b}$ and $X_{T,d,b}$ for $T = \{0^n, a\}$, $r = 3$.

a	$W_{T,b}$	$X_{T,d,b}$
$(1,0,0)$, $(0,1,0)$	0	0
$(1,1,0)$	0	0
$(0,0,1)$	$|L|^2/2$	$|L|^2/4$
$(1,0,1)$, $(0,1,1)$	$|L|^2/2$	$|L|^2/4$
$(1,1,1)$	$-|L|^2/2$	$|L|^2/4$

The following theorem follows from Corollary 3.

Theorem 8. *Let $q \in B_n$ be quadratic rank 3. Let $\chi(A) = (-1)^{q(dA)}$. If $d \in V_n$*

$$
\Gamma^\chi_S(f) = \frac{N}{(2^n-1)(2^n-2)(2^n-4)} (|L|-2)(|L|-4)(|L|Z_{d+L^\perp}(F) - Z_{V_n}(F)).
$$

Table 2. Values of $W_{T,b}$ and $X_{T,d,b}$ for $T = \{0^n, a\}^\perp$, $r = 3$.

a	$W_{T,b}$	$X_{T,d,b}$
$(1,0,0), (0,1,0)$	$|L|$	$|L|$
$(1,1,0)$	0	0
$(0,0,1)$	0	0
$(1,0,1), (0,1,1)$	0	0
$(1,1,1)$	$|L|$	$|L|$

Corollary 4. *If $q \in B_n$ is a rank 3 quadratic form, $f \in B_n$ is q-bent, and $L \subseteq V_n$ is a linear subspace of dimension k, then for all $d \in V_n$ we have $2^{n/2-6}|I_{d+L^\perp}(f)$.*

6   The q-Transform of q with q Quadratic

In this section we consider the q-transform of q. When q is linear, this is equivalent to the WHT of a linear function, which has one large peak and is otherwise zero. The q-transform of q seems to be hard to compute in general, but we have the following when s is a rank r quadratic form $q(b)$ on V_n. If $A \in GL_n$, let
$$q^A(b) = q(b) + q(bA)$$

Theorem 9. *Let $q \in B_n$ be quadratic with rank r. Every even integer between 0 and $\min(2r,n)$ occurs as the rank of qA for some $A \in GL_n$. Thus for every even r' with $0 \le r' \le \min(2r,n)$ there is a matrix $A \in GL_n$ with $|W_q(q)(A)| = 2^{n-r'/2}$.*

7   Questions

Perhaps the biggest open problem is how to construct q-bent functions for even the simplest nonlinear $q \in B_n$, such as a rank 3 quadratic function. A computer search showed that if $n = 4$, then there are no q-bent functions for any nonlinear q. We further wonder what the right definition of q-bent is if q is not balanced. Two possibilities are (1) functions f whose distance from S_q is maximal and (2) functions f such that $|\bar{W}_q(f)(A)|$ is constant. We also plan to explore notions of cryptographic security, such as resilience, in the context of q-transforms.

References

1. Carlet, C.: Boolean functions for cryptography and error correcting codes. In: Crama, Y., Hammer, P. (eds.) Boolean Methods and Models, pp. 257–397. Cambridge University Press, Cambridge (2010). http://www.math.univ-paris13.fr/~carlet/pubs.html

2. Courtois, N.T.: Fast algebraic attacks on stream ciphers with linear feedback. In: Boneh, D. (ed.) CRYPTO 2003. LNCS, vol. 2729, pp. 176–194. Springer, Heidelberg (2003)
3. Cusick, T., Stănică, P.: Cryptographic Boolean Functions and Applications. Academic Press, Amsterdam (2009)
4. Goresky, M., Klapper, A.: Algebraic Shift Register Sequences. Cambridge University Press, Cambridge (2012)
5. van Lint, J.H., Wilson, R.M.: A Course in Combinatorics. Cambridge University Press, Cambridge (1992)

Constructing Hyper-Bent Functions from Boolean Functions with the Walsh Spectrum Taking the Same Value Twice

Chunming Tang[1] and Yanfeng Qi[2]($\boxtimes$)

[1] School of Mathematics and Information,
China West Normal University, Nanchong 637002, Sichuan, China
tangchunmingmath@163.com
[2] School of Science, Hangzhou Dianzi University,
Hangzhou 310018, Zhejiang, China
qiyanfeng07@163.com

Abstract. Hyper-bent functions as a subclass of bent functions attract much interest and it is elusive to completely characterize hyper-bent functions. Most of known hyper-bent functions are Boolean functions with Dillon exponents and they are often characterized by special values of Kloosterman sums. In this paper, we present a method for characterizing hyper-bent functions with Dillon exponents. A class of hyper-bent functions with Dillon exponents over $\mathbb{F}_{2^{2m}}$ can be characterized by a Boolean function over $\mathbb{F}_{2^m}$, whose Walsh spectrum takes the same value twice. Further, we show several classes of hyper-bent functions with Dillon exponents characterized by Kloosterman sum identities and the Walsh spectra of some common Boolean functions.

Keywords: Bent function · Hyper-bent function · Dillon exponents · Walsh-Hadamard transform · Kloosterman sums

1 Introduction

Bent functions are maximally nonlinear Boolean functions with even numbers of variables whose Hamming distance to the set of all affine functions equals $2^{n-1} \pm 2^{\frac{n}{2}-1}$. These functions introduced by Rothaus [22] as interesting combinatorial objects have been extensively studied for their applications not only in cryptography, but also in coding theory [3,18] and combinatorial design. A bent function can be considered as a Boolean function defined over $\mathbb{F}_2^n$, $\mathbb{F}_{2^m} \times \mathbb{F}_{2^m}$ ($n = 2m$) or $\mathbb{F}_{2^n}$. Thanks to good structures and properties of the finite field $\mathbb{F}_{2^n}$, bent functions can be well studied. Much research on bent functions on $\mathbb{F}_{2^n}$ can be found in [2,4,5,7–9,12,14,17–20]. Youssef and Gong [26] introduced a class of bent functions called hyper-bent functions, which achieve the maximal

Y. Qi – Part of this work was done when he was a postdoctor in Peking University and Aisino Corporation Inc.

© Springer International Publishing Switzerland 2014
K.-U. Schmidt and A. Winterhof (Eds.): SETA 2014, LNCS 8865, pp. 60–71, 2014.
DOI: 10.1007/978-3-319-12325-7_5

minimum distance to all the coordinate functions of all bijective monomials (i.e., functions of the form $\mathrm{Tr}_1^n(ax^i) + \epsilon$, $\gcd(i, 2^n - 1) = 1$). However, the definition of hyper-bent functions was given by Gong and Golomb [13] by a property of the extended Hadamard transform of Boolean functions. Hyper-bent functions as special bent functions with strong properties are hard to characterize and many related problems are open. Much research give the precise characterization of hyper-bent functions in certain forms, such as hyper-bent functions with Dillon exponents and hyper-bent functions with Niho exponents.

Charpin and Gong [4] studied the hyper-bent functions with multiple trace terms of the form

$$f(x) = \sum_{r \in R} \mathrm{Tr}_1^n(a_r x^{r(2^m - 1)}),$$

where $n = 2m$, R is a set of representations of the cyclotomic cosets modulo $2^m + 1$ of full size n and $a_r \in \mathbb{F}_{2^m}$. The characterization of these hyper-bent functions was presented by the character sums on $\mathbb{F}_{2^m}$. Lisonek [15] presented another characterization of Charpin and Gong's hyper-bent functions in terms of the number of rational points on certain hyperelliptic curves. And they proved that there exists an algorithm for determining such hyper-bent functions with time complexity and space complexity $O(r_{max}^a m^b)$, where r_{max} is the biggest element in R, and a, b are some positive constants irrelevant to r_{max} and m. In particular, when $R = r$ and $(r, 2^m + 1) = 1$, these hyper-bent function are monomial functions via Dillon-like exponents. Dillon [7] proved that $Tr_1^n(ax^{r(2^m - 1)})$ $(a \in \mathbb{F}_{2^m})$ is hyper-bent if and only if $K_m(a) = 0$.

Mesnager [18] generalized Charpin and Gong's hyper-bent functions and presented the characterization of hyper-bent functions of the form

$$f(x) = \sum_{r \in R} \mathrm{Tr}_1^n(a_r x^{r(2^m - 1)}) + Tr_1^2(bx^{\frac{2^n - 1}{3}}),$$

where $b \in \mathbb{F}_4$ and $a_r \in \mathbb{F}_{2^m}$. In the case $\#R = 1$, explicit characterization in [17] by Mesnager is presented. With the similar approach, Wang et al. [25] characterized the hyper-bentness of a class of Boolean functions of the form

$$f(x) = \sum_{r \in R} \mathrm{Tr}_1^n(a_r x^{r(2^m - 1)}) + Tr_1^4(bx^{\frac{2^n - 1}{5}}),$$

where $b \in \mathbb{F}_{16}$ and $a_r \in \mathbb{F}_{2^m}$. In [23,24], explicit characterization for the case $\#R = 1$ is given. When r_{max} is small, Flori and Mesnager [10,11] used the number of rational points on hyper-elliptic curves to determine those classes of Wang et al.'s hyper-bent functions. Mesnager and Flori [21] generalized the above results and characterized the hyper-bentness of Boolean functions of the form

$$f(x) = \sum_{r \in R} \mathrm{Tr}_1^n(a_r x^{r(2^m - 1)}) + Tr_1^t(bx^{s(2^m - 1)}),$$

where $s|(2^m+1)$, $t = o(s(2^m-1))$, i.e., t is the size of the cyclotomic coset of s modulo 2^m+1, $a_r \in \mathbb{F}_{2^m}$, and $b \in \mathbb{F}_{2^t}$.

Li et al. [16] considered a class of Boolean functions of the form

$$f(x) = \sum_{i=0}^{q-1} Tr_1^n(ax^{(ri+s)(q-1)}) + Tr_1^2(bx^{\frac{q^2-1}{3}}),$$

where $n = 2m$, $q = 2^m$, m is odd, $gcd(r, q+1) = 1$, $a \in \mathbb{F}_{q^2}$, and $b \in \mathbb{F}_4$. The hyper-bentness of these functions is characterized by Kloosterman sums.

This paper characterizes hyper-bent functions with Dillon exponents $c(2^m-1)$ with a new method. A hyper-bent function with Dillon exponents over $\mathbb{F}_{2^{2m}}$ can be characterized by two elements in $\mathbb{F}_{2^m}$, which take the same Walsh-Hadamard coefficient of a Boolean function over $\mathbb{F}_{2^m}$. Further, Kloosterman sum identities and the Walsh spectra of some common Boolean functions are used to characterize several classes of hyper-bent functions.

This paper is organized as follows: Sect. 2 introduces some notations, hyper-bent functions, and results of exponential sums. Section 3 presents our main method for characterizing hyper-bent functions over $\mathbb{F}_{2^{2m}}$ from Boolean functions over $\mathbb{F}_{2^m}$. Then we give several classes of hyper-bent functions from some common Boolean functions over $\mathbb{F}_{2^m}$. Kloosterman sum identities and the Walsh spectra of some common Boolean functions are of use in the characterization of these hyper-bent functions. Section 4 makes a conclusion for this paper.

2   Preliminaries

2.1   Boolean Functions and Bent Functions

Let n be a positive integer, $n = 2m$, and $q = 2^m$. Let $\mathbb{F}_{2^n}$ be a finite field with 2^n elements and $\mathbb{F}_{2^n}^*$ the multiplicative group of $\mathbb{F}_{2^n}$. Let α be a primitive element of $\mathbb{F}_{2^n}$. Let U be a subgroup of $\mathbb{F}_{2^n}^*$ generated by $\xi = \alpha^{q-1}$. Then U is a cyclic group of $q+1$ elements.

Let $\mathbb{F}_{2^k}$ be a subfield of $\mathbb{F}_{2^n}$. The trace function from $\mathbb{F}_{2^n}$ to $\mathbb{F}_{2^k}$, denoted by $Tr_k^n(x)$, is a map defined as $Tr_k^n(x) := x + x^{2^k} + x^{2^{2k}} + \cdots + x^{2^{n-k}}$.

A Boolean function f over $\mathbb{F}_{2^n}$ is an $\mathbb{F}_2$-valued function. The "sign" function of f is defined by $\chi(f) := (-1)^f$. The Walsh-Hadamard transform of f is the discrete Fourier transform of χ_f, whose value at $\omega \in \mathbb{F}_{2^n}$ is defined by

$$\widehat{\chi_f}(w) := \sum_{x \in \mathbb{F}_{2^n}} (-1)^{f(x)+Tr_1^n(wx)},$$

where $w \in \mathbb{F}_{2^n}$. Then we can define the bent functions.

Definition 1. *A Boolean function* $f : \mathbb{F}_{2^n} \to \mathbb{F}_2$ *is called a bent function, if* $\widehat{\chi_f}(w) = \pm 2^{\frac{n}{2}}$ $(\forall w \in \mathbb{F}_{2^n})$.

If f is a bent function, n must be even. Further, $\deg(f) \le \frac{n}{2}$ [2]. Hyper-bent functions as an important subclass of bent functions are defined below.

Definition 2. *A bent function $f : \mathbb{F}_{2^n} \to \mathbb{F}_2$ is called a hyper-bent function, if, for any i satisfying $(i, 2^n - 1) = 1$, $f(x^i)$ is also a bent function.*

Many hyper-bent Boolean functions are with Dillon exponents. A Boolean function is with Dillon exponents if the exponents of the trace representation of this function have the form $c(q - 1)$, where c is a positive integer. Such functions satisfies that for any $y \in \mathbb{F}_q^*$ and $x \in \mathbb{F}_{2^n}$, $f(yx) = f(x)$. The characterization of hyper-bent functions with Dillon exponents is given in the following proposition [16,17].

Proposition 1. *Let $f(x)$ be a Boolean function with Dillon exponents defined over $\mathbb{F}_{2^{2m}}$. Then $f(x)$ is hyper-bent if and only if $\Lambda_f = \sum_{u \in U}(-1)^{f(u)} = (-1)^{f(0)}$.*

2.2   Exponential Sums

In this subsection, we introduce some results for special exponential sums.

Definition 3. *The binary Kloosterman sums associated with a on finite field $\mathbb{F}_{2^m}$ are*

$$K_m(a) = \sum_{x \in \mathbb{F}_{2^m}} (-1)^{Tr_1^m(\frac{1}{x}+ax)}, a \in \mathbb{F}_{2^m}.$$

Note that $\frac{1}{0} = 0$ for $x = 0$.

Definition 4. *The cubic sums on $\mathbb{F}_{2^m}$ are*

$$C_m(a, b) = \sum_{x \in \mathbb{F}_{2^m}} (-1)^{Tr_1^m(ax^3+bx)}, a \in \mathbb{F}_{2^m}^*, b \in \mathbb{F}_{2^m}.$$

Carlitz computed the exact values of the cubic sums in the following two propositions [1].

Proposition 2. *Let m be an odd integer. Then*
   (1) $C_m(1, 1) = (-1)^{(m^2-1)/8} 2^{(m+1)/2}$.
   (2) *If $Tr_1^m(c) = 0$, then $C_m(1, c) = 0$.*
   (3) *If $Tr_1^m(c) = 1$ and $c \neq 1$, then $C_m(1, c) = (-1)^{Tr_1^m(\gamma^3+\gamma)}(\frac{2}{m})2^{(m+1)/2}$,*
where $c = \gamma^4 + \gamma + 1, \gamma \in \mathbb{F}_{2^m}$, and $(\frac{2}{m})$ is the Jacobi symbol.

Proposition 3. *Let m be an even integer. Then,*
   (1) $C_m(1, 0) = (-1)^{\frac{m}{2}+1} 2^{\frac{m}{2}+1}$;
   (2) $C_m(1, \lambda) = \begin{cases} (-1)^{Tr_1^m(\gamma^3)}(-1)^{\frac{m}{2}+1} 2^{\frac{m}{2}+1}, & Tr_2^m(\lambda) = 0 \\ 0, & Tr_2^m(\lambda) \neq 0 \end{cases}$, *where γ is a*
solution of $\gamma^4 + \gamma = \lambda^2$.

3   A Class of Hyper-Bent Functions with Dillon Exponents

Let n be a positive integer, $n = 2m$, and $q = 2^m$. In this section, we present our new method for characterizing hyper-bent functions over $\mathbb{F}_{2^n}$ by a Boolean function over $\mathbb{F}_q$, whose Walsh spectrum takes the same value twice.

Note that $\frac{1}{0} = 0$. Let $g(y)$ be a Boolean function defined over $\mathbb{F}_q$. Then we define a Boolean function over $\mathbb{F}_{q^2}$ of the form

$$f(x) = g\left(\frac{1}{\lambda_1 + \lambda_2} \cdot \frac{1}{x^{q-1} + x^{-(q-1)}}\right) + Tr_1^m\left(\frac{\lambda_i}{\lambda_1 + \lambda_2} \cdot \frac{1}{x^{q-1} + x^{-(q-1)}}\right) \quad (1)$$

where $\lambda_i \in \mathbb{F}_q$ ($i = 1$ or 2) and $\lambda_1 \neq \lambda_2$. Note that $x^{q-1} + x^{-(q-1)} \in \mathbb{F}_q$. Then $f(x)$ is well defined. The hyper-bentness of $f(x)$ is characterized by the same Walsh-Hadamard coefficient of $g(y)$ in the following theorem.

Theorem 1. *Let $f(x)$ be defined in (1). Let $g(0) = 0$. Then $f(x)$ is hyper-bent if and only if $\widehat{\chi}_g(\lambda_1) = \widehat{\chi}_g(\lambda_2)$, where $\widehat{\chi}_g(\lambda)$ is the Walsh-Hadamard transform of $g(y)$.*

Proof. Note that $f(x)$ is a function with Dillon exponents $c(q - 1)$. When $y \neq 0$ and $Tr_1^n(y) = 1$, the equation $\frac{1}{u+u^{-1}} = y$ has two solutions. Then $u \mapsto \frac{1}{u+u^{-1}}$ is a 2-to-1 map from $U \setminus \{1\}$ to $\{y \in \mathbb{F}_q : Tr_1^n(y) = 1\}$ [17]. The map $u \mapsto u^{q-1}$ is a permutation of U. Then

$$\Lambda_f = \sum_{u \in U} (-1)^{g\left(\frac{1}{\lambda_1+\lambda_2} \cdot \frac{1}{u+u^{-1}}\right) + Tr_1^m\left(\frac{\lambda_i}{\lambda_1+\lambda_2} \cdot \frac{1}{u+u^{-1}}\right)}$$

$$= (-1)^{g(0)} + 2 \sum_{y \in \mathbb{F}_q, Tr_1^m(y)=1} (-1)^{g\left(\frac{y}{\lambda_1+\lambda_2}\right) + Tr_1^m\left(\frac{\lambda_i}{\lambda_1+\lambda_2}y\right)}.$$

Further, we have

$$\Lambda_f = (-1)^{g(0)} + \sum_{y \in \mathbb{F}_q} (-1)^{g\left(\frac{y}{\lambda_1+\lambda_2}\right) + Tr_1^m\left(\frac{\lambda_i}{\lambda_1+\lambda_2}y\right)}$$

$$- \sum_{y \in \mathbb{F}_q} (-1)^{g\left(\frac{y}{\lambda_1+\lambda_2}\right) + Tr_1^m\left(\frac{\lambda_i}{\lambda_1+\lambda_2}y\right) + Tr_1^m(y)}$$

$$= (-1)^{g(0)} + \sum_{y \in \mathbb{F}_q} (-1)^{g\left(\frac{y}{\lambda_1+\lambda_2}\right) + Tr_1^m\left(\frac{\lambda_i}{\lambda_1+\lambda_2}y\right)} - \sum_{y \in \mathbb{F}_q} (-1)^{g\left(\frac{y}{\lambda_1+\lambda_2}\right) + Tr_1^m\left(\frac{\lambda_{3-i}}{\lambda_1+\lambda_2}y\right)}.$$

Note that $y \mapsto \frac{y}{\lambda_1+\lambda_2}$ is a permutation of $\mathbb{F}_q$ and $g(0) = 0$. Then $\Lambda_f = 1 + \sum_{y \in \mathbb{F}_q} (-1)^{g(y) + Tr_1^m(\lambda_i y)} - \sum_{y \in \mathbb{F}_q} (-1)^{g(y) + Tr_1^m(\lambda_{3-i}y)}$. From Proposition 1, $f(x)$ is hyper-bent if and only if $\sum_{y \in \mathbb{F}_q} (-1)^{g(y) + Tr_1^m(\lambda_i y)} = \sum_{y \in \mathbb{F}_q} (-1)^{g(y) + Tr_1^m(\lambda_{3-i}y)}$, i.e., $\widehat{\chi}_g(\lambda_1) = \widehat{\chi}_g(\lambda_2)$. Hence, this theorem follows.

Theorem 1 offers a new method to present hyper-bent funtions of the form (1). On the Walsh spectra of $g(y)$, there are many exisiting results, which can be used to find two different elements λ_1 and λ_2 satisfying $\widehat{\chi}_g(\lambda_1) = \widehat{\chi}_g(\lambda_2)$. From the proper choice of a Boolean function $g(y)$, λ_1, and λ_2, a lot of hyper-bent functions $f(x)$ can be given.

For further consideration, we give the following lemma.

Lemma 1. *Let* $x \in \mathbb{F}_{q^2}$, $u = x^{q-1}$, $\lambda \in \mathbb{F}_q$, *and* $m \geq t \geq 1$. *Then*

(1) $\frac{1}{u+u^{-1}} = \sum_{i=1}^{2^{m-2}} (u^{2(2i-1)} + u^{-2(2i-1)})$;

(2) $Tr_1^m(\lambda \frac{1}{x^{q-1}+x^{-(q-1)}}) = \sum_{i=1}^{2^{m-2}} Tr_1^n(\lambda^{2^{m-1}} x^{(2i-1)(q-1)})$;

(3) $(\frac{1}{u+u^{-1}})^{2^{t-1}-1} = \sum_{i=1}^{2^{m-t}} (u^{2^{t-1}(2i-1)} + u^{-2^{t-1}(2i-1)})$;

(4) $Tr_1^m(\lambda(\frac{1}{x^{q-1}+x^{-(q-1)}})^{2^{t-1}-1}) = \sum_{i=1}^{2^{m-t}} Tr_1^n(\lambda^{2^{m-t+1}} x^{(2i-1)(q-1)})$;

(5) $(u+u^{-1})^{2^t-1} = \sum_{i=1}^{2^{t-1}} (u^{2i-1} + u^{-(2i-1)})$;

(6) $Tr_1^m(\lambda(x^{q-1}+x^{-(q-1)})^{2^t-1}) = \sum_{i=1}^{2^{t-1}} Tr_1^n(\lambda x^{(2i-1)(q-1)})$;

(7) $(u+u^{-1})^{2^t+1} = u^{2^t-1} + u^{-(2^t-1)} + u^{2^t+1} + u^{-(2^t+1)}$;

(8) $Tr_1^m(\lambda(x^{q-1}+x^{-(q-1)})^{2^t+1}) = Tr_1^n(\lambda(x^{(2^t-1)(q-1)} + x^{(2^t+1)(q-1)}))$.

Proof. This lemma can be easily verified.

In the rest of this section, some common classes of Boolean functions over $\mathbb{F}_q$ are used to characterize hyper-bent functions over $\mathbb{F}_{2^n}$. Kloosterman sum identities and cubic sums are linked with the characterization of hyper-bent functions.

3.1   Hyper-Bent Functions from $g(y) = Tr_1^m(ay^{-d})$

From Theorem 1, we have the following proposition.

Proposition 4. *Let* d *be an odd integer such that* $q - 3 \geq d \geq 1$ *and* $\gcd(d, q-1) = e > 1$. *Let* $a \in \mathbb{F}_q$, $\rho \in \mathbb{F}_q^*$, $\rho^e = 1$, *and* $\rho \neq 1$. *Then, the Boolean function*

$f(x) = \sum_{j=0}^{\frac{d-1}{2}} \binom{d}{j} Tr_1^n(ax^{(d-2j)(q-1)}) + \sum_{j=1}^{2^{m-2}} Tr_1^n(\frac{\rho^i}{1+\rho} x^{(2j-1)(q-1)}) \in \mathbb{F}_2[x]$ *is hyper-bent, where* $i = 0$ *or* $i = 1$.

Proof. Let $g(y) = Tr_1^m(ay^{-d})$. For any $\lambda \in \mathbb{F}_q^*$, we have

$$\widehat{\chi}_g(\lambda) = \sum_{y \in \mathbb{F}_q} (-1)^{Tr_1^m(ay^{-d}+\lambda y)}$$

$$= \sum_{y \in \mathbb{F}_q} (-1)^{Tr_1^m(a(\rho y)^{-d}+\lambda(\rho y))} = \sum_{y \in \mathbb{F}_q} (-1)^{Tr_1^m(ay^{-d}+\lambda\rho y)},$$

i.e., $\widehat{\chi}_g(\lambda) = \widehat{\chi}_g(\lambda\rho)$. From Theorem 1, we have the hyper-bent function

$$f(x) = Tr_1^m(a\lambda^d(1+\rho)^d(x^{q-1} + x^{-(q-1)})^d) + Tr_1^m(\frac{\rho^i}{1+\rho} \frac{1}{x^{q-1}+x^{-(q-1)}}).$$

From Result (2) in Lemma 1, we have

$$f(x) = \sum_{j=0}^{d} Tr_1^m(a\lambda^d(1+\rho)^d \binom{d}{j} x^{(2j-d)(q-1)})$$

$$+ \sum_{j=1}^{2^{m-2}} Tr_1^n((\frac{\rho^i}{1+\rho})^{2^{m-1}} x^{(2j-1)(q-1)}),$$

$$= \sum_{j=0}^{\frac{d-1}{2}} Tr_1^m(a\lambda^d(1+\rho)^d \binom{d}{j} (x^{(2j-d)(q-1)} + x^{(d-2j)(q-1)}))$$

$$+ \sum_{j=1}^{2^{m-2}} Tr_1^n((\frac{\rho^i}{1+\rho})^{2^{m-1}} x^{(2j-1)(q-1)}),$$

$$= \sum_{j=0}^{\frac{d-1}{2}} \binom{d}{j} Tr_1^n(a\lambda^d(1+\rho)^d x^{(d-2j)(q-1)})$$

$$+ \sum_{j=1}^{2^{m-2}} Tr_1^n((\frac{\rho^i}{1+\rho})^{2^{m-1}} x^{(2j-1)(q-1)}).$$

We can replace a by $\frac{a}{\lambda^d(1+\rho)^d}$ and ρ by $\rho^{2^{m-1}}$ and get that $f(x)$ is still hyper-bent. Hence, this proposition holds.

The coefficient $\binom{d}{j}$ mod 2 can be determined by Lucas's theorem. We will give the hyper-bent function $f(x)$ for cases $d = 2^s - 1$ and $d = 2^s + 1$ correspondingly in the following corollary.

Corollary 1. *Let $a \in \mathbb{F}_q$ and s be a positive integer.*
   (1) *Let $gcd(m,s) > 1$, $e = 2^{gcd(m,s)} - 1$, $\rho \in \mathbb{F}_q \setminus \mathbb{F}_2$, $\rho^e = 1$, and $i \in \{0,1\}$. Then the Boolean function $f(x) = \sum_{j=0}^{2^{s-1}} Tr_1^n(ax^{(2j-1)(q-1)}) + \sum_{j=1}^{2^{m-2}} Tr_1^n(\frac{\rho^i}{1+\rho}x^{(2j-1)(q-1)})$ is hyper-bent.*
   (2) *Let $\frac{m}{gcd(m,s)}$ be even, $e = 2^{gcd(m,s)} + 1$, $\rho \in \mathbb{F}_q \setminus \mathbb{F}_2$, $\rho^e = 1$, and $i \in \{0,1\}$. Then the Boolean function $f(x) = Tr_1^n(a(x^{(2^s-1)(q-1)} + x^{(2^s+1)(q-1)})) + \sum_{j=1}^{2^{m-2}} Tr_1^n(\frac{\rho^i}{1+\rho}x^{(2j-1)(q-1)})$ is hyper-bent.*

Proof. Take $d = 2^s - 1$. Then $e = 2^{gcd(m,s)} - 1 = gcd(d, q-1)$. From Proposition 4, we have the hyper-bent function

$$f(x) = \sum_{j=0}^{2^{s-1}-1} \binom{2^s-1}{j} Tr_1^n(ax^{(d-2j)(q-1)}) + \sum_{j=1}^{2^{m-2}} Tr_1^n(\frac{\rho^i}{1+\rho}x^{(2j-1)(q-1)}).$$

From Lucas's Theorem, when $2^{s-1} - 1 \geq j \geq 0$, $\binom{2^s-1}{j} \equiv 1 \mod 2$. We have the hyper-bent function

$$f(x) = \sum_{j=1}^{2^{s-1}} Tr_1^n(ax^{(2j-1)(q-1)}) + \sum_{j=1}^{2^{m-2}} Tr_1^n(\frac{\rho^i}{1+\rho}x^{(2j-1)(q-1)}).$$

Result (1) holds.

Take $d = 2^s + 1$. Since $\frac{m}{gcd(m,s)}$ is even, $e = 2^{gcd(m,s)} + 1 = gcd(d, q-1)$. From Proposition 4, we have the hyper-bent function

$$f(x) = \sum_{j=0}^{2^{s-1}} \binom{2^s+1}{j} Tr_1^n(ax^{(d-2j)(q-1)}) + \sum_{j=1}^{2^{m-2}} Tr_1^n(\frac{\rho^i}{1+\rho}x^{(2j-1)(q-1)}).$$

From Lucas's Theorem, when $2^{s-1} \geq j \geq 0$, $\binom{2^s+1}{j} \equiv 1 \mod 2$ holds only for $j = 0, 1$. Then we have the hyper-bent function

$$f(x) = Tr_1^n(a(x^{(2^s-1)(q-1)} + x^{(2^s+1)(q-1)})) + \sum_{j=1}^{2^{m-2}} Tr_1^n(\frac{\rho^i}{1+\rho}x^{(2j-1)(q-1)}).$$

Result (2) holds.

3.2 Hyper-Bent Functions from $g(y) = Tr_1^m(y)$

Take $g(y) = Tr_1^m(y)$. Note that $\sum_{y \in \mathbb{F}_q} (-1)^{Tr_1^m(\mu y)} = 0$ ($\mu \neq 0$). Thus, for any $\lambda \in \mathbb{F}_q \setminus \mathbb{F}_2$, we have $\widehat{\chi_g}(0) = \widehat{\chi_g}(\lambda) = 0$. From Theorem 1, we have the following hyper-bent function $f(x) = Tr_1^m(\frac{1}{\lambda} \cdot \frac{1}{x^{q-1}+x^{-(q-1)}})$. Further, from Lemma 1, we have the following hyper-bent function

$$f(x) = \sum_{i=1}^{2^{m-2}} Tr_1^n(\frac{1}{\lambda^{2^{m-1}}}x^{(2i-1)(q-1)}).$$

Remark 1. Note that $\{\frac{1}{\lambda^{2^{m-1}}} : \lambda \in \mathbb{F}_q \setminus \mathbb{F}_2\} = \mathbb{F}_q \setminus \mathbb{F}_2$. Then, the Boolean function $f(x) = \sum_{i=1}^{2^{m-2}} Tr_1^n(\lambda x^{(2i-1)(q-1)})$ is hyper-bent if and only if $\lambda \notin \mathbb{F}_2$. This hyper-bent function has been characterized in Corollary 4 in [16].

3.3 Hyper-Bent Functions from $g(y) = Tr_1^m(\frac{1}{y})$

Take $g(y) = Tr_1^m(\frac{1}{y})$, $\lambda_i \in \mathbb{F}_q$ ($i = 1, 2$), and $\lambda_1 \neq \lambda_2$. The Boolean function defined in (1) over $\mathbb{F}_{q^2}$ is

$$f(x) = Tr_1^m((\lambda_1 + \lambda_2)(x^{q-1} + x^{-(q-1)})) + Tr_1^m(\frac{\lambda_i}{\lambda_1 + \lambda_2}\frac{1}{x^{q-1} + x^{-(q-1)}})$$

$$= Tr_1^n((\lambda_1 + \lambda_2)x^{q-1}) + Tr_1^m(\frac{\lambda_i}{\lambda_1 + \lambda_2}\frac{1}{x^{q-1} + x^{-(q-1)}})$$

$$= Tr_1^n((\lambda_1 + \lambda_2)x^{q-1}) + \sum_{j=1}^{2^{m-2}} Tr_1^n((\frac{\lambda_i}{\lambda_1 + \lambda_2})^{2^{m-1}}x^{(2j-1)(q-1)}).$$

Note that $\widehat{\chi}_g(\lambda_i) = K_m(\lambda_i)$ $(i = 1, 2)$. Hence, from Theorem 1, we have the following theorem.

Theorem 2. *Let $\lambda_i \in \mathbb{F}_q$ $(i = 1, 2)$ and $\lambda_1 \neq \lambda_2$. The following conditions are equivalent.*

(1) $f_1(x) = Tr_1^n((\lambda_1 + \lambda_2)x^{q-1}) + \sum_{i=1}^{2^{m-2}} Tr_1^n((\frac{\lambda_1}{\lambda_1+\lambda_2})^{2^{m-1}} x^{(2i-1)(q-1)})$ *is hyper-bent.*

(2) $f_1(x) = Tr_1^n((\lambda_1 + \lambda_2)x^{q-1}) + \sum_{i=1}^{2^{m-2}} Tr_1^n((\frac{\lambda_2}{\lambda_1+\lambda_2})^{2^{m-1}} x^{(2i-1)(q-1)})$ *is hyper-bent.*

(3) $K_m(\lambda_1) = K_m(\lambda_2)$.

Usually, special values of Kloosterman sums are used to characterize hyper-bent functions. From Theorem 2, we can characterize hyper-bent functions from two distinct elements, which have the same evaluation of Kloosterman sums. Known results on Kloosterman sum identities are of use. From known Kloosterman sum identities, several hyper-bent functions can be given immediately.

Corollary 2. *Let $b \in \mathbb{F}_q$ and $\epsilon \in \mathbb{F}_2$. The following Boolean functions $Tr_1^n((b^2 + b)x^{q-1}) + \sum_{i=1}^{2^{m-2}} Tr_1^n((b+\epsilon)x^{(2i-1)(q-1)})$, $(b \notin \mathbb{F}_2)$, $Tr_1^n((b^2 + b)x^{q-1}) + \sum_{i=1}^{2^{m-2}} Tr_1^n((b^2 + \epsilon)x^{(2i-1)(q-1)})$, $(b \notin \mathbb{F}_2)$, and $Tr_1^n((b^4 + b)x^{q-1}) + \sum_{i=1}^{2^{m-2}} Tr_1^n((b^4 + \epsilon)x^{(2i-1)(q-1)})$, $(b \notin \mathbb{F}_4)$ are all hyper-bent.*

Proof. From [6], when $b \in \mathbb{F}_q \setminus \mathbb{F}_2$, we have the following Kloosterman sum identities: $K_m(b^3(1 + b)) = K_m((1 + b)^3 b)$, $K_m(b^5(1 + b)) = K_m((1 + b)^5 b)$, and $K_m(b^8(b^4 + b)) = K_m((1 + b)^8(b^4 + b))$. Consider the following three cases:

(1) $\lambda_1 = b^3(1 + b)$ and $\lambda_2 = (1 + b)^3 b$, where $b \in \mathbb{F}_q \setminus \mathbb{F}_2$. Then $\lambda_1 \neq \lambda_2$;

(2) $\lambda_1 = b^5(1 + b)$ and $\lambda_2 = (1 + b)^5 b$, where $b \in \mathbb{F}_q \setminus \mathbb{F}_2$. Then $\lambda_1 \neq \lambda_2$;

(3) $\lambda_1 = b^8(b^4 + b)$ and $\lambda_2 = (1+b)^8(b^4+b)$, where $b \in \mathbb{F}_q \setminus \mathbb{F}_4$. Then $\lambda_1 \neq \lambda_2$;

From Theorem 2, this corollary can be obtained immediately.

3.4 Hyper-Bent Functions from $g(y) = Tr_1^m(y^{2^{t-1}-1})$

Take $g(y) = Tr_1^m(y^{2^{t-1}-1})$, $t \geq 1$, $\lambda_i \in \mathbb{F}_q$ $(i = 1, 2)$, and $\lambda_1 \neq \lambda_2$. From Result (2) and Result (4) in Lemma 1, the Boolean function defined in (1) over $\mathbb{F}_{q^2}$ is

$$f(x) = \sum_{j=1}^{2^{m-t}} Tr_1^n((\lambda_1 + \lambda_2)^{2^{m-t+1}-1} x^{(2j-1)(q-1)})$$

$$+ \sum_{j=1}^{2^{m-2}} Tr_1^n((\frac{\lambda_i}{\lambda_1 + \lambda_2})^{2^{m-1}} x^{(2j-1)(q-1)}). \tag{2}$$

From Theorem 1, we have the following theorem.

Theorem 3. *Let $f(x)$ be defined in (2). Then $f(x)$ is hyper-bent if and only if $\sum_{y \in \mathbb{F}_q}(-1)^{Tr_1^m(y^{2^{t-1}-1}+\lambda_1 y)} = \sum_{y \in \mathbb{F}_q}(-1)^{Tr_1^m(y^{2^{t-1}-1}+\lambda_2 y)}$.*

If $gcd(t - 1, m) = 1$, then $gcd(2^{t-1} - 1, 2^m - 1) = 1$ and $y \mapsto y^{2^{t-1}-1}$ is a permutation of $\mathbb{F}_q$, and $\sum_{y \in \mathbb{F}_q}(-1)^{Tr_1^m(y^{2^{t-1}-1})} = 0$. Hence, we have the following corollary.

Corollary 3. *Let $gcd(t - 1, m) = 1$, $\lambda \in \mathbb{F}_q^*$, and $\epsilon \in \mathbb{F}_2$. The Boolean function*

$$f(x) = \sum_{j=1}^{2^{m-t}} Tr_1^n(\lambda^{2^{m-t+1}-1}x^{(2j-1)(q-1)}) + \epsilon \sum_{j=1}^{2^{m-2}} Tr_1^n(x^{(2j-1)(q-1)})$$

is hyper-bent if and only if $\sum_{y \in \mathbb{F}_q}(-1)^{Tr_1^m(y^{2^{t-1}-1}+\lambda y)} = 0$.

This corollary generalizes Theorem 6 in [16]. It is easy to verify that when $t = 1, 2$, the hyper-bent function defined in (2) is just the hyper-bent function in Remark 1. In the following subsection, we discuss the case $t = 3$. When $t = 3$, $\widehat{\chi_g}(\lambda)$ is just the cubic sum $C_m(1, \lambda)$.

When m is odd, from Proposition 2, we have $\widehat{\chi_g}(\lambda) \in \{0, \pm(\frac{2}{m})2^{(m+1)/2}\}$. Define $H_{1,0} = \{\lambda \in \mathbb{F}_q : \widehat{\chi_g}(\lambda) = 0\}$, $H_{1,1} = \{\lambda \in \mathbb{F}_q : \widehat{\chi_g}(\lambda) = (\frac{2}{m})2^{(m+1)/2}\}$, and $H_{1,-1} = \{\lambda \in \mathbb{F}_q : \widehat{\chi_g}(\lambda) = -(\frac{2}{m})2^{(m+1)/2}\}$. Further, from Proposition 2, we have $H_{1,0} = \{\lambda \in \mathbb{F}_q : Tr_1^m(\lambda) = 0\}$, $H_{1,1} = \{\gamma^4+\gamma+1 : Tr_1^m(\gamma^3+\gamma) = 0\} \cup \{1\}$, and $H_{1,-1} = \{\gamma^4 + \gamma + 1 : Tr_1^m(\gamma^3 + \gamma) = 1\}$.

From Theorem 1, we have the following corollary.

Corollary 4. *Let m be odd, $\lambda_i \in \mathbb{F}_q (i = 1, 2)$, and $\lambda_1 \neq \lambda_2$. Then, the Boolean function*

$$f(x) = \sum_{j=1}^{2^{m-3}} Tr_1^n((\lambda_1 + \lambda_2)^{2^{m-2}-1}x^{(2j-1)(q-1)}) + \sum_{j=1}^{2^{m-2}} Tr_1^n((\frac{\lambda_i}{\lambda_1 + \lambda_2})^{2^{m-1}}x^{(2j-1)(q-1)})$$

is hyper-bent if and only if there exists $j \in \{0, 1, -1\}$ such that $\lambda_1, \lambda_2 \in H_{1,j}$.

Remark 2. Note that the cardinality of $\{\widehat{\chi_g}(\lambda)|\lambda \in \mathbb{F}_q\}$ is 3. If we suppose $q = 2^m > 3$ and take four elements in $\mathbb{F}_q$, then there exists two elements $\lambda_1, \lambda_2 \in \mathbb{F}_q$ lying in some $H_{1,j}$. Hence we can get a corresponding hyper-bent function.

Note that $0 \in H_{1,0}$. Then we have the following corollary.

Corollary 5. *Let m be odd, $\lambda \in \mathbb{F}_q^*$, and $\epsilon \in \mathbb{F}_2$. The Boolean function $f(x) = \sum_{j=1}^{2^{m-3}} Tr_1^n(\lambda^{2^{m-2}-1}x^{(2j-1)(q-1)}) + \epsilon \sum_{j=1}^{2^{m-2}} Tr_1^n(x^{(2j-1)(q-1)})$ is hyper-bent if and only if $Tr_1^m(\lambda) = 0, \lambda \neq 0$.*

These corollaries generalize Result (3) in Corollary 6 in [16].

When m is even, from Proposition 3, $\widehat{\chi_g}(\lambda) \in \{0, \pm(-1)^{\frac{m}{2}+1}2^{\frac{m}{2}+1}\}$. Define $H_{0,0} = \{\lambda \in \mathbb{F}_q : \widehat{\chi_g}(\lambda) = 0\}$, $H_{0,1} = \{\lambda \in \mathbb{F}_q : \widehat{\chi_g}(\lambda) = (-1)^{\frac{m}{2}+1}2^{\frac{m}{2}+1}\}$, and $H_{0,-1} = \{\lambda \in \mathbb{F}_q : \widehat{\chi_g}(\lambda) = -(-1)^{\frac{m}{2}+1}2^{\frac{m}{2}+1}\}$. From Proposition 3, we have $H_{0,0} = \{\lambda \in \mathbb{F}_q : Tr_2^m(\lambda) \neq 0\}$, $H_{0,1} = \{(\gamma^4 + \gamma)^{2^{m-1}} : \gamma \in \mathbb{F}_q, Tr_1^m(\gamma^3) = 0\}$, and $H_{0,-1} = \{(\gamma^4 + \gamma)^{2^{m-1}} : \gamma \in \mathbb{F}_q, Tr_1^m(\gamma^3) = 1\}$. Obviously, $0 \in H_{0,1}$.

Lemma 2. $1 \in H_{0,1}$ *if and only if $8|m$.*

Proof. From the definition of $H_{0,1}$, we have $1 \in H_{0,1}$ if and only if there exists $\gamma \in \mathbb{F}_q$ satisfying $\gamma^4 + \gamma + 1 = 0$ and $Tr_1^m(\gamma^3) = 0$. It is easy to verify that $\gamma^4 + \gamma + 1 = 0$ is irreducible over $\mathbb{F}_2$. Thus, $4 \mid m$. Further, $Tr_1^m(\gamma^3) = Tr_1^4(Tr_4^m(\gamma^3)) = \frac{m}{4} = 0$. Hence, this theorem follows.

From Theorem 1, we have the following corollary.

Corollary 6. *Let m be even, $\lambda_i \in \mathbb{F}_q (i = 1, 2)$, and $\lambda_1 \neq \lambda_2$. The Boolean function*

$$f(x) = \sum_{j=1}^{2^{m-3}} Tr_1^n((\lambda_1 + \lambda_2)^{2^{m-2}-1} x^{(2j-1)(q-1)})$$

$$+ \sum_{j=1}^{2^{m-2}} Tr_1^n((\frac{\lambda_i}{\lambda_1 + \lambda_2})^{2^{m-1}} x^{(2j-1)(q-1)})$$

is hyper-bent if and only if there exists $j \in \{0, 1, -1\}$ satisfying $\lambda_1, \lambda_2 \in H_{0,j}$.

When $8 \mid m$, from Lemma 2, we have $0, 1 \in H_{0,1}$. Hence, we have the following hyper-bent functions: $f_0(x) = \sum_{j=1}^{2^{m-3}} Tr_1^n(x^{(2j-1)(q-1)})$ and $f_1(x) = \sum_{j=2^{m-3}+1}^{2^{m-2}} Tr_1^n(x^{(2j-1)(q-1)})$.

4 Conclusion

In this paper, we characterize hyper-bent functions from Boolean functions with the Walsh spectrum taking the same value twice. From our method, many results on exponential sums can be used in the characterization of hyper-bent functions. We use some Kloosterman sum identities and the Walsh spectra of some common Boolean functions to characterize several classes of hyper-bent functions.

Acknowledgements. This work was supported by the Natural Science Foundation of China (Grant No.10990011, 11401480 & No. 61272499).

References

1. Carlitz, L.: Explicit evaluation of certain exponential sums. Math. Scand. **44**, 5–16 (1979)
2. Carlet, C.: Boolean functions for cryptography and error correcting codes. In: Crama, Y., Hammer, P.L. (eds.) Chapter of the Monography Boolean Models and Method in Mathematics, Computer Science, and Engineering, pp. 257–397. Cambridge University Press, Cambridge (2010)
3. Carlet, C., Gaborit, P.: Hyperbent functions and cyclic codes. J Combin. Theory Ser. A **113**(3), 466–482 (2006)
4. Charpin, P., Gong, G.: Hyperbent functions, Kloosterman sums and Dickson polynomials. IEEE Trans. Inf. Theory **9**(54), 4230–4238 (2008)

5. Charpin, P., Kyureghyan, G.: Cubic monomial bent functions: a subclass of $\mathcal{M}$. SIAM J. Discr. Math. **22**(2), 650–665 (2008)
6. Cao, X., Hollmann, H.D.L., Xiang, Q.: New Kloosterman sum identities and equalities over finite fields. Finite Fields Appl. **14**, 823–833 (2008)
7. Dillon, J.: Elementary Hadamard difference sets. Ph.D., University of Maryland (1974)
8. Dobbertin, H., Leander, G.: A survey of some recent results on bent functions. In: Helleseth, T., Sarwate, D., Song, H.-Y., Yang, K. (eds.) SETA 2004. LNCS, vol. 3486, pp. 1–29. Springer, Heidelberg (2005)
9. Dobbertin, H., Leander, G., Canteaut, A., Carlet, C., Felke, P., Gaborit, P.: Construction of bent functions via Niho power functions. J. Combin. Theory Ser. A **113**, 779–798 (2006)
10. Flori, J.P., Mesnager, S.: An efficient characterization of a family of hyper-bent functions with multiple trace terms. J. Math. Crypt. **7**(1), 43–68 (2013)
11. Flori, J.-P., Mesnager, S.: Dickson polynomials, hyperelliptic curves and hyper-bent functions. In: Helleseth, T., Jedwab, J. (eds.) SETA 2012. LNCS, vol. 7280, pp. 40–52. Springer, Heidelberg (2012)
12. Gold, R.: Maximal recursive sequences with 3-valued recursive crosscorrelation functions. IEEE Trans. Inf. Theory **14**(1), 154–156 (1968)
13. Gong, G., Golomb, S.W.: Transform domain analysis of DES. IEEE Trans. Inf. Theory **45**(6), 2065–2073 (1999)
14. Leander, G.: Monomial bent functions. IEEE Trans. Inf. Theory **2**(52), 738–743 (2006)
15. Lisonek, P.: An efficient characterization of a family of hyperbent functions. IEEE Trans. Inf. Theory **57**(9), 6010–6014 (2011)
16. Li, N., Helleseth, T., Tang, X., Kholosha, A.: Several new classes of bent functions from Dillon exponents. IEEE Trans. Inf. Theory **59**(3), 1818–1831 (2013)
17. Mesnager, S.: A new class of bent and hyper-bent Boolean functions in polynomial forms. Des. Codes Crypt. **59**(1–3), 265–279 (2011)
18. Mesnager, S.: Bent and hyper-bent functions in polynomial form and their link with some exponential sums and Dickson polynomials. IEEE Trans. Inf. Theory **57**(9), 5996–6009 (2011)
19. Mesnager, S.: Hyper-bent Boolean functions with multiple trace terms. In: Hasan, M.A., Helleseth, T. (eds.) WAIFI 2010. LNCS, vol. 6087, pp. 97–113. Springer, Heidelberg (2010)
20. Mesnager, S.: A new family of hyper-bent Boolean functions in polynomial form. In: Parker, M.G. (ed.) Cryptography and Coding 2009. LNCS, vol. 5921, pp. 402–417. Springer, Heidelberg (2009)
21. Mesnager, S., Flori, J.P.: Hyper-bent functions via Dillon-like exponents. IEEE Trans. Inf. Theory. **59**(5), 3215–3232 (2013)
22. Rothaus, O.S.: On bent functions. J. Combin. Theory Ser. A **20**, 300–305 (1976)
23. Wang, B., Tang, C., Qi, Y., Yang, Y.: A generalization of the class of hyper-bent Boolean functions in binomial forms. Cryptology ePrint Archive, Report 2011/698 (2011). http://eprint.iacr.org/
24. Wang, B., Tang, C., Qi, Y., Yang, Y., Xu, M.: A new class of hyper-bent Boolean functions in binomial forms. CoRR, abs/1112.0062 (2011)
25. Wang, B., Tang, C., Qi, Y., Yang, Y., Xu, M.: A new class of hyper-bent Boolean functions with multiple trace terms. Cryptology ePrint Archive, Report 2011/600 (2011). http://eprint.iacr.org/
26. Youssef, A.M., Gong, G.: Hyper-bent functions. In: Pfitzmann, B. (ed.) EUROCRYPT 2001. LNCS, vol. 2045, pp. 406–419. Springer, Heidelberg (2001)

Characterizations of Plateaued and Bent Functions in Characteristic p

Sihem Mesnager[1,2][✉]

[1] Department of Mathematics, University of Paris VIII, Saint-Denis, France
[2] LAGA, UMR 7539, CNRS, University of Paris XIII, Villetaneuse, France
smesnager@univ-paris8.fr

Abstract. We characterize bent functions and plateaued functions in terms of moments of their Walsh transforms. We introduce in any characteristic the notion of directional difference and establish a link between the fourth moment and that notion. We show that this link allows to identify bent elements of particular families. Notably, we characterize bent functions of algebraic degree 3.

1 Introduction

Binary bent functions are usually called Boolean bent functions. These functions were first introduced by Rothaus in [12]. Bent functions are closely related to other combinatorial and algebraic objects such as Hadamard difference sets, relative difference sets, planar functions and commutative semi-fields. Later, this notion has been generalized to that of p-ary bent functions [11]. Several studies on p-ary bent functions have been performed (a non exhaustive list is [5,7–10,13]). Most of them concern constructions of bent functions or studies of their properties. Another important family of binary functions is that of plateaued functions [3]. Like the notion of bent function, the notion of plateaued function can be generalized to p-ary plateaued functions (see [4] for instance). In this paper, we establish characterizations of bent functions and plateaued functions in terms of sums of powers of the Walsh transform (Theorems 1 and 3). We also introduce the notion of directional difference for p-ary functions, generalizing the directional derivative of Boolean functions (Definition 1). We then show that one can establish identities linking sums of fourth-powers of the Walsh transform and directional derivatives of a p-ary function (Proposition 1). We then deduce from our characterizations of all bent p-ary functions of algebraic degree 3 when p is odd (Theorem 4). We finally establish a link between the bentness of all elements of a family of p-ary functions and counting zeros of their directional differences (Theorem 6 and Corollary 2).

2 Notation and Preliminaries

Let p be a prime integer, $n \geq 1$ be an integer. We will denote $\mathbb{F}_{p^n}$ the finite field of size p^n and $\mathbb{F}_{p^n}^*$ the set of nonzero elements of $\mathbb{F}_{p^n}$. Let ξ_p be a primitive

© Springer International Publishing Switzerland 2014
K.-U. Schmidt and A. Winterhof (Eds.): SETA 2014, LNCS 8865, pp. 72–82, 2014.
DOI: 10.1007/978-3-319-12325-7_6

pth-root of unity and set $\chi_p(a) = \xi_p^a$. Let f be a function from $\mathbb{F}_{p^n}$ to $\mathbb{F}_p$. The Walsh transform of f at $w \in \mathbb{F}_{p^n}$ is defined as

$$\widehat{\chi_f}(w) = \sum_{x \in \mathbb{F}_{p^n}} \chi_p\Big(f(x) - Tr_p^{p^n}(wx)\Big).$$

Then f is bent if and only if $\big|Waf(w)\big|^2 = p^n$ for every $w \in \mathbb{F}_{p^n}$. It is said to be *regular bent* if there exists $f^* : \mathbb{F}_{p^n} \to \mathbb{F}_p$ such that $\widehat{\chi_f}(w) = \chi_p(f^*(w))p^{\frac{n}{2}}$ for all $w \in \mathbb{F}_{p^n}$. The function f^* is called the *dual function* of f (in characteristic 2, all bent functions are regular bent; when p is odd, regular bent functions can exist only if $p \equiv 1 \mod 4$). A function $f : \mathbb{F}_{p^n} \to \mathbb{F}_p$ is said to be *weakly regular bent* if, for all $w \in \mathbb{F}_{p^n}$, we have $\widehat{\chi_f}(w) = \epsilon\chi_p(f^*(w))p^{\frac{n}{2}}$ for some complex number with $|\epsilon| = 1$ (in fact ϵ can only be ± 1 or $\pm i$). For every function f from $\mathbb{F}_{p^n}$ to $\mathbb{F}_p$, we have

$$\sum_{w \in \mathbb{F}_{p^n}} \widehat{\chi_f}(w) = p^n\chi_p(f(0)). \tag{1}$$

Set $|z|^2 = z\bar{z}$ where $\bar{z}$ stands for the conjugate of z. Then

$$\sum_{w \in \mathbb{F}_{p^n}} \big|\widehat{\chi_f}(w)\big|^2 = p^{2n}. \tag{2}$$

In the sequel, we shall refer to (2) as the *Parseval identity*. If $\big|\widehat{\chi_f}(w)\big| \in \{0, p^{\frac{n+s}{2}}\}$ for some nonnegative integer s then f is said to be s-plateaued. With this definition, bent functions are 0-plateaued functions (in the case where $s = 0$, $|\widehat{\chi_f}(w)| \in \{0, p^{\frac{n}{2}}\}$ is equivalent to $|\widehat{\chi_f}(w)| = p^{\frac{n}{2}}$). The Parseval identity allows to compute the multiplicity of each value of the Walsh transform (when $p = 2$, a more precise statement has been shown in [2]).

Lemma 1. *Let $f : \mathbb{F}_{p^n} \to \mathbb{F}_p$ be s-plateaued. Then the absolute value of the Walsh transform $\widehat{\chi_f}$ takes p^{n-s} times the value $p^{\frac{n+s}{2}}$ and $p^n - p^{n-s}$ times the value 0.*

Proof. If N denotes the number of $w \in \mathbb{F}_{p^n}$ such that $|\widehat{\chi_f}(w)| = p^{\frac{n+s}{2}}$, then $\sum_{w \in \mathbb{F}_{p^n}} |\widehat{\chi_f}(w)|^2 = p^{n+s}N$. Now, according to Eq. (2), one must have that $p^{n+s}N = p^{2n}$, that is, $N = p^{n-s}$. The result follows.

A map F from $\mathbb{F}_{p^n}$ to $\mathbb{F}_{p^n}$ is said to be planar if and only if the function from $\mathbb{F}_{p^n}$ to $\mathbb{F}_{p^n}$ induced by the polynomial $F(X + a) - F(x) - F(a)$ is bijective for every $a \in \mathbb{F}_{p^n}^*$. We finally introduce the directional difference.

Definition 1. *Let $f : \mathbb{F}_{p^n} \to \mathbb{F}_p$. The directional difference of f at $a \in \mathbb{F}_{p^n}$ is the map $D_a f$ from $\mathbb{F}_{p^n}$ to $\mathbb{F}_p$ defined by*

$$\forall x \in \mathbb{F}_{p^n}, \quad D_a f(x) = f(x + a) - f(x).$$

3   New Characterizations of Plateaued Functions

Let p be a positive prime integer. For any nonnegative integer k, we set

$$S_k(f) = \sum_{w \in \mathbb{F}_{p^n}} \left| \widehat{\chi_f}(w) \right|^{2k} \text{ and } T_k(f) = \frac{S_{k+1}(f)}{S_k(f)}$$

with the convention regarding $k = 0$ that $S_0(f) = p^n$ (in this case, $T_0(f) = \frac{S_1(f)}{S_0(f)} = p^n$). Let us make a preliminary but important remark : for every integer A and every positive integer k, it holds

$$\sum_{w \in \mathbb{F}_{p^n}} \left(\left| \widehat{\chi_f}(w) \right|^2 - A \right)^2 \left| \widehat{\chi_f}(w) \right|^{2(k-1)}$$

$$= S_{k+1}(f) - 2A S_k(f) + A^2 S_{k-1}(f). \tag{3}$$

We are now going to deduce from (3) a characterization of plateaued functions in terms of moments of the Walsh transform (in Sect. 4, we shall specialize our characterization to bent functions, see Theorem 3).

Theorem 1. *Let n and k be two positive integers. Let f be a function from $\mathbb{F}_{p^n}$ to $\mathbb{F}_p$. Then, the two following assertions are equivalent.*

1. *f is plateaued, that is, there exists a nonnegative integer s such that f is s-plateaued.*
2. *$T_{k+1}(f) = T_k(f)$.*

Proof.

1. Suppose that f is s-plateaued for some nonnegative integer s, that is, $\left| \widehat{\chi_f}(w) \right| \in \{0, p^{\frac{n+s}{2}}\}$. Then, by Lemma 1,

$$S_k(f) = \sum_{w \in \mathbb{F}_{p^n}} \left| \widehat{\chi_f}(w) \right|^{2k} = p^{n-s} \times p^{k(n+s)} = p^{(k+1)n+(k-1)s}$$

$$S_{k+1}(f) = p^{n-s} \times p^{(k+1)(n+s)} = p^{(k+2)n+ks}$$

$$S_{k+2}(f) = p^{n-s} \times p^{(k+2)(n+s)} = p^{(k+3)n+(k+1)s}.$$

Therefore

$$T_k(f) = \frac{p^{(k+2)n+ks}}{p^{(k+1)n+(k-1)s}} = p^{n+s}$$

and

$$T_{k+1}(f) = \frac{p^{(k+3)n+(k+1)s}}{p^{(k+2)n+ks}} = p^{n+s} = T_k(f).$$

2. Suppose $T_{k+1}(f) = T_k(f)$. According to (3)

$$\sum_{w \in \mathbb{F}_{p^n}} \left(\left| \widehat{\chi_f}(w) \right|^2 - T_k(f) \right)^2 \left| \widehat{\chi_f}(w) \right|^{2k}$$

$$= S_{k+2}(f) - 2T_k(f) S_{k+1}(f) + T_k^2(f) S_k(f)$$

$$= S_{k+1}(f) \left(T_{k+1}(f) - 2T_k(f) + T_k(f) \right) = 0$$

proving that $|\widehat{\chi_f}(w)| \in \{0, \sqrt{T_k(f)}\}$ for every $w \in \mathbb{F}_{p^n}$. Thus,

$$\sum_{w \in \mathbb{F}_{p^n}} |\widehat{\chi_f}(w)|^2 = T_k(f) \#\{w \in \mathbb{F}_{p^n} \mid |\widehat{\chi_f}(w)| = \sqrt{T_k(f)}\}.$$

Now, the Parseval identity (2) states that

$$\sum_{w \in \mathbb{F}_{p^n}} |\widehat{\chi_f}(w)|^2 = p^{2n}.$$

Therefore $T_k(f)$ divides p^{2n} proving that $T_k(f) = p^\rho$ for some positive integer ρ. Now, one has $\#\{w \in \mathbb{F}_{p^n} \mid |\widehat{\chi_f}(w)| = \sqrt{T_k(f)}\} = p^{2n-\rho} \leq p^n$ which implies that $\rho \geq n$, that is, $\rho = n + s$ for some nonnegative integer s.

Remark 1. Specializing Theorem 1 to the case where $k = 1$, we get that f is plateaued if and only if $T_2(f) = T_1(f)$, that is

$$S_3(f)S_1(f) - S_2^2(f) = p^{2n}S_3(f) - S_2^2(f) = 0.$$

Remark 2. In the proof, we have shown more than the sole equivalence between (1) and (2). Indeed, we have shown that if (2) holds then f is s-plateaued and $|\widehat{\chi_f}(w)| \in \{0, \sqrt{T_k(f)}\}$.

In Theorem 1, we have considered the ratio of two consecutive sums $S_k(f)$. In fact, one can get a more general result than Theorem 1. Indeed, for every positive integer k and every nonnegative integer l, we have

$$\sum_{w \in \mathbb{F}_{p^n}} \left(|\widehat{\chi_f}(w)|^{2l} - A\right)^2 |\widehat{\chi_f}(w)|^{2(k-1)} \tag{4}$$

$$= S_{k+2l-1}(f) - 2AS_{k+l-1}(f) + A^2 S_{k-1}(f).$$

Then, one can make the same kind of proof as that of Theorem 1 but with (4) in place of (3) (the proof being very similar, we omit it).

Theorem 2. *Let n, k and l be positive integers and $f : \mathbb{F}_{p^n} \to \mathbb{F}_p$. Then, the two following assertions are equivalent*

1. *f is plateaued, that is, there exists a nonnegative integer s such that f is s-plateaued.*
2. *$\frac{S_{k+2l}(f)}{S_{k+l}(f)} = \frac{S_{k+l}(f)}{S_k(f)}$.*

4   The Case of Bent Functions

In this section, we shall specialize our study to bent functions and suppose that p is a positive prime integer. In the whole section, n is a positive integer. In Theorem 1, we have excluded the possibility to for the integer k to be equal to 0 because it does concern both plateaued functions and bent functions. In fact, if we aim to characterize only bent functions, we are going to show that it follows from comparing $T_1(f) = \frac{S_2(f)}{S_1(f)} = \frac{S_2(f)}{p^{2n}}$ to $T_0(f) = \frac{S_1(f)}{S_0(f)} = p^n$.

Theorem 3. *Let n be a positive integer. Let f be a function from $\mathbb{F}_{p^n}$ to $\mathbb{F}_p$. Then*

$$S_2(f) = \sum_{w \in \mathbb{F}_{p^n}} \left| \widehat{\chi_f}(w) \right|^4 \geq p^{3n}$$

and f is bent if and only if $S_2(f) = p^{3n}$.

Proof. If we apply (3) with $A = p^n$ at $k = 1$, we get that

$$\sum_{w \in \mathbb{F}_{p^n}} \left(\left| \widehat{\chi_f}(w) \right|^2 - p^n \right)^2 = S_2(f) - 2p^n S_1(f) + p^{2n} S_0(f).$$

Now, $S_0(f) = p^n$ and $S_1(f) = p^{2n}$ (Parseval identity, Eq. 2). Hence

$$\sum_{w \in \mathbb{F}_{p^n}} \left(\left| \widehat{\chi_f}(w) \right|^2 - p^n \right)^2 = S_2(f) - p^{3n}. \tag{5}$$

Since $\left(\left| \widehat{\chi_f}(w) \right|^2 - p^n \right)^2 \geq 0$ for every $w \in \mathbb{F}_{p^n}$, it implies that $S_2(f) \geq p^{3n}$. Now, f is bent if and only if $\left| \widehat{\chi_f}(w) \right|^2 = p^n$ for every $w \in \mathbb{F}_{p^n}$. Therefore, f is bent if and only if the left-hand side of Eq. (5) vanishes, that is, if and only if $S_2(f) = p^{3n}$.

In characteristic 2, identities have been established involving the Walsh transform of a Boolean function and its directional derivatives (see [1,3]). For instance, for every Boolean function f, $S_2(f)$ and the second-order derivatives of f have been linked. We now show that one can link $S_2(f)$ and the directional difference defined in Definition 1.

Proposition 1. *Let n be a positive integer. Let f be a function from $\mathbb{F}_{p^n}$ to $\mathbb{F}_p$. Then*

$$\sum_{w \in \mathbb{F}_{p^n}} \left| \widehat{\chi_f}(w) \right|^4 = p^n \sum_{(a,b,x) \in \mathbb{F}_{p^n}^3} \chi_p(D_a D_b f(x)). \tag{6}$$

Proof. Since $|z|^4 = z^2 \bar{z}^2$ where $\bar{z}$ stands for the conjugate of z and $\overline{\xi_p} = \xi_p^{-1}$, we have

$$\sum_{w \in \mathbb{F}_{p^n}} \left| \widehat{\chi_f}(w) \right|^4$$

$$= \sum_{w \in \mathbb{F}_{p^n}} \sum_{(x_1,x_2,x_3,x_4) \in \mathbb{F}_{p^n}^4} \chi_p\big(f(x_1) - f(x_2) + f(x_3) - f(x_4)$$

$$- Tr_p^{p^n}\left(w(x_1 - x_2 + x_3 - x_4)\right)\big).$$

Now,

$$\sum_{w \in \mathbb{F}_{p^n}} \chi_p\big(-Tr_p^{p^n}\left(w(x_1 - x_2 + x_3 - x_4)\right)\big) = \begin{cases} p^n & \text{if } x_1 - x_2 + x_3 - x_4 = 0 \\ 0 & \text{otherwise.} \end{cases}$$

Hence,

$$\sum_{w \in \mathbb{F}_{p^n}} \left| \widehat{\chi_f}(w) \right|^4 = p^n \sum_{(x_1, x_2, x_3) \in \mathbb{F}_{p^n}^3} \chi_p\big(f(x_1) - f(x_2) + f(x_3) - f(x_1 - x_2 + x_3)\big).$$

Now note that

$$D_{x_2 - x_1} D_{x_3 - x_2} f(x_1) = f(x_1) + f(x_3) - f(x_2) - f(x_1 + x_3 - x_2).$$

Then, since $(x_1, x_2, x_3) \mapsto (x_1, x_2 - x_1, x_3 - x_2)$ is a permutation of $\mathbb{F}_{p^n}^3$, we get

$$\sum_{w \in \mathbb{F}_{p^n}} \left| \widehat{\chi_f}(w) \right|^4 = p^n \sum_{(a,b,x) \in \mathbb{F}_{p^n}^3} \chi_p\big(D_a D_b f(x)\big).$$

Remark 3. In odd characteristic p, when f is a quadratic form over $\mathbb{F}_{p^n}$, that is, $f(x) = \phi(x, x)$ for some symmetric bilinear map ϕ from $\mathbb{F}_{p^n} \times \mathbb{F}_{p^n}$ to $\mathbb{F}_{p^n}$, then, $f(x+y) = f(x) + f(y) + 2\phi(x, y)$. Let us now compute the directional differences of f at $(a, b) \in \mathbb{F}_{p^n}$:

$$D_b f(x) = f(x + b) - f(x) = f(b) + 2\phi(b, x)$$
$$D_a D_b f(x) = 2\phi(b, x + a) - 2\phi(b, x) = 2\phi(b, a).$$

According to Proposition 1, one has

$$S_2(f) = p^n \sum_{(a,b,x) \in \mathbb{F}_{p^n}^3} \chi_p(2\phi(b, a))$$

$$= p^{2n} \sum_{b \in \mathbb{F}_{p^n}} \sum_{a \in \mathbb{F}_{p^n}} \chi_p(2\phi(b, a)).$$

Now, classical results about character sums over finite abelian groups say that

$$\sum_{a \in \mathbb{F}_{p^n}} \chi_p(2\phi(b, a)) = \begin{cases} p^n & \text{if } \phi(b, \bullet) = 0 \\ 0 & \text{otherwise.} \end{cases}$$

Hence,

$$S_2(f) = p^{3n} \#\mathfrak{rad}(\phi)$$

where $\mathfrak{rad}(\phi)$ stands for the radical of ϕ : $\mathfrak{rad}(\phi) = \{b \in \mathbb{F}_{p^n} \mid \phi(b, \bullet) = 0\}$. One can then conclude thanks to Theorem 3 that f is bent if and only if $\mathfrak{rad}(\phi) = \{0\}$.

Suppose that p is odd and consider now functions of the form

$$f(x) = Tr_p^{p^n} \left(\sum_{\substack{i,j,k=0 \\ i \neq j, j \neq k, k \neq i}}^{n-1} a_{ijk} x^{p^i + p^j + p^k} + \sum_{\substack{i,j=0 \\ i \neq j}}^{n-1} b_{ij} x^{p^i + p^j} \right). \tag{7}$$

We are going to characterize bent functions of that form thanks to Theorem 3 and Proposition 1. But before, let us note that we can rewrite the expression of f as follows

$$f(x) = Tr_p^{p^n}\left(\sum_{\substack{i,j,k=0\\i\neq j,j\neq k,k\neq i}}^{n-1} a_{ijk}x^{p^i+p^j+p^k}\right) + Tr_p^{p^n}\left(\sum_{\substack{i,j=0\\i\neq j}}^{n-1} b_{ij}x^{p^i+p^j}\right)$$

$$= Tr_p^{p^n}\left(\sum_{\substack{i,j,k=0\\i\neq j,j\neq k,k\neq i}}^{n-1} a_{ijk}^{p^{-i}}x^{1+p^{j-i}+p^{k-i}}\right) + Tr_p^{p^n}\left(\sum_{\substack{i,j=0\\i\neq j}}^{n-1} b_{ij}x^{p^i+p^j}\right)$$

$$= Tr_p^{p^n}\left(x\sum_{\substack{i,j,k=0\\i\neq j,j\neq k,k\neq i}}^{n-1} a_{ijk}^{p^{-i}}x^{p^{j-i}+p^{k-i}}\right) + Tr_p^{p^n}\left(\sum_{\substack{i,j=0\\i\neq j}}^{n-1} b_{ij}x^{p^i+p^j}\right).$$

In the second equality, we have used the fact that $Tr_p^{p^n}$ is invariant under the Frobenius map $x \mapsto x^p$. Set

$$\psi(x,y) = \frac{1}{2}\sum_{\substack{i,j,k=0\\i\neq j,j\neq k,k\neq i}}^{n-1} a_{ijk}^{p^{-i}}\left(x^{p^{j-i}}y^{p^{k-i}} + x^{p^{k-i}}y^{p^{j-i}}\right)$$

$$\phi(x,y) = \frac{1}{2}Tr_p^{p^n}\left(\sum_{\substack{i,j=0\\i\neq j}}^{n-1} b_{ij}(x^{p^i}y^{p^j} + x^{p^j}y^{p^i})\right),$$

Therefore, a function f of the form (7) can be written

$$f(x) = Tr_p^{p^n}\left(x\psi(x,x)\right) + \phi(x,x) \tag{8}$$

where $\psi : \mathbb{F}_{p^n} \to \mathbb{F}_{p^n}$ is a symmetric bilinear map and $\phi : \mathbb{F}_{p^n} \to \mathbb{F}_{p^n}$ is a symmetric bilinear form. We can now state our characterization.

Theorem 4. *Suppose that p is odd. Let ϕ be a symmetric bilinear form over $\mathbb{F}_{p^n} \times \mathbb{F}_{p^n}$ and ψ be a symmetric bilinear map from $\mathbb{F}_{p^n} \times \mathbb{F}_{p^n}$ to $\mathbb{F}_{p^n}$. Define $f : \mathbb{F}_{p^n} \to \mathbb{F}_p$ by $f(x) = Tr_p^{p^n}(x\psi(x,x)) + \phi(x,x)$ for $x \in \mathbb{F}_{p^n}$. For $(a,b) \in \mathbb{F}_{p^n}$, set $\ell_{a,b}(x) = Tr_p^{p^n}(\psi(a,b)x + a\psi(b,x) + b\psi(a,x))$. For every $a \in \mathbb{F}_{p^n}$, define the vector space $\mathfrak{K}_a = \{b \in \mathbb{F}_{p^n} \mid \ell_{a,b} = 0\}$. Then f is bent if and only if $\{a \in \mathbb{F}_{p^n}, \phi(a, \bullet)|_{\mathfrak{K}_a} = 0\} = \{0\}$.*

Proof. According to Theorem 3 and Proposition 1, f is bent if and only if

$$\sum_{(a,b,x)\in\mathbb{F}_{p^n}^3} \chi_p(D_bD_af(x)) = p^{2n}. \tag{9}$$

Now, for $(a, b) \in \mathbb{F}_{p^n}^2$,

$$
\begin{aligned}
D_a f(x) &= Tr_p^{p^n} \left((x + a)\psi(a + x, a + x) - x\psi(x, x) \right) \\
&\quad + \phi(x + a, x + a) - \phi(x, x) \\
&= Tr_p^{p^n} \left(a\psi(x, x) + 2x\psi(a, x) + 2a\psi(a, x) + x\psi(a, a) + a\psi(a, a) \right) \\
&\quad + 2\phi(a, x) + \phi(a, a). \\
D_b D_a f(x) &= Tr_p^{p^n} \left(2a\psi(b, x) + a\psi(b, b) + 2b\psi(a, x) + 2x\psi(a, b) + 2b\psi(a, b) \right. \\
&\quad \left. + 2a\psi(a, b) + b\psi(a, a) \right) + 2\phi(a, b)) \\
&= 2\ell_{a,b}(x) + Tr_p^{p^n} \left(a\psi(b, b) + b\psi(a, a) + 2(a + b)\psi(a, b) \right) + 2\phi(a, b).
\end{aligned}
$$

Note that, $\ell_{a,b}$ is a linear map from $\mathbb{F}_{p^n}$ to $\mathbb{F}_{p^n}$. Furthermore, for any $a \in \mathbb{F}_{p^n}$ and $b \in \mathfrak{K}_a$, one has

$$
\ell_{a,b}(a) = Tr_p^{p^n} \left(\psi(a, b)a + a\psi(b, a) + b\psi(a, a) \right) = 0,
$$
$$
\ell_{a,b}(b) = Tr_p^{p^n} \left(\psi(a, b)b + a\psi(b, b) + b\psi(a, b) \right) = 0
$$

which implies, summing those two equations, that

$$
Tr_p^{p^n} \left(a\psi(b, b) + b\psi(a, a) + 2(a + b)\psi(a, b) \right) = 0.
$$

Hence,

$$
\begin{aligned}
\sum_{(a,b,x) \in \mathbb{F}_{p^n}^3} \chi_p(D_b D_a f(x)) &= \sum_{(a,b) \in \mathbb{F}_{p^n}^3} \chi_p(2\phi(a, b)) \sum_{x \in \mathbb{F}_{p^n}} \chi_p(2\ell_{a,b}(x)) \\
&= p^n \sum_{a \in \mathbb{F}_{p^n}} \sum_{b \in \mathfrak{K}_a} \chi_p(2\phi(a, b)).
\end{aligned}
$$

Now, for every $a \in \mathbb{F}_{p^n}$, the map $b \in \mathfrak{K}_a \mapsto \phi(a, b)$ is linear over $\mathfrak{K}_a$. Therefore

$$
\sum_{b \in \mathfrak{K}_a} \chi_p(2\phi(a, b)) = \begin{cases} \#\mathfrak{K}_a & \text{if } \phi(a, \bullet)\big|_{\mathfrak{K}_a} = 0 \\ 0 & \text{otherwise} \end{cases}
$$

Hence, according to (9), f is bent if and only if

$$
\sum_{(a,b,x) \in \mathbb{F}_{p^n}^3} \chi_p(D_a D_b f(x)) = p^n \sum_{a \in \mathbb{F}_{p^n}, \, \phi(a,\bullet)\big|_{\mathfrak{K}_a} = 0} \#\mathfrak{K}_a = p^{2n},
$$

that is, if and only if,

$$
\sum_{a \in \mathbb{F}_{p^n}, \, \phi(a,\bullet)\big|_{\mathfrak{K}_a} = 0} \#\mathfrak{K}_a = p^n.
$$

Now, if $a = 0$, then $\mathfrak{K}_0 = \mathbb{F}_{p^n}$ because $\ell_{0,b} = 0$ for every $b \in \mathbb{F}_{p^n}$. Therefore, f is bent if and only if

$$
\sum_{a \in \mathbb{F}_{p^n}^*, \, \phi(a,\bullet)\big|_{\mathfrak{K}_a} = 0} \#\mathfrak{K}_a = 0
$$

which is equivalent to $\#\mathfrak{K}_a = 0$ for every $a \in \mathbb{F}_{p^n}^*$ such that $\phi(a, \bullet)\big|_{\mathfrak{K}_a} = 0$.

We now turn our attention towards maps from $\mathbb{F}_{p^n}$ to $\mathbb{F}_{p^m}$. Let us extend the notion of bentness to those maps as follows.

Definition 2. *Let F be a Boolean map from $\mathbb{F}_{p^n}$ to $\mathbb{F}_{p^m}$. For every $\lambda \in \mathbb{F}_{p^m}^\star$, define $f_\lambda : \mathbb{F}_{p^n} \to \mathbb{F}_p$ as : $f_\lambda(x) = Tr_p^{p^m}(\lambda F(x))$ for every $x \in \mathbb{F}_{p^n}$. Then F is said to be bent if and only if f_λ is bent for every $\lambda \in \mathbb{F}_{p^n}^\star$.*

Theorem 3 implies

Theorem 5. *Let F be a map from $\mathbb{F}_{p^n}$ to $\mathbb{F}_{p^m}$. Then, F is bent if and only if*

$$\sum_{\lambda \in \mathbb{F}_{p^m}^\star} S_2(f_\lambda) = p^{3n}(p^m - 1). \tag{10}$$

Proof. According to Theorem 3, for every $\lambda \in \mathbb{F}_{p^m}^\star$, f_λ is bent if and only if $S_2(f_\lambda) = p^{3n}$ which gives (10). Conversely, suppose that (10) holds. Theorem 3 states that $S_2(f_\lambda) \geq p^{3n}$ for every $\lambda \in \mathbb{F}_{p^m}^\star$. Thus, one has necessarily, for every $\lambda \in \mathbb{F}_{p^n}^\star$, $S_2(f_\lambda) = p^{3n}$ implying that f_λ is bent for every $\lambda \in \mathbb{F}_{p^n}$, proving that F is bent.

We now show that one can compute the left-hand side of (10) by counting the zeros of the second-order directional differences.

Proposition 2. *Let F be a Boolean map from $\mathbb{F}_{p^n}$ to $\mathbb{F}_{p^m}$. Then*

$$\sum_{\lambda \in \mathbb{F}_{p^m}^\star} S_2(f_\lambda) = p^{n+m}\mathfrak{N}(F) - p^{4n}$$

where $\mathfrak{N}(F)$ is the number of elements of $\{(a, b, x) \in \mathbb{F}_{p^n}^3 \mid D_a D_b F(x) = 0\}$.

Proof. According to Proposition 1, we have

$$\sum_{\lambda \in \mathbb{F}_{p^m}^\star} S_2(f_\lambda) = p^n \sum_{\lambda \in \mathbb{F}_{p^m}^\star} \sum_{a,b,x \in \mathbb{F}_{p^n}} \chi_p\big(D_a D_b f_\lambda(x)\big).$$

Next, $D_a D_b f_\lambda = Tr_p^{p^m}(\lambda D_a D_b F)$. Therefore

$$\sum_{\lambda \in \mathbb{F}_{p^m}^\star} S_2(f_\lambda) = p^n \sum_{a,b,x \in \mathbb{F}_{p^n}} \sum_{\lambda \in \mathbb{F}_{p^m}^\star} \chi_p\big(Tr_p^{p^m}(\lambda D_a D_b F(x))\big).$$

That is

$$\sum_{\lambda \in \mathbb{F}_{p^m}^\star} S_2(f_\lambda) = p^n \sum_{a,b,x \in \mathbb{F}_{p^n}} \Big(\sum_{\lambda \in \mathbb{F}_{p^m}} \chi_p\big(Tr_p^{p^m}(\lambda D_a D_b F(x))\big)\Big) - p^{4n}.$$

We finally get the result from

$$\sum_{\lambda \in \mathbb{F}_{p^m}} \chi_p\big(Tr_p^{p^m}(\lambda D_a D_b F(x))\big) = \begin{cases} 0 & \text{if } D_a D_b F(x) \neq 0 \\ p^m & \text{if } D_a D_b F(x) = 0 \end{cases}$$

We then deduce from Theorem 3 a characterization of bentness in terms of zeros of the second-order directional differences.

Theorem 6. *Let F be a map from $\mathbb{F}_{p^n}$ to $\mathbb{F}_{p^m}$. Then F is bent if and only if* $\mathfrak{N}(F) = p^{3n-m} + p^{2n} - p^{2n-m}$.

Proof. F is bent if and only if all the functions f_λ, $\lambda \in \mathbb{F}_{p^m}^*$, are bent. Therefore, according to Proposition 3, if F is bent then

$$\sum_{\lambda \in \mathbb{F}_{p^m}^*} S_2(f_\lambda) = (p^m - 1)p^{3n}.$$

Now, according to Proposition 2, one has

$$\sum_{\lambda \in \mathbb{F}_{p^m}^*} S_2(f_\lambda) = p^{n+m}\mathfrak{N}(F) - p^{4n}.$$

We deduce from the two above equalities that

$$\mathfrak{N}(F) = p^{-n-m}(p^{4n} + (p^m - 1)p^{3n})$$
$$= p^{3n-m} + p^{2n} - p^{2n-m}.$$

Conversely, suppose that $\mathfrak{N}(F) = p^{3n-m} + p^{2n} - p^{2n-m}$. Then

$$\sum_{\lambda \subset \mathbb{F}_{p^m}^*} S_2(f_\lambda) = p^{n+m}\mathfrak{N}(F) - p^{4n} = p^{4n} + p^{3n+m} - p^{3n} - p^{4n} = p^{3n}(p^m - 1).$$

We then conclude by Theorem 5 that F is bent.

Note that when $a = 0$ or $b = 0$, $D_a D_b F$ is trivially equal to 0. We state below a slightly different version of Theorem 6 to exclude those trivial cases to characterize the bentness of F.

Corollary 1. *Let F be a map from $\mathbb{F}_{p^n}$ to $\mathbb{F}_{p^m}$. Then F is bent if and only if $\mathfrak{N}^*(F) = p^n(p^n - 1)(p^{n-m} - 1)$ where $\mathfrak{N}^*(F)$ is the number of elements of $\{(a, b, x) \in \mathbb{F}_{p^n}^* \times \mathbb{F}_{p^n}^* \times \mathbb{F}_{p^n} \mid D_a D_b F(x) = 0\}$.*

Proof. It follows from Proposition 2 by noting that $\{(a, b, x) \in \mathbb{F}_{p^n}^3 \mid D_a D_b F(x) = 0\}$ contains the set $\{(a, 0, x), a, x \in \mathbb{F}_{p^n}, \} \cup \{(0, a, x), a, x \in \mathbb{F}_{p^n}\}$ whose cardinality equals $p^n(1 + 2(p^n - 1)) = 2p^{2n} - p^n$. Hence, the cardinality of $\mathfrak{N}^*(F)$ equals $p^{3n-m} + p^{2n} - p^{2n-m} - (2p^{2n} - p^n) = p^{3n-m} - p^{2n-m} + p^n - p^{2n} = p^{2n-m}(p^n - 1) + p^n(1 - p^n) = p^n(p^n - 1)(p^{n-m} - 1)$.

In the particular case of planar functions, Theorem 1 rewrites as follows

Corollary 2. *Let $F : \mathbb{F}_{p^n} \to \mathbb{F}_{p^n}$. Then, F is planar if and only if, $D_a D_b F$ does not vanish on $\mathbb{F}_{p^n}$ for every $(a, b) \in \mathbb{F}_{p^n}^* \times \mathbb{F}_{p^n}^*$.*

Proof. F is planar if and only if F is bent ([6, Lemma 1.1]). Hence, according to Corollary 1, F is planar if and only if $\mathfrak{N}^*(F) = 0$ proving the result.

References

1. Canteaut, A., Carlet, C., Charpin, P., Fontaine, C.: Propagation characteristics and correlation-immunity of highly nonlinear boolean functions. In: Preneel, B. (ed.) EUROCRYPT 2000. LNCS, vol. 1807, pp. 507–522. Springer, Heidelberg (2000)
2. Canteaut, A., Charpin, P.: Decomposing bent functions. IEEE Trans. Inf. Theory **49**(8), 2004–2019 (2003)
3. Carlet, C.: Boolean functions for cryptography and error correcting codes. In: Crama, Y., Hammer, P.L. (eds.) Boolean Models and Methods in Mathematics, Computer Science, and Engineering, pp. 257–397. Cambridge University Press, Cambridge (2010)
4. Cesmelioglu, A., Meidl, W.: A construction of bent functions from plateaued functions. Des. Codes Crypt. **66**(1–3), 231–242 (2013)
5. Coulter, R.S., Matthews, R.W.: Planar functions and planes of Lenz-Barlotti class II. Des. Codes Crypt. **10**(2), 167–184 (1997)
6. Helleseth, T., Hollmann, H., Kholosha, A., Wang, Z., Xiang, Q.: Proofs of two conjectures on ternary weakly regular bent functions. IEEE Trans. Inf. Theory **55**(11), 5272–5283 (2009)
7. Helleseth, T., Kholosha, A.: Monomial and quadratic bent functions over the finite fields of odd characteristic. IEEE Trans. Inf. Theory **52**(5), 2018–2032 (2006)
8. Helleseth, T., Kholosha, A.: On the dual of monomial quadratic p-ary bent functions. In: Golomb, S.W., Gong, G., Helleseth, T., Song, H.-Y. (eds.) SSC 2007. LNCS, vol. 4893, pp. 50–61. Springer, Heidelberg (2007)
9. Hou, X.-D.: p-ary and q-ary versions of certain results about bent functions and resilient functions. Finite Fields Appl. **10**(4), 566–582 (2004)
10. Hou, X.-D.: On the dual of a Coulter-Matthews bent function. Finite Fields Appl. **14**(2), 505–514 (2008)
11. Kumar, P.V., Scholtz, R.A., Welch, L.R.: Generalized bent functions and their properties. J. Comb. Theory, Ser. A **40**(1), 90–107 (1985)
12. Rothaus, O.S.: On "bent" functions. J. Comb. Theory, Ser. A **20**(3), 300–305 (1976)
13. Tan, Y., Yang, J., Zhang, X.: A recursive construction of p-ary bent functions which are not weakly regular. In: IEEE International Conference on Information Theory and Information Security (ICITIS), pp. 156–159 (2010)

Perfect Sequences

A Method of Optimisation of the Exhaustive Computer Search for Perfect Sequences

Oleg Kuznetsov[✉]

School of Mathematical Sciences, Monash University,
Wellington Road, Clayton, Victoria 3800, Australia
oleg.kuznetsov@monash.edu

Abstract. We have obtained an equality which expresses the absolute value of the real part of autocorrelation values of a sequence over the quaternions as a fraction of the sums of the norms of its particular elements. Based on this result, we obtained a condition necessary for perfection of a sequence over the quaternions. This condition becomes necessary and sufficient for perfection when applied to a symmetric sequence. Our result also allows increasing efficiency of the exhaustive search for perfect sequences. During exhaustive search experiments, we have attained up to 6 times reduction in computer time required for completion of the exhaustive search for perfect sequences, in comparison with the traditional method involving direct calculation of autocorrelation values. While we focused our study mainly on sequences over the quaternions, all results are equally applicable for sequences over the complex numbers.

Keywords: Quaternion · Perfect sequence · Left perfect · Right perfect

1 Introduction

Perfect sequences have many applications in communication systems. Traditionally, perfect sequences are considered over commutative alphabets, usually complex roots of unity. Since complex numbers are special cases of the quaternions, perfect sequences over the quaternions can be considered as a natural generalization of perfect sequences over the complex numbers. Studying the structure and properties of quaternion perfect sequences may provide for advances in understanding perfect sequences over the complex numbers.

Some papers in electronic communications stated the importance of signal design over the quaternions for polarization based systems. Isaeva and Sarytchev [3] suggested that the polarization state of an electromagnetic wave can be modelled by means of quaternions, whereby two complex signals $z_1 = x_0 + x_1 i$ and $z_2 = y_0 + y_1 i$ are represented by the single quaternion number $q = x_0 + x_1 i + y_0 j + y_1 k$. Wysocki et al. [11] described a signal transmitted between two dual-polarized antennas using a quaternion notation. The channel can then be modelled by a single quaternion gain, instead of a matrix of four complex gains as in the classic approach.

© Springer International Publishing Switzerland 2014
K.-U. Schmidt and A. Winterhof (Eds.): SETA 2014, LNCS 8865, pp. 85–96, 2014.
DOI: 10.1007/978-3-319-12325-7_7

We continue the study of perfect sequences over the quaternions initiated by Kuznetsov [5]. The perfect sequence results presented in this paper have the potential to be used as synchronization preambles and channel sounding sequences in polarization based communication systems.

Unlike complex numbers, the quaternion algebra is non-commutative. The non-commutative nature of the quaternions calls for the definition of two different autocorrelation functions: right and left autocorrelations, which, in general, have non-equal values. Kuznetsov [5] defined the right and the left periodic autocorrelation functions and gives corresponding definitions of right and left perfect sequences. It has been proved that left and right perfection are equivalent over the quaternions, that is, every right perfect sequence is also left perfect, and vice versa.

Acevedo and Hall [1] found that there exist arbitrary long perfect sequences over the quaternions. They have shown that any Lee sequence, which are proved to exist for unbounded lengths [6–8], can be converted into a sequence over the basic quaternions, that is, the elements of the quaternion group Q_8 formed by unit quaternions $\{\pm 1, \pm i, \pm j, \pm k\}$ [10]. Since elements of the group Q_8 are, in fact, the quaternionic 4-roots of unity, this result clearly indicates that the Mows conjecture [9] stating the upper limit for the length of a perfect sequence over complex roots of unity does not hold for sequences over the quaternionic roots of unity.

At the present time, there exist no universal algorithms for constructing perfect sequences over the quaternions, except the conversion of Lee sequences into sequences over the quaternions. However, this method can only deliver perfect sequences of a very special form: they contain only one true quaternion, which is not a complex number, with all other elements being complex 4-roots of unity $\{\pm 1, \pm i\}$. For finding perfect sequences of a general form, the exhaustive computer search, when every possible sequence of a given length over a certain alphabet is taken and checked for perfection, is the only available method. Since each arithmetical operation with quaternions involves many operations with real numbers, the exhaustive computer search is a very inefficient method of finding new sequences over quaternion alphabets. With the most advanced modern computers, even connected in grids, only sequences of lengths in the order of tens can be accessed.

In the present paper, we introduce a new formula which allows accelerating the exhaustive computer search for perfect sequences by 3–6 times in comparison with the traditional method involving direct calculation of autocorrelation values. Our new optimized method of the exhaustive search is a result of application of the new equality, introduced in the present paper, which relates the absolute value of the real part of autocorrelation functions of a sequence with the quotient of the sums of the norms of its individual elements.

It is worth noting that, since complex numbers are quaternions of a special form, all our results are equally valid for sequences over the complex numbers.

2   Notations and Definitions

In this paper, quaternions and sequences of quaternions are denoted by **bold fonts** and ordinary fonts are reserved for real and complex numbers. All summations in indices are assumed *modulo n*.

Definition 1. *An ordered n-tuple* $\boldsymbol{x} = [\boldsymbol{x}_0, \boldsymbol{x}_1, \ldots, \boldsymbol{x}_{n-1}]$ *of elements from a set A is called a sequence. The set A is called an alphabet. The number n is called the length of the sequence* $\boldsymbol{x}$.

Definition 2. *The sequence* $\boldsymbol{x} = [\boldsymbol{x}_0, \boldsymbol{x}_1, \ldots, \boldsymbol{x}_{n-1}]$ *is called symmetric if there exists an integer s so that* $\boldsymbol{x}_{s+m} = \boldsymbol{x}_{s-m}$ *for every integer m.*

The non-commutative quaternion algebra $\mathbb{H}$ over the real field $\mathbb{R}$ is generated by two elements $\boldsymbol{i}$ and $\boldsymbol{j}$, which satisfy $\boldsymbol{i}^2 = \boldsymbol{j}^2 = -1$ and $\boldsymbol{ij} + \boldsymbol{ji} = 0$. This algebra has a conjugation, commonly denoted by the star '$*$', $\boldsymbol{i}^* = -\boldsymbol{i}$, $\boldsymbol{j}^* = -\boldsymbol{j}$, $(\boldsymbol{ij})^* = \boldsymbol{j}^*\boldsymbol{i}^* = -\boldsymbol{ij}$. The elements of the algebra $\mathbb{H}$ are real linear combinations of $1, \boldsymbol{i}, \boldsymbol{j}$ and $\boldsymbol{k} = \boldsymbol{ij}$:

$$\boldsymbol{q} = q_0 + q_1\boldsymbol{i} + q_2\boldsymbol{j} + q_3\boldsymbol{k}. \tag{1}$$

The addition of quaternions is defined by the component-wise addition rule. That is, for the quaternions $\boldsymbol{p} = p_0 + p_1\boldsymbol{i} + p_2\boldsymbol{j} + p_3\boldsymbol{k}$ and $\boldsymbol{q} = q_0 + q_1\boldsymbol{i} + q_2\boldsymbol{j} + q_3\boldsymbol{k}$, the sum is

$$\boldsymbol{p} + \boldsymbol{q} = (p_0 + q_0) + (p_1 + q_1)\boldsymbol{i} + (p_2 + q_2)\boldsymbol{j} + (p_3 + q_3)\boldsymbol{k}. \tag{2}$$

The quaternion addition and multiplication satisfy the distributive and associative laws, implying the following multiplication formula:

$$\begin{aligned}\boldsymbol{pq} = &(p_0q_0 - p_1q_1 - p_2q_2 - p_3q_3) + (p_0q_1 + p_1q_0 + p_2q_3 - p_3q_2)\boldsymbol{i} \\ &+ (p_0q_2 + p_2q_0 + p_3q_1 - p_1q_3)\boldsymbol{j} + (p_0q_3 + p_3q_0 + p_1q_2 - p_2q_1)\boldsymbol{k},\end{aligned} \tag{3}$$

for any real quaternions $\boldsymbol{p} = p_0 + p_1\boldsymbol{i} + p_2\boldsymbol{j} + p_3\boldsymbol{k}$ and $\boldsymbol{q} = q_0 + q_1\boldsymbol{i} + q_2\boldsymbol{j} + q_3\boldsymbol{k}$.

Definition 3. *The (reduced) norm of a quaternion* $\boldsymbol{q}$, *denoted by* $\|\boldsymbol{q}\|_r$, *is defined by* $\|\boldsymbol{q}\|_r = \boldsymbol{qq}^*$. *A quaternion of norm 1 is called a unit quaternion.*

It is easy to see that $\|\boldsymbol{q}\|_r = \boldsymbol{qq}^* = \boldsymbol{q}^*\boldsymbol{q}$ and $\|\boldsymbol{q}\|_r = q_0^2 + q_1^2 + q_2^2 + q_3^2$.

Definition 4. *The left and right (periodic) autocorrelation functions of a sequence* $\boldsymbol{x} = [\boldsymbol{x}_0, \boldsymbol{x}_1, \ldots, \boldsymbol{x}_{n-1}]$ *are defined as*

$$ACF_{\boldsymbol{x}}^L(m) = \frac{1}{\|\sum_{t=0}^{n-1} \boldsymbol{x}_t\|_r} \sum_{t=0}^{n-1} \boldsymbol{x}_t^* \boldsymbol{x}_{t+m} \tag{4}$$

and

$$ACF_{\boldsymbol{x}}^R(m) = \frac{1}{\|\sum_{t=0}^{n-1} \boldsymbol{x}_t\|_r} \sum_{t=0}^{n-1} \boldsymbol{x}_t \boldsymbol{x}_{t+m}^* \tag{5}$$

respectively, for any integer m.

Definition 5. *A sequence* $x = [x_0, x_1, \ldots, x_{n-1}]$ *over an arbitrary quaternion alphabet is called left (right) perfect if its left (right) periodic autocorrelation function is equal to zero for all* m, $1 \leq m \leq n-1$.

It is known [5] that, over the quaternions, a sequence $x = [x_0, x_1, \ldots, x_{n-1}]$ is left perfect if and only if it is right perfect. Therefore, the designations 'left' and 'right' perfect are redundant for quaternion sequences and can be omitted. Any left or right perfect sequence can be simply called 'perfect'.

3   The Absolute Value of the Real Part of Autocorrelation Coefficients of a Sequence Over the Quaternions

In this section we introduce the equality which relates the absolute value of the real part of an arbitrary sequence over the quaternions to the quotient of the sums of the norms of its particular elements.

Proposition 1. *Let* $x = [x_0, x_1, \ldots, x_{n-1}]$ *be a sequence over the algebra of real quaternions* $\mathbb{H}$ *and let* m *be an integer,* $1 \leq m \leq n-1$. *Then*

$$|1 + Re(ACF_x^L(m)| = \frac{\sum_{t=0}^{n-1} \|x_t + x_{t+m}\|_r}{2 \sum_{t=0}^{n-1} \|x_t\|_r}. \tag{6}$$

Before we prove Proposition 1, we prove some lemmas.

Lemma 1. *Let* $x = [x_0, x_1, \ldots, x_{n-1}]$ *be a sequence over the algebra of real quaternions* $\mathbb{H}$. *Then*

$$\sum_{s=0}^{n-1} \sum_{t=0}^{n-1} \|x_s x_t\|_r = \left(\sum_{t=0}^{n-1} \|x_t\|_r \right)^2. \tag{7}$$

Proof.

$$\left(\sum_{t=0}^{n-1} \|x_t\|_r \right)^2 = \sum_{s=0}^{n-1} \sum_{t=0}^{n-1} \|x_s\|_r \|x_t\|_r = \sum_{s=0}^{n-1} \sum_{t=0}^{n-1} \|x_s x_t\|_r \tag{8}$$

Lemma 2. *Let* $x = [x_0, x_1, \ldots, x_{n-1}]$ *be a sequence over the algebra of real quaternions* $\mathbb{H}$. *Then*

$$\sum_{m=0}^{n-1} \|ACF_x^L(m)\|_r = \left(\frac{1}{\|\sum_{t=0}^{n-1} x_t\|_r} \right)^2 \sum_{t=0}^{n-1} \sum_{s=0}^{n-1} \sum_{m=0}^{n-1} x_t^* x_s x_{s+m}^* x_{t+m}. \tag{9}$$

and

$$\sum_{m=0}^{n-1} \|ACF_x^R(m)\|_r = \left(\frac{1}{\|\sum_{t=0}^{n-1} x_t\|_r} \right)^2 \sum_{t=0}^{n-1} \sum_{s=0}^{n-1} \sum_{m=0}^{n-1} x_t x_s^* x_{s+m} x_{t+m}^*. \tag{10}$$

Proof. As shown in [5],

$$\sum_{m=0}^{n-1} \|ACF_x^L(m)\|_r = \frac{1}{\|\sum_{t=0}^{n-1} x_t\|_r} \sum_{t=0}^{n-1}\sum_{s=0}^{n-1} x_t^* ACF_x^R(s-t)x_s \qquad (11)$$

Hence,

$$\sum_{m=0}^{n-1} \|ACF_x^L(m)\|_r = \frac{1}{\|\sum_{t=0}^{n-1} x_t\|_r} \sum_{m=0}^{n-1}\sum_{t=0}^{n-1} x_t^* ACF_x^R(m)x_{t+m}$$
$$= \frac{1}{\|\sum_{t=0}^{n-1} x_t\|_r} \sum_{t=0}^{n-1}\sum_{s=0}^{n-1}\sum_{m=0}^{n-1} x_t^* x_s x_{s+m}^* x_{t+m}. \qquad (12)$$

The second identity is proved similarly.

Lemma 3. *Let* $x = [x_0, x_1, \ldots, x_{n-1}]$ *be a sequence over the algebra of real quaternions* $\mathbb{H}$. *Then*

$$\sum_{m=0}^{n-1} ACF_x^L(m)^2 = \left(\frac{1}{\|\sum_{t=0}^{n-1} x_t\|_r}\right)^2 \sum_{t=0}^{n-1}\sum_{s=0}^{n-1}\sum_{m=0}^{n-1} x_t^* x_{s+m} x_s^* x_{t+m}. \qquad (13)$$

and

$$\sum_{m=0}^{n-1} ACF_x^R(m)^2 = \left(\frac{1}{\|\sum_{t=0}^{n-1} x_t\|_r}\right)^2 \sum_{t=0}^{n-1}\sum_{s=0}^{n-1}\sum_{m=0}^{n-1} x_t x_{s+m}^* x_s x_{t+m}^*. \qquad (14)$$

Proof.

$$\sum_{m=0}^{n-1} ACF_x^L(m)^2 = \left(\frac{1}{\|\sum_{t=0}^{n-1} x_t\|_r}\right)^2 \sum_{m=0}^{n-1}\sum_{t=0}^{n-1} x_t^* x_{t+m} \sum_{s=0}^{n-1} x_s^* x_{s+m}$$
$$= \left(\frac{1}{\|\sum_{t=0}^{n-1} x_t\|_r}\right)^2 \sum_{m=0}^{n-1}\sum_{t=0}^{n-1}\sum_{s=0}^{n-1} x_t^* x_{t+m} x_s^* x_{s+m} \qquad (15)$$

The second identity is proved similarly.

Proof (Proposition 1).
    First, note that

$$\sum_{s=0}^{n-1}\sum_{t=0}^{n-1} \|(x_s + x_{s+m})(x_t + x_{t+m})\|_r = \left(\sum_{s=0}^{n-1} \|x_s + x_{s+m}\|_r\right)^2 \qquad (16)$$

The norm $\|(x_s + x_{s+m})(x_t + x_{t+m})\|_r$ can be expanded as follows:

$$\|(x_s + x_{s+m})(x_t + x_{t+m})\|_r = \|(x_s^* + x_{s+m}^*)(x_t + x_{t+m})\|_r$$
$$= (x_t^* + x_{t+m}^*)(x_s + x_{s+m})(x_s^* + x_{s+m}^*)(x_t + x_{t+m})$$
$$= (\|x_s x_t\|_r + \|x_{s+m} x_t\|_r + \|x_s x_{t+m}\|_r + \|x_{s+m} x_{t+m}\|_r)$$
$$+ (\|x_s\|_r + \|x_{s+m}\|_r)(x_t^* x_{t+m} + x_{t+m}^* x_t)$$
$$+ (x_t^*(x_s x_{s+m}^* + x_{s+m} x_s^*)x_t + x_{t+m}^*(x_s x_{s+m}^* + x_{s+m} x_s^*)x_{t+m})$$
$$+ (x_t^* x_s x_{s+m}^* x_{t+m} + x_t^* x_{s+m} x_s^* x_{t+m} + x_{t+m}^* x_s x_{s+m}^* x_t + x_{t+m}^* x_{s+m} x_s^* x_t) \qquad (17)$$

Then,

$$\sum_{t=0}^{n-1}\sum_{s=0}^{n-1}\|(\boldsymbol{x}_s + \boldsymbol{x}_{s+m})(\boldsymbol{x}_t + \boldsymbol{x}_{t+m})\|_r$$

$$= \sum_{t=0}^{n-1}\sum_{s=0}^{n-1}(\|\boldsymbol{x}_s\boldsymbol{x}_t\|_r + \|\boldsymbol{x}_{s+m}\boldsymbol{x}_t\|_r + \|\boldsymbol{x}_s\boldsymbol{x}_{t+m}\|_r + \|\boldsymbol{x}_{s+m}\boldsymbol{x}_{t+m}\|_r)$$

$$+ \sum_{t=0}^{n-1}\sum_{s=0}^{n-1}((\|\boldsymbol{x}_s\|_r + \|\boldsymbol{x}_{s+m}\|_r)(\boldsymbol{x}_t^*\boldsymbol{x}_{t+m} + \boldsymbol{x}_{t+m}^*\boldsymbol{x}_t))$$

$$+ \sum_{t=0}^{n-1}\sum_{s=0}^{n-1}(\boldsymbol{x}_t^*(\boldsymbol{x}_s\boldsymbol{x}_{s+m}^* + \boldsymbol{x}_{s+m}\boldsymbol{x}_s^*)\boldsymbol{x}_t + \boldsymbol{x}_{t+m}^*(\boldsymbol{x}_s\boldsymbol{x}_{s+m}^* + \boldsymbol{x}_{s+m}\boldsymbol{x}_s^*)\boldsymbol{x}_{t+m})$$

$$+ \sum_{t=0}^{n-1}\sum_{s=0}^{n-1}(\boldsymbol{x}_t^*\boldsymbol{x}_s\boldsymbol{x}_{s+m}^*\boldsymbol{x}_{t+m} + \boldsymbol{x}_t^*\boldsymbol{x}_{s+m}\boldsymbol{x}_s^*\boldsymbol{x}_{t+m} + \boldsymbol{x}_{t+m}^*\boldsymbol{x}_s\boldsymbol{x}_{s+m}^*\boldsymbol{x}_t + \boldsymbol{x}_{t+m}^*\boldsymbol{x}_{s+m}\boldsymbol{x}_s^*\boldsymbol{x}_t)$$

$$(18)$$

Consider each of the four terms of the above sum separately. By Lemma 1, we have

$$\sum_{t=0}^{n-1}\sum_{s=0}^{n-1}(\|\boldsymbol{x}_s\boldsymbol{x}_t\|_r + \|\boldsymbol{x}_{s+m}\boldsymbol{x}_t\|_r + \|\boldsymbol{x}_s\boldsymbol{x}_{t+m}\|_r + \|\boldsymbol{x}_{s+m}\boldsymbol{x}_{t+m}\|_r)$$

$$= 4\left(\sum_{t=0}^{n-1}\|\boldsymbol{x}_t\|_r\right)^2$$

$$(19)$$

$$\sum_{t=0}^{n-1}\sum_{s=0}^{n-1}((\|\boldsymbol{x}_s\|_r + \|\boldsymbol{x}_{s+m}\|_r)(\boldsymbol{x}_t^*\boldsymbol{x}_{t+m} + \boldsymbol{x}_{t+m}^*\boldsymbol{x}_t))$$

$$= \left(\sum_{t=0}^{n-1}\|\boldsymbol{x}_t\|_r\right)\sum_{s=0}^{n-1}((\|\boldsymbol{x}_s\|_r + \|\boldsymbol{x}_{s+m}\|_r)(ACF_{\boldsymbol{x}}^L(m) + ACF_{\boldsymbol{x}}^L(-m)))$$

$$= 2\left(\sum_{t=0}^{n-1}\|\boldsymbol{x}_t\|_r\right)\sum_{s=0}^{n-1}((\|\boldsymbol{x}_s\|_r + \|\boldsymbol{x}_{s+m}\|_r)Re(ACF_{\boldsymbol{x}}^L(m)))$$

$$= 4\left(\sum_{t=0}^{n-1}\|\boldsymbol{x}_t\|_r\right)^2 Re(ACF_{\boldsymbol{x}}^L(m))$$

$$(20)$$

$$\sum_{t=0}^{n-1}\sum_{s=0}^{n-1}(\boldsymbol{x}_t^*(\boldsymbol{x}_s\boldsymbol{x}_{s+m}^* + \boldsymbol{x}_{s+m}\boldsymbol{x}_s^*)\boldsymbol{x}_t + \boldsymbol{x}_{t+m}^*(\boldsymbol{x}_s\boldsymbol{x}_{s+m}^* + \boldsymbol{x}_{s+m}\boldsymbol{x}_s^*)\boldsymbol{x}_{t+m})$$

$$= 2\left(\sum_{t=0}^{n-1}\|\boldsymbol{x}_t\|_r\right)\sum_{t=0}^{n-1}\boldsymbol{x}_t^*((ACF_{\boldsymbol{x}}^R(m)) + ACF_{\boldsymbol{x}}^R(-m))\boldsymbol{x}_t$$

$$= 2\left(\sum_{t=0}^{n-1}\|\boldsymbol{x}_t\|_r\right)\sum_{t=0}^{n-1}\boldsymbol{x}_t^*((ACF_{\boldsymbol{x}}^R(m)) + ACF_{\boldsymbol{x}}^R(m)^*)\boldsymbol{x}_t$$

$$= 4\left(\sum_{t=0}^{n-1}\|\boldsymbol{x}_t\|_r\right)Re(ACF_{\boldsymbol{x}}^R(m))\sum_{t=0}^{n-1}\boldsymbol{x}_t^*\boldsymbol{x}_t = 4\left(\sum_{t=0}^{n-1}\|\boldsymbol{x}_t\|_r\right)^2 Re(ACF_{\boldsymbol{x}}^R(m))$$

$$(21)$$

The fourth term, by lemmas 2 and 3, satisfies the equations

$$
\sum_{t=0}^{n-1}\sum_{s=0}^{n-1}(x_t^* x_s x_{s+m}^* x_{t+m} + x_t^* x_{s+m} x_s^* x_{t+m}
$$
$$
+x_{t+m}^* x_s x_{s+m}^* x_t + x_{t+m}^* x_{s+m} x_s^* x_t)
$$
$$
= \left(\sum_{t=0}^{n-1} \|x_t\|_r\right)^2 (\|ACF_x^L(m)\|_r + ACF_x^L(m)^2
$$
$$
+ACF_x^L(-m)^2 + \|ACF_x^L(-m)\|_r)
$$
$$
= \left(\sum_{t=0}^{n-1} \|x_t\|_r\right)^2 (ACF_x^L(m)^2 + ACF_x^L(m)^* ACF_x^L(m) \tag{22}
$$
$$
+ACF_x^L(m)ACF_x^L(m)^* + (ACF_x^L(m)^*)^2)
$$
$$
= \left(\sum_{t=0}^{n-1} \|x_t\|_r\right)^2 (ACF_x^L(m) + ACF_x^L(m)^*)^2
$$
$$
= 4 \left(\sum_{t=0}^{n-1} \|x_t\|_r\right)^2 Re(ACF_x^L(m))^2
$$

Then,

$$
\left(\sum_{s=0}^{n-1} \|x_s + x_{s+m}\|_r\right)^2 = \sum_{t=0}^{n-1}\sum_{s=0}^{n-1} \|(x_s + x_{s\,|\,m})(x_t + x_{t+m})\|_r
$$
$$
= 4 \left(\sum_{t=0}^{n-1} \|x_t\|_r\right)^2 + 4 \left(\sum_{t=0}^{n-1} \|x_t\|_r\right)^2 Re(ACF_x^I(m))
$$
$$
+4 \left(\sum_{t=0}^{n-1} \|x_t\|_r\right)^2 Re(ACF_x^R(m)) + 4 \left(\sum_{t=0}^{n-1} \|x_t\|_r\right)^2 Re(ACF_x^L(m))^2
$$
$$
= 4 \left(\sum_{t=0}^{n-1} \|x_t\|_r\right)^2 + 8 \left(\sum_{t=0}^{n-1} \|x_t\|_r\right)^2 Re(ACF_x^L(m)) \tag{23}
$$
$$
+4 \left(\sum_{t=0}^{n-1} \|x_t\|_r\right)^2 Re(ACF_x^L(m))^2
$$
$$
= 4 \left(\sum_{t=0}^{n-1} \|x_t\|_r\right)^2 (1 + 2Re(ACF_x^L(m)) + Re(ACF_x^L(m))^2)
$$
$$
= 4 \left(\sum_{t=0}^{n-1} \|x_t\|_r\right)^2 (1 + Re(ACF_x^L(m)))^2
$$

Thus,

$$
(1 + Re(ACF_x^L(m)))^2 = \frac{(\sum_{t=0}^{n-1} \|x_t + x_{t+m}\|_r)^2}{4(\sum_{t=0}^{n-1} \|x_t\|_r)^2}, \tag{24}
$$

or

$$|1 + Re(ACF_x^L(m)| = \frac{\sum_{t=0}^{n-1} \|x_t + x_{t+m}\|_r}{2\sum_{t=0}^{n-1} \|x_t\|_r}, \tag{25}$$

which proves the proposition.

Remark 1. A proposition similar to 1, but involving the right autocorrelation values $ACF_x^R(m)$, can be proved in the same way. However, it is known [4] that $Re(ACF_x^L(m)) = Re(ACF_x^R(m))$, for every m. Therefore, such a proof would be redundant here.

While cumbersome at first glance, the formula of Proposition 1 is greatly simplified when applied to a perfect sequence. Since all out-of-phase autocorrelation values of a perfect sequence are equal to zero, the following corollary follows from Proposition 1.

Corollary 1. *Let $x = [x_0, x_1, \ldots, x_{n-1}]$ be a perfect sequence over the algebra of real quaternions $\mathbb{H}$ and let m be an integer, $1 \le m \le n - 1$. Then*

$$2\sum_{t=0}^{n-1} \|x_t\|_r = \sum_{t=0}^{n-1} \|x_t + x_{t+m}\|_r. \tag{26}$$

4   Necessary and Sufficient Condition for Perfection of a Symmetric Sequence

In this part, we introduce a necessary and sufficient condition for perfection of a symmetric sequence over the quaternions. We start with Lemma 4, which states an important property of a symmetric sequence over the quaternions.

Lemma 4. *Let $a = [a_0, a_1, \ldots, a_{n-1}]$ be a symmetric sequence over the algebra of real quaternions $\mathbb{H}$. Then all its left (right) autocorrelation values are real numbers.*

Proof. Let $a = [a_0, a_1, \ldots, a_{n-1}]$ be a symmetric sequence, and let s be the integer for which $a_{s+m} = a_{s-m}$, for every integer m. Then

$$
\begin{aligned}
ACF_a^L(m)^* &= \frac{1}{\|\sum_{t=0}^{n-1} a_t\|_r} \left(\sum_{t=0}^{n-1} a_{s+t}^* a_{s+t+m} \right)^* \\
&= \frac{1}{\|\sum_{t=0}^{n-1} a_t\|_r} \sum_{t=0}^{n-1} a_{s+t+m}^* a_{s+t} = \frac{1}{\|\sum_{t=0}^{n-1} a_t\|_r} \sum_{t=0}^{n-1} a_{s+t}^* a_{s+t-m} \\
&= \frac{1}{\|\sum_{t=0}^{n-1} a_t\|_r} \sum_{t=0}^{n-1} a_{s-t}^* a_{s-t+m} = ACF_a^L(m)
\end{aligned} \tag{27}
$$

Since $ACF_a^L(m) = ACF_a^L(m)^*$, it is a real number.

Remark 2. Since all autocorrelation values of a symmetric sequence are real numbers, the left and the right autocorrelation values of a symmetric sequence are equal, $ACF_a^L(m) = ACF_a^R(m)$, for every integer m.

From Proposition 1, it is clear that, for any sequence $x = [x_0, x_1, \ldots, x_{n-1}]$, $Re(ACF_x^L(m)) = 0$ if and only if $2\sum_{t=0}^{n-1} \|x_t\|_r = \sum_{t=0}^{n-1} \|x_t + x_{t+m}\|_r$. Since all autocorrelation values of a symmetric sequence are real numbers (that is, quaternions with the imaginary part equal to zero), the following corollary is true:

Corollary 2 (necessary and sufficient condition for perfection of a symmetric sequence). *Let $a = [a_0, a_1, \ldots, a_{n-1}]$ be a symmetric sequence over the algebra of real quaternions $\mathbb{H}$. Then a is perfect if and only if, for all integers m, $1 \leq m \leq n - 1$,*

$$2\sum_{t=0}^{n-1} \|a_t\|_r = \sum_{t=0}^{n-1} \|a_t + a_{t \mid m}\|_r. \tag{28}$$

Remark 3 A symmetric sequence over the complex numbers is perfect if and only if the same condition is satisfied.

5 An Improved Method of the Exhaustive Computer Search for Perfect Sequences

While it is proved that perfect sequences of unbounded lengths over the quaternions exist [1], there is no known universal algorithm for finding long perfect sequences of all possible lengths. In most cases, the exhaustive computer search is the only available method for finding examples of perfect sequences. When implementing the exhaustive computer search over symmetric sequences, the formula of Corollary 2 may be used for a great reduction in computer time required for checking perfection of each possible sequence. Indeed, checking perfection of a symmetric unimodular (that is, with the property of having all elements of equal norm) sequence of length n by direct calculation of its (left) autocorrelation values by the formula $ACF_a^L(m) = \frac{1}{\|\sum_{t=0}^{n-1} a_t\|_r} \sum_{t=0}^{n-1} a_t^* a_{t+m} = 0$ would require to complete n operations of taking conjugate (3 operations of taking negatives over real numbers each), plus n operations of multiplication of one quaternion by another (16 multiplications and 12 summation of real numbers each), plus n summations of quaternions (4 operations over real numbers each), which results in $35n$ operations over real numbers for each calculation of each autocorrelation value.

In the contrast, checking perfection by calculation of $\sum_{t=0}^{n-1} \|a_t + a_{t+m}\|_r$ (note that $\sum_{t=0}^{n-1} \|a_t\|_r = n$ is a constant value for an unimodular sequence) takes n summations of quaternions (4 operations over real numbers each), plus n operations of calculating quaternion norm (4 multiplications and 3 summations of real numbers each), plus n summations of real numbers, which only comes to $12n$ operations over real numbers for each m. Thus, checking perfection of a symmetric sequence by means of the formula of Corollary 2 is 3 times more time efficient than by direct calculation of autocorrelation values.

6 Experimental Results of the Exhaustive Computer Search

We have performed some experimentation with finding perfect sequences over the quaternions by the exhaustive search with our ordinary desktop computer (Intel Pentium 4, 2.80 GHz, 1.00 GB of RAM) running the computational software package Magma [2] under Windows XP. We conducted the exhaustive search over symmetric sequences of the form $a = [a_0, a_1, \ldots, a_{n-2}, a_{n-1}, a_{n-2}, \ldots, a_1]$ for different lengths $2(n-1)$ by using both methods: direct calculations of each autocorrelation value and using the formula of Corollary 2. We considered sequences with elements in two finite quaternion groups [10]: the *double pyramid group* $\mathbb{Q}_8$ of order 8, formed by the unit quaternions $\{\pm 1, \pm i, \pm j, \pm k\}$, and the *double tetrahedron group* $\mathbb{Q}_{24}$ of order 24, which is a group generated by two unit quaternions, i and $\frac{1+i+j+k}{2}$, elements of which are unit quaternions $\{\pm 1, \pm i, \pm j, \pm k\}$ along with unit quaternions $\frac{\pm 1 \pm i \pm j \pm k}{2}$, for all possible combinations of the plus and minus signs. Results of our experiments were compiled in Tables 1 and 2 below for comparison.

It is worth noting that Proposition 1 can also be used for improvement of the exhaustive computer search for perfect sequences in general (non-symmetric) form. Since, by Proposition 1, $Re(ACF_x^L(m)) = 0$ if and only if $2\sum_{t=0}^{n-1} \|x_t\|_r = \sum_{t=0}^{n-1} \|x_t + x_{t+m}\|_r$, checking the condition $2\sum_{t=0}^{n-1} \|x_t\|_r = \sum_{t=0}^{n-1} \|x_t + x_{t+m}\|_r$ for different m during the exhaustive search would eliminate many sequences whose autocorrelation values have non-zero real part. The remaining sequences can then be checked for perfection by direct calculation of autocorrelation values. This way, a reduction in computer time can still be attained. Tables 3 and 4 below lists the results of our exhaustive search experiments for sequences over $\mathbb{Q}_8$ and $\mathbb{Q}_{24}$.

It is clear from Tables 1, 2, 3 and 4 that use of the improved method offers a great reduction in computer time required for completion of the exhaustive computer search for both symmetric and general perfect sequences, especially over larger quaternion alphabets.

Table 1. Computer time (in seconds) required for completion of the exhaustive search for perfect sequences over the set of all symmetric sequences of the form $a = [a_0, a_1, \ldots, a_{n-2}, a_{n-1}, a_{n-2}, \ldots, a_1]$ with elements from the group $\mathbb{Q}_8$

Length $2(n-1)$	Total number of perfect sequences	Time for completion of search by direct calculation	Time for completion of search by use of Corollary 2
6	384	0.988	0.609
8	384	9.391	5.656
10	1152	88.406	52.594

Table 2. Computer time (in seconds) required for completion of the exhaustive search for perfect sequences over the set of all symmetric sequences of the form $a = [a_0, a_1, \ldots, a_{n-2}, a_{n-1}, a_{n-2}, \ldots, a_1]$ with elements from the group $\mathbb{Q}_{24}$

Length $2(n-1)$	Total number of perfect sequences	Time for completion of search by direct calculation	Time for completion of search by use of Corollary 2
6	3456	518.594	159.88
8	3456	15952.906	4608.938
10	35712	467773.344	139080.609

Table 3. Computer time (in seconds) required for completion of the exhaustive search for perfect sequences of the general form $x = [x_0, x_1, \ldots, x_{n-1}]$ with elements from the group $\mathbb{Q}_8$

Length n	Total number of perfect sequences	Time for completion of search by direct calculation	Time for completion of search by use of Corollary 2
6	1152	39.453	26.438
8	1536	1690.438	1401.313

Table 4. Computer time (in seconds) required for completion of the exhaustive search for perfect sequences of the general form $x = [x_0, x_1, \ldots, x_{n-1}]$ with elements from the group $\mathbb{Q}_{24}$

Length n	Total number of perfect sequences	Time for completion of search by direct calculation	Time for completion of search by use of Corollary 2
4	2304	284.828	57.359
6	10368	233364.609	34179.516

7   Conclusion

We have introduced a new equality, which holds for sequences over the quaternions and relates the absolute value of the real part of its autocorrelation values to the fraction of the sums of the norms of its elements. Applying this equality to perfect sequences over the quaternions, we have found a new condition necessary for perfection of an arbitrary sequence and necessary and sufficient for perfection of a symmetric sequence. Checking this condition during implementation of the exhaustive computer search for perfect sequences provides for great reduction (up to 6 times reduction) in computer time, in comparison with the exhaustive search involving direct calculation of autocorrelation values.

As a side result, we have shown that symmetric sequences have real (left and right) autocorrelation values.

References

1. Acevedo, S.B., Hall, T.E.: Perfect sequences of unbounded lengths over the basic quaternions. In: Helleseth, T., Jedwab, J. (eds.) SETA 2012. LNCS, vol. 7280, pp. 159–167. Springer, Heidelberg (2012)
2. Bosma, W., Cannon, J., Playoust, C.: The Magma algebra system I: the user language. J. Symb. Comput. **24**, 235–265 (1997)
3. Isaeva, O., Sarytchev, V.: Quaternion presentations polarization state. In: IEEE Proceedings of the Second Topical Symposium on Combined Optical-Microwave Earth and Atmosphere Sensing, pp. 195–196 (1995)
4. Kuznetsov, O.: On some invariances of left and right autocorrelation functions of sequences over the quaternions. Int. J. Inf. Coding Theory **2**(2/3), 105–114 (2013)
5. Kuznetsov, O.: Perfect sequences over the real quaternions. In: IEEE Proceedings of the 4th International Workshop on Signal Design and Its Applications in Communications, pp. 8–11 (2009)
6. Lee, C.E.: Perfect q-ary sequences from multiplicative characters over GF(p). Electron. Lett. **28**(9), 833–835 (1992)
7. Luke, H.D.: BTP transform and perfect sequences with small phase alphabet. IEEE Trans. Aerosp. Electron. Syst. **32**(1), 497–499 (1996)
8. Luke, H.D.: Mismatched filtering of periodic quadriphase and 8-phase sequences. IEEE Trans. Commun. **51**(7), 1061–1063 (2003)
9. Mow, W.H.: A unified construction of perfect polyphase sequences. In: IEEE Proceedings of International Symposium on Information Theory, p. 459 (1995)
10. Stringham, W.I.: Determination of the finite quaternion groups. Am. J. Math. **4**(1), 345–357 (1881)
11. Wysocki, B., Wysocki, T., Seberry, J.: Modeling dual polarization wireless fading channels using quaternions. In: IEEE Proceedings of the Joint IST Workshop on Mobile Future and Symposium on Trends in Communications, pp. 68–71 (2006)

Almost Six-Phase Sequences with Perfect Periodic Autocorrelation Function

Vladimir E. Gantmakher and Mikhail V. Zaleshin$^{(\boxtimes)}$

Yaroslav-the-Wise Novgorod State University, Veliky Novgorod, Russia
mikhailzaleshin@gmail.com

Abstract. A new family of almost six-phase sequences with perfect periodic autocorrelation function is obtained. The distinctive features of these sequences are: fairly frequent grid of periods, quasi-perfect periodic cross-correlation function, and nearly unit peak-factor.

Keywords: Almost six-phase sequence · Perfect PACF · M-sequence · Galois field

1 Introduction

Sequences with perfect periodic autocorrelation function (PACF) are required in a variety of systems (communication, navigation, radiolocation, etc.) [1]. For a long time the developers were interested in sequences with a small phase alphabet (bi-phase, three-phase, and quadriphase). The generation and processing of such sequences with a long length were not a problem [2]. The development of digital electronic components and signal processors was followed by the appearance of demand on multiphase sequences (MPS). The large families of MPS with perfect PACF were synthesized by Frank [3], Chu [4] and Milewski [5]. In [6] Mow offered the generalized methodology for generation of MPS with perfect PACF and obtained the estimation for the total amount of such sequences. Based on this estimation, he assumed that all possible MPS with perfect PACF and unit peak-factor $pf = N/W$ are already synthesized. Here N is the period of the sequence and W – its weight (the number of nonzero symbols on the period).

Further expansion of the array of sequences with perfect PACF is related to the synthesis of MPS with zero symbols on the period, which are frequently called almost MPS [7]. Such sequences include ternary Ipatov [8] and Hoholdt-Justesen [9] sequences, quadriphase Lee [10] and eight-phase Lüke [11] sequences with one zero symbol, as well as sequences presented in the papers [12–14]. All the listed sequences are generated over extended Galois fields. However, the need for almost MPS is not satisfied both for a variety of properties and for a density of periods' grid.

The aim of this paper is the synthesis and study of one more almost MPS family: almost six-phase sequences with perfect PACF.

The paper is prepared with financial support of the Ministry of Education and Science of the Russian Federation within the basic part of the government assignment.

© Springer International Publishing Switzerland 2014
K.-U. Schmidt and A. Winterhof (Eds.): SETA 2014, LNCS 8865, pp. 97–103, 2014.
DOI: 10.1007/978-3-319-12325-7_8

2   Preliminaries

Let the generation of a new almost MPS family be implemented over an extended Galois field. Further, we use the following notation:

- $GF(q^m)$ – an extended Galois field, where $q = p^s$; p is a prime number; $m \geq 2$ and $s \geq 1$ – natural numbers;
- θ – a primitive root of the field $GF(q)$;
- $\{d_n\}$ – q-ary M-sequence with the period q^{m-1};
- $h = (q^m - 1) / (q - 1)$ – the length of M-sequence's train.

It is possible to sort nonzero elements of the field $GF(q)$ in ascending order by the power of the primitive root θ^n, $n = 0, 1, \ldots, q - 2$. If $q \equiv 1 \pmod 6$, then all nonzero elements of the field may be divided into six sets:

$$H_r = \left\{ \theta^{6k+r} \,\middle|\, k = 0, 1, \ldots, \frac{q-1}{6} - 1 \right\}, \text{where } r = 0, 1, \ldots, 5.$$

Based on each set, we construct six binary sequences defined by the rule

$$x_n^{(r)} = \begin{cases} 1, \, d_n \in H_r; \\ 0, \, d_n \notin H_r. \end{cases}$$

We will refer to the binary sequences $\left\{ x_n^{(r)} \right\}$ as "structural sequences" (SS) because they determine the cyclic structure of generated sequences. Let us note several properties of SS $\left\{ x_n^{(r)} \right\}$, which are defined by the known properties of M-sequences [15].

Property 2.1. Every SS has the same period $6h$.

Property 2.2. Every SS has the same cyclic structure.

Property 2.3. Every SS has the same PACF:

$$R_x(\tau) = q^{m-2} \begin{cases} q, & \tau \equiv 0 \pmod{6h}; \\ 0, & \tau \equiv 0 \pmod{h}; \\ 1, & \text{else.} \end{cases}$$

Property 2.4. Every pair of SS $\left\{ x_n^{(j)} \right\}$ and $\left\{ x_n^{(l)} \right\}$ has the same, with an accuracy up to a cyclic shift, periodic cross-correlation function (PCCF):

$$R_{j,l}(\tau) = R_{0,l-j}(\tau) = R_x \left(\tau - (l - j) h \right) \ .$$

Now we associate one of the symbols from the alphabet

$$Z = \left\{ \exp\left(\frac{\pi i k}{3} \right) \,\middle|\, k = 0, 1, \ldots, 5 \right\}$$

with each single symbol of every SS. The almost MPS is generated by the following rule:

$$y_n = (-1)^n \sum_{r=0}^{5} z_r x_n^{(r)}, \text{ where } z_r \in Z. \tag{1}$$

We may now find necessary and sufficient conditions for the almost MPS to have a perfect PACF with respect to the coding rule (1).

Necessary conditions

$$\begin{aligned} z_{r+3} &= -z_r; \\ q, m &- \text{ odd numbers}; \end{aligned} \tag{2}$$

follow from the Properties of SS 2.1 – 2.4 and the coding rule (1).

Sufficient conditions follow from the comparison of two nonisomorphic almost MPS

$$a_n = (-1)^n \left[x_n^{(0)} - x_n^{(1)} + \exp\left(\frac{4\pi i}{3}\right) x_n^{(2)} - x_n^{(3)} + x_n^{(4)} + \exp\left(\frac{\pi i}{3}\right) x_n^{(5)} \right] \text{ and}$$

$$b_n = (-1)^n \left[x_n^{(0)} - x_n^{(1)} + \exp\left(\frac{2\pi i}{3}\right) x_n^{(2)} - x_n^{(3)} + x_n^{(4)} + \exp\left(\frac{5\pi i}{3}\right) x_n^{(5)} \right],$$

which have a perfect PACF only if the following conditions take place:

$$\begin{cases} z_2 = \exp(\frac{4\pi i}{3}); \\ z_5 = \exp(\frac{\pi i}{3}). \end{cases} \text{ or } \begin{cases} z_2 = \exp(\frac{2\pi i}{3}); \\ z_5 = \exp(\frac{5\pi i}{3}). \end{cases} \tag{3}$$

Theorem 1. *The sequence $\{a_n\}$ has a perfect PACF.*

Proof. The PACF of the sequence $\{a_n\}$ is defined by the formula:

$$R_a(\tau) = (-1)^\tau \left[\left(R_x(\tau) - R_{0,1}(\tau) + \left[\exp\left(\frac{4\pi i}{3}\right) \right]^* R_{0,2}(\tau) - R_{0,3}(\tau) + \right. \right.$$

$$\left. + R_{0,4}(\tau) + \left[\exp\left(\frac{\pi i}{3}\right) \right]^* R_{0,5}(\tau) \right) + \left(R_x(\tau) - \left[\exp\left(\frac{4\pi i}{3}\right) \right]^* R_{0,1}(\tau) + \right.$$

$$\left. \left. + R_{0,2}(\tau) - R_{0,3}(\tau) - \left[\exp\left(\frac{\pi i}{3}\right) \right]^* R_{0,4}(\tau) - R_{0,5}(\tau) \right) + \dots \right].$$

It follows from the properties of SS 2.3 and 2.4 that:

$$R_a(\tau) = (-1)^\tau 6 \left[R_x(\tau) - R_x(\tau - 3h) \right] =$$

$$= (-1)^\tau 6 q^{m-1} \begin{cases} 1, & \tau \equiv 0 \pmod{6h}; \\ -1, & \tau \equiv 0 \pmod{3h}; \\ 0, & \text{else}. \end{cases}$$

Hence we obtain that the PACF $R_a(\tau)$ is perfect, if $(-1)^\tau = -1$ for $\tau \equiv 0$ (mod $3h$). This condition holds when q and m are odd numbers. At the same time, the PACF becomes:

$$R_a(\tau) = 3q^{m-1} \begin{cases} 1, & \tau \equiv 0 \quad (\text{mod } 3h); \\ 0, & \text{else.} \end{cases}$$

This completes the proof.

$\square$

In the similar way we can prove that the sequence $\{b_n\}$ has a perfect PACF.

Thus, there are only two coding rules of almost six-phase sequences that satisfy the necessary existence conditions (2) and the sufficient existence conditions (3). That is to say, for $z_0 = z_4 = 1$, $z_1 = z_3 = -1$, as well as for z_2 and z_5, which satisfy (3), the sequences generated by (1) have a perfect PACF. Moreover, the parameters of the extended Galois field q and m should be odd numbers. This result does not depend on choice of initial values of M-sequence.

3   Properties of Almost Six-Phase MPS

Properties 2.3 and 2.4 of structural sequences determine the following property of the almost MPS:

Property 3.1. The following formula defines the absolute value of the PCCF of two sequences $\{a_n\}$ and $\{b_n\}$ corresponding to the same M-sequence:

$$|R_{a,b}(\tau)| = \begin{cases} \sqrt{3}q^{m-1}, & \tau \equiv 0 \quad (\text{mod } h); \\ 0, & \text{else.} \end{cases}$$

The proof of the Property 3.1 is similar to the proof of the theorem about the perfect PACF of $\{a_n\}$.

From the cyclic properties of M-sequences we obtain the following property.

Property 3.2. The period of almost six-phase sequences with perfect PACF equals to $3\left(q^m - 1\right)/\left(q - 1\right)$.

The proof of the Property 3.2 follows from the definition of the coding rule (1) and the properties of SS.

The peak-factor of the almost MPS is:

$$pf = \frac{3h}{3q^{m-1}} = \frac{q}{q-1} - \frac{1}{q^{m-1}\left(q-1\right)} \; .$$

Hence:

Property 3.3. The limit of the peak-factor, as the characteristic of the Galois field over which the almost MPS was generated approaches infinity, is one.

Property 3.4. The number of cyclically distinct almost six-phase MPS with perfect PACF is defined by the number of cyclically distinct corresponding M-sequences and equals to:

$$\frac{\phi(q^m - 1)}{m}, \text{ where } \phi(n) - \text{Euler's totient function}.$$

It follows from the Property 3.4 that the sequences with perfect PACF generated by (1) based on nonisomorphic M-sequences are also nonisomorphic.

4   Examples

Now we illustrate the generation principles and properties of almost six-phase MPS by several examples.

Example 1. Here we generate almost six-phase MPS over the Galois field $GF(7^3)$ and consider its properties.

1. Sets H_r over the prime Galois field $GF(7)$ with the primitive root $\theta = 3$ are:

$$H_0 = \{1\}; H_1 = \{3\}; H_2 = \{2\}; H_3 = \{6\}; H_4 = \{4\}; H_5 = \{5\}.$$

2. The coding rules of $\{a_n\}$ and $\{b_n\}$ are:

$$a_n = (-1)^n \begin{cases} 1, & d_n \in \{1, 4\}; \\ -1, & d_n \in \{3, 6\}; \\ \exp(\frac{4\pi i}{3}), & d_n = 2; \\ \exp(\frac{\pi i}{3}), & d_n = 5; \\ 0, & \text{else.} \end{cases}$$

$$b_n = (-1)^n \begin{cases} 1, & d_n \in \{1, 4\}; \\ -1, & d_n \in \{3, 6\}; \\ \exp(\frac{2\pi i}{3}), & d_n = 2; \\ \exp(\frac{5\pi i}{3}), & d_n = 5; \\ 0, & \text{else.} \end{cases}$$

3. The PACFs of the almost MPS are:

$$R_a(\tau) = R_b(\tau) = \begin{cases} 147, & \tau \equiv 0 \pmod{171}; \\ 0, & \text{else.} \end{cases}$$

4. The absolute value of the PCCF is:

$$|R_{a,b}(\tau)| = \begin{cases} 49\sqrt{3}, & \tau \equiv 0 \pmod{57}; \\ 0, & \text{else.} \end{cases}$$

5. The peak-factor of the almost MPS equals to $pf \approx 1.16$.

Example 2. Here we generate almost six-phase MPS over the Galois field $GF(25^3)$ and consider its properties.

1. Sets H_r over the extended Galois field $GF(5^2)$ with the primitive root $\theta = x$ and the primitive polynomial $f(x) = x^2 + x + 2$ are:

$$H_0 = \{1, 2, 3, 4\}\, ; H_1 = \{x, 2x, 3x, 4x\}\, ;$$
$$H_2 = \{x + 2, 2x + 4, 3x + 1, 4x + 3\}\, ;$$
$$H_3 = \{x + 3, 2x + 1, 3x + 4, 4x + 2\}\, ;$$
$$H_4 = \{x + 4, 2x + 3, 3x + 2, 4x + 1\}\, ;$$
$$H_5 = \{x + 1, 2x + 2, 3x + 3, 4x + 4\}\, .$$

2. The coding rules of $\{a_n\}$ and $\{b_n\}$ are:

$$a_n = (-1)^n \begin{cases} 1, & d_n \in \{1, 2, 3, 4, x + 4, 2x + 3, 3x + 2, 4x + 1\}\, ; \\ -1, & d_n \in \{x, x + 3, 2x, 2x + 1, 3x, 3x + 4, 4x, 4x + 2\}\, ; \\ \exp(\frac{4\pi i}{3}), & d_n \in \{x + 2, 2x + 4, 3x + 1, 4x + 3\}\, ; \\ \exp(\frac{\pi i}{3}), & d_n \in \{x + 1, 2x + 2, 3x + 3, 4x + 4\}\, ; \\ 0, & \text{else.} \end{cases}$$

$$b_n = (-1)^n \begin{cases} 1, & d_n \in \{1, 2, 3, 4, x + 4, 2x + 3, 3x + 2, 4x + 1\}\, ; \\ -1, & d_n \in \{x, x + 3, 2x, 2x + 1, 3x, 3x + 4, 4x, 4x + 2\}\, ; \\ \exp(\frac{2\pi i}{3}), & d_n \in \{x + 2, 2x + 4, 3x + 1, 4x + 3\}\, ; \\ \exp(\frac{5\pi i}{3}), & d_n \in \{x + 1, 2x + 2, 3x + 3, 4x + 4\}\, ; \\ 0, & \text{else.} \end{cases}$$

3. The PACFs of the almost MPS are:

$$R_a(\tau) = R_b(\tau) = \begin{cases} 1875, & \tau \equiv 0 \pmod{1953}; \\ 0, & \text{else.} \end{cases}$$

4. The absolute value of the PCCF is:

$$|R_{a,b}(\tau)| = \begin{cases} 625\sqrt{3}, & \tau \equiv 0 \pmod{651}; \\ 0, & \text{else.} \end{cases}$$

5. The peak-factor of the almost MPS equals to $pf \approx 1.04$.

5   Conclusions

In this paper, we offered a new family of almost six-phase sequences with perfect PACF. Their distinctive features are the following:

1. The grid of periods is fairly frequent: $N = 3\,(q^m - 1) / (q - 1)$;

2. The periodic cross-correlation function is quasi-perfect:

$$|R_{a,b}(\tau)| = \begin{cases} \sqrt{3}q^{m-1}, & \tau \equiv 0 \pmod{h}; \\ 0, & \text{else}; \end{cases}$$

3. The peak-factor nearly equals to one:

$$pf = \frac{q}{q-1} - \frac{1}{q^{m-1}(q-1)} \; ;$$

4. The number of cyclically distinct almost MPS is determined by the number of cyclically distinct corresponding M-sequences and equals to $\phi(q^m - 1)/m$.

The validity of this paper is verified by numerous examples, and the achieved results are modeled on the computer.

References

1. Golomb, S.W., Gong, G.: Signal Design for Good Correlation: for Wireless Communication, Cryptography and Radar, p. 438. Cambridge University Press, Cambridge (2005)
2. Fan, P., Darnell, M.: Sequence Design for Communications Applications, p. 493. Research Studies Press Ltd., London (1996)
3. Frank, R.L.: Phase coded communication system. U.S. Patent 3,099,795, 30 July 1963
4. Chu, D.C.: Polyphase codes with good periodic correlation properties. IEEE Trans. Inf. Theor. **IT–18**, 531–533 (1972)
5. Milewski, A.: Periodic sequences with optimal properties for channel estimation and fast start-up equalization. IBM J. Res. Dev. **27**(5), 425–431 (1983)
6. Mow, W.H.: A new unified construction of perfect root-of-unity sequences. In: Proceedings of International Symposium on Spread Spectrum Techniques and its Applications (ISSSTA'96), Mainz, Germany, pp. 955–959 (1996)
7. Lüke, H.D., Schotten, H.D., Hadinejad-Mahram, H.: Binary and quadriphase sequences with optimal autocorrelation properties: a survey. IEEE Trans. Inf. Theor. **IT–49**(12), 3271–3282 (2003)
8. Ipatov, V.P.: Periodic discrete signals with optimal correlation properties. Moscow, "Radio i svyaz", p. 152 (1992)
9. Hoholdt, T., Justesen, J.: Ternary sequences with perfect periodic auto-correlation. IEEE Trans. Inf. Theor. **IT–29**(4), 597–600 (1983)
10. Lee, C.E.: Perfect q-ary sequences from multiplicative characters over GF(p). Electron. Lett. **3628**(9), 833–835 (1992)
11. Lüke, H.D.: BTP-transform and perfect sequences with small phase alphabet. IEEE Trans. Aerosp. Syst. **32**, 497–499 (1996)
12. Krengel, E.I.: Some constructions of almost-perfect, odd-perfect and perfect polyphase and almost-polyphase sequences. In: Carlet, C., Pott, A. (eds.) SETA 2010. LNCS, vol. 6338, pp. 387–398. Springer, Heidelberg (2010)
13. Schotten, H.D., Lüke, H.D.: New perfect and w-cyclic-perfect sequences. In: Proceedings of 1996 IEEE International Symposium on Information Theory, pp. 82–85 (1996)
14. Krengel, E.I.: A method of construction of perfect sequences. Radiotekhnika **11**, 15–21 (2009)
15. Zierler, N.: Linear recurring sequences. J. Soc. Ind. Appl. Math. **7**, 31–48 (1959)

A Construction for Perfect Periodic Autocorrelation Sequences

Samuel T. Blake and Andrew Z. Tirkel[(✉)]

School of Mathematical Sciences, Monash University, Melbourne, Australia
atirkel@bigpond.net.au

Abstract. We introduce a construction for perfect periodic autocorrelation sequences over roots of unity. The sequences share similarities to the perfect periodic sequence constructions of Liu, Frank, and Milewski.

Perfect periodic autocorrelation sequences see applications in many areas, including spread spectrum communications [14], channel estimation and fast start-up equalization [12], pulse compression radars [3], sonar systems [17], CDMA systems [8], system identication [16], and watermarking [15].

There exists a number of known constructions for perfect periodic autocorrelation sequences over roots of unity. These include Frank sequences of length n^2 over n roots of unity [4,6], Chu sequences of length n over n roots of unity for n odd and length n over $2n$ roots of unity for n even [2], Milewski sequences of length m^{2k+1} over m^{k+1} roots of unity [12], Liu-Fan sequences of length n over n roots of unity for n even [11]. Other sequence constructions exist [1,5,7,9,10,13].

The periodic cross-correlation of the sequences, $\mathbf{a} = [a_0, a_1, \cdots, a_{n-1}]$ and $\mathbf{b} = [b_0, b_1, \cdots, b_{n-1}]$, for shift τ is given by

$$\theta_{\mathbf{a},\mathbf{b}}(\tau) = \sum_{i=0}^{n-1} a_i b_{i+\tau}^*,$$

where $i + \tau$ is computed modulo n. The periodic autocorrelation of a sequence, $\mathbf{s}$ for shift τ is given by $\theta_{\mathbf{s}}(\tau) = \theta_{\mathbf{s},\mathbf{s}}(\tau)$. For $\tau \neq 0 \bmod n$, $\theta_{\mathbf{s}}(\tau)$ is called an *off-peak* autocorrelation. A sequence is *perfect* if all off-peak autocorrelation values are zero.

The periodic autocorrelation of a sequence, $\mathbf{s} = [s_0, s_1, \cdots, s_{ld^2-1}]$, can be expressed in terms of the autocorrelation and cross-correlation of an array *associated* with $\mathbf{s}$ [4,6,13]. The sequence $\mathbf{s}$ has the *array orthogonality property* (AOP) for the *divisor d*, if the array $\mathbf{S}$ associated with $\mathbf{s}$ has the following two properties:

1. For all τ, the periodic cross-correlation of any two distinct columns of $\mathbf{S}$ is zero.
2. For $\tau \neq 0$, the sum of the periodic autocorrelation of all columns of $\mathbf{S}$ is zero.

Any sequence with the AOP is perfect [13].

© Springer International Publishing Switzerland 2014
K.-U. Schmidt and A. Winterhof (Eds.): SETA 2014, LNCS 8865, pp. 104–108, 2014.
DOI: 10.1007/978-3-319-12325-7_9

In most perfect sequence constructions, one proves the sequence has perfect autocorrelation by reducing the autocorrelation to a Gaussian sum. A Gaussian sum is given by $\sum_{k=0}^{n-1} \omega^{qk}$, where $\omega = e^{2\pi\sqrt{-1}/n}$ and $q \in \mathbb{Z}$. If $q \neq 0 \bmod n$, then the sum is zero.

We present a construction for perfect sequences over roots of unity. Let $\mathbf{s}$ be a sequence of length $4mn^{k+1}$ over $2mn^k$ roots of unity, where $n, m, k \in \mathbb{N}$. Construct a $2mn^{k+1} \times 2$ array, $\mathbf{S}$, over $2mn^k$ roots of unity, where $\mathbf{S} = [S_{i,j}] = \omega^{\lfloor i(i+j)/n \rfloor}$ and $\omega = e^{2\pi\sqrt{-1}/(2mn^k)}$. The sequence $\mathbf{s}$ is constructed by enumerating, row-by-row, the array $\mathbf{S}$.

We now show that $\mathbf{s}$ has perfect periodic autocorrelation. We show $\mathbf{s}$ is perfect by showing that it has the array orthogonality property (AOP) for the divisor 2. First, we show that the cross-correlation of the two columns of $\mathbf{S}$ is zero for every non-zero shift.

$$\theta_{S_{i,0},S_{i,1}}(\kappa) = \sum_{i=0}^{2mn^{k+1}-1} S_{i,0} S_{i+\kappa,1}^* \tag{1}$$

Let $i = qn + r$, $(r < n)$, and $\kappa = q'n + r'$, $(r' < n)$, then (1) becomes

$$\theta_{S_{qn+r,0},S_{qn+r,1}}(q'n + r') = \sum_{q=0}^{2mn^k-1} \sum_{r=0}^{n-1} S_{qn+r,0} S_{(q+q')n+r+r',1}^*$$

$$= \sum_{q=0}^{2mn^k-1} \sum_{r=0}^{n-1} \omega^{\left\lfloor \frac{(qn+r)^2}{n} \right\rfloor} \omega^{-\left\lfloor \frac{((q+q')n+r+r')^2+(q+q')n+r+r'}{n} \right\rfloor}$$

$$= \sum_{q=0}^{2mn^k-1} \sum_{r=0}^{n-1} \omega^{-(2nq'+2r'+1)q-2q'r+\left\lfloor \frac{r^2}{n} \right\rfloor - \left\lfloor \frac{(r+r')(r+r'+1)}{n} \right\rfloor}$$

$$= \omega^{-2q'r'-q'} \left(\sum_{q=0}^{2mn^k-1} \omega^{-(2nq'+2r'+1)q} \right) \times$$

$$\left(\sum_{r=0}^{n-1} \omega^{-2q'r+\left\lfloor \frac{r^2}{n} \right\rfloor - \left\lfloor \frac{(r+r')(r+r'+1)}{n} \right\rfloor} \right).$$

The leftmost summation above is zero, as $-2nq' - 2r' - 1 \neq 0 \bmod 2mn^k$ (since $-2nq' - 2r' - 1$ is odd for all n, q', r', whereas $2mn^k$ is even for all n, q', r'). Thus $\mathbf{s}$ satisfies the first condition of the AOP.

Now we show that $\mathbf{s}$ satisfies the second condition of the AOP. That is, for all non-zero shifts, we show that the sum of the periodic autocorrelations of both columns of $\mathbf{S}$ sums to zero.

$$\theta_{S_{i,0}}(\kappa) + \theta_{S_{i,1}}(\kappa) = \sum_{i=0}^{2mn^{k+1}-1} S_{i,0}\, S^*_{i+\kappa,0} + \sum_{i=0}^{2mn^{k+1}-1} S_{i,1}\, S^*_{i+\kappa,1}$$

$$= \sum_{i=0}^{2mn^{k+1}-1} \omega^{\left\lfloor \frac{i^2}{n} \right\rfloor} \omega^{-\left\lfloor \frac{(i+\kappa)^2}{n} \right\rfloor}$$

$$+ \sum_{i=0}^{2mn^{k+1}-1} \omega^{\left\lfloor \frac{i^2+i}{n} \right\rfloor} \omega^{-\left\lfloor \frac{(i+\kappa)^2+i+\kappa}{n} \right\rfloor} \qquad ((2)+(3))$$

Let $i = qn + r$, $(r < n)$, and $\kappa = q'n + r'$, $(r' < n)$, then (2) becomes

$$\omega^{-2q'r'-nq'^2} \sum_{q=0}^{2mn^k-1}\sum_{r=0}^{n-1} \omega^{-2(nq'+r')q-2q'r+\left\lfloor \frac{r^2}{n} \right\rfloor - \left\lfloor \frac{r^2+r'^2+2r'r}{n} \right\rfloor}$$

$$= \omega^{-2q'r'-nq'^2} \left(\sum_{q=0}^{2mn^k-1} \omega^{-2(nq'+r')q} \right) \left(\sum_{r=0}^{n-1} \omega^{-2q'r+\left\lfloor \frac{r^2}{n} \right\rfloor - \left\lfloor \frac{r^2+r'^2+2r'r}{n} \right\rfloor} \right). \qquad (4)$$

Similarly, (3) becomes:

$$\omega^{-2q'r'-nq'^2-q'} \sum_{q=0}^{2mn^k-1}\sum_{r=0}^{n-1} \omega^{-2(nq'+r')q-2q'r+\left\lfloor \frac{r^2+r}{n} \right\rfloor - \left\lfloor \frac{r^2+2r'r+r+r'^2+r'}{n} \right\rfloor}$$

$$= \omega^{-2q'r'-nq'^2-q'} \left(\sum_{q=0}^{2mn^k-1} \omega^{-2(nq'+r')q} \right)$$

$$\times \left(\sum_{r=0}^{n-1} \omega^{-2q'r+\left\lfloor \frac{r^2+r}{n} \right\rfloor - \left\lfloor \frac{r^2+2r'r+r+r'^2+r'}{n} \right\rfloor} \right). \qquad (5)$$

Then $\theta_{S_{i,0}}(\kappa) + \theta_{S_{i,1}}(\kappa) = (4) + (5)$ is given by

$$\omega^{-2q'r'-nq'^2} \left(\sum_{q=0}^{2mn^k-1} \omega^{-2(nq'+r')q} \right) \left(\sum_{r=0}^{n-1} \omega^{-2q'r+\left\lfloor \frac{r^2}{n} \right\rfloor - \left\lfloor \frac{r^2+r'^2+2r'r}{n} \right\rfloor} + \right.$$

$$\left. \omega^{-q'} \sum_{r=0}^{n-1} \omega^{-2q'r+\left\lfloor \frac{r^2+r}{n} \right\rfloor - \left\lfloor \frac{r^2+2r'r+r+r'^2+r'}{n} \right\rfloor} \right).$$

The summation $\sum_{q=0}^{2mn^k-1} \omega^{-2(nq'+r')q}$ is non-zero when $-2(nq' + r') = 0 \bmod 2mn^k$, which is when $q' = -mn^{k-1}$, $r' = 0$ (excluding $q' = r' = 0$ as we only consider off-peak autocorrelations). In which case we have

$$\sum_{r=0}^{n-1} \omega^{-2q'r+\left\lfloor\frac{r^2}{n}\right\rfloor-\left\lfloor\frac{r^2+r'^2+2r'r}{n}\right\rfloor} = \sum_{r=0}^{n-1} \omega^{-2q'r+\left\lfloor\frac{r^2+r}{n}\right\rfloor-\left\lfloor\frac{r^2+2r'r+r+r'^2+r'}{n}\right\rfloor}$$

$$= \sum_{r=0}^{n-1} \omega^{-2q'r} = \sum_{r=0}^{n-1} e^{\left(\frac{2\pi\sqrt{-1}}{n}\right)r} = 0.$$

Thus, $\theta_{S_{i,0}}(\kappa) + \theta_{S_{i,1}}(\kappa) = 0$, so **s** satisfies the second condition of the AOP. It follows that **s** is a perfect sequence.

We note that the array, **S**, also has perfect periodic autocorrelation. The proof follows from the sequence, **s**, having the AOP.

In terms of the ratio of the sequence length to the number of phases, this construction sits below the construction of Milewski and above the constructions of Chu and Liu.

References

1. Alltop, W.O.: Complex sequences with low periodic correlations. IEEE Trans. Inform. Theor. **26**(3), 350–354 (1980)
2. Chu, D.C.: Polyphase codes with good periodic correlation properties. IEEE Trans. Inf. theor. **18**(4), 531–532 (1972)
3. Farnett, E.C., et al.: Pulse compression radar. In: Skolnik, M. (ed.) Radar Handbook, 2nd edn. McGraw-Hill, New York (1990)
4. Frank, R.L., Zadoff, S.A., Heimiller, R.: Phase shift pulse codes with good periodic correlation properties. IRE Trans. Inf. Theor. **8**(6), 381–382 (1961)
5. Gabidulin, E.M.: Non-binary sequences with perfect periodic auto-correlation and with optimal periodic cross-correlation. In: Proceedings of the IEEE International Symposium on Informational Theory, San Antonio, USA, pp. 412, January 1993
6. Heimiller, R.C.: Phase shift pulse codes with good periodic correlation properties. IRE Trans. Inf. Theor. **7**(4), 254–257 (1961)
7. Ipatov, V.P.: Ternary sequences with ideal autocorrelation properties. Radio Eng. Electron. Phys. **24**, 75–79 (1979)
8. Ipatov, V.P.: Spread Spectrum and CDMA: Principles and Applications. Wiley, Chichester (2005)
9. Kumar, P.V., Scholtz, R.A., Welch, L.R.: Generalized Bent functions and their properties. J. Combinat. Theor. Ser. A **40**(1), 90–107 (1985)
10. Lewis, B.L., Kretschmer, F.F.: Linear frequency modulation derived polyphase pulse compression. IEEE Trans. AES **18**(5), 637–641 (1982)
11. Liu, Y., Fan, P.: Modified Chu sequences with smaller alphabet size. Electron. Lett. **40**(10), 598–599 (2004)
12. Milewski, A.: Periodic sequences with optimal properties for channel estimation and fast start-up equalization. IBM J. Res. Dev. **27**(5), 426–431 (1983)
13. Mow, W.H.: A study of correlation of sequences. Ph.D. Department of Information Engineering, The Chinese University of Hong Kong (1993)
14. Simon, M.K.: Spread Spectrum Communications, vol. 1. Computer Science Press, The University of Michigan (1985)
15. Tirkel, A.Z., Rankin, G.A., Van Schyndel, R.M., Ho, W.J., Mee, N.R.A., Osborne, C.F.: Electronic water mark. In: DICTA 93, Macquarie University, pp. 666–673 (1993)

16. Van Schyndel, R.G.: Using phase-modulated probe signals to recover delays from higher order non-linear systems. In: Biomedical Research in 2001 IEEE Engineering in Medicine and Biology, pp. 94–97 (2001)
17. Xu, L.: Phase coded waveform design for sonar sensor network. In: 2011 6th International Conference on Communications and Networking in China (CHINACOM) ICST, pp. 251–256, August 2011

A Simple Construction of Almost Perfect Quinary ASK and QAM Sequences

Guang Gong[1](✉) and Solomon Golomb[2]

[1] Department of Electrical and Computer Engineering,
University of Waterloo, Waterloo, ON, Canada
ggong@uwaterloo.ca
[2] Department of Electrical Engineering,
University of Southern California, Los Angels, CA, USA
sgolomb@usc.edu

Abstract. In this paper, we give a simple construction of almost perfect quinary sequences for quinary amplitude shift keying (ASK) modulation schemes in digital communication, and using those almost perfect quinary sequences, we derive almost perfect quinary sequences for quinary quadrature amplitude modulation with a correlation receiver. Those sequences are constructed from the sequences with the two-tuple balance property over a finite field with 5 elements where the field elements are presented in a symmetric way.

Keywords: Almost perfect sequences · Quinary sequences · ASK and QAM sequences · 2.-tuple balance

1 Introduction

Sequences have been widely used in communications such as spread-spectrum modulation including code-division multiple access (CDMA), frequency hopping, and ultra wide-band (UWB) communications; orthogonal frequency division multiplexing (OFDM) transmission; channel estimation and synchronization; radar distance range and deep water detection; and compressing sensing.

We consider a sequence $\mathbf{a} = \{a_t\}$ with period N where a_t is a complex number. The autocorrelation of $\mathbf{a}$ is defined as $C_{\mathbf{a}}(\tau) = \sum_{i=0}^{N-1} a_i \bar{a}_{i+\tau}$ where $\bar{x}$ is the complex conjugate of x. In the applications of continuous wave (CW) radar systems [21] or channel estimation, autocorrelation detection is used on the returning reflected signal to determine the round-trip delay time, and thus the range, to the target for former, and to determine the channel condition for the later. In such an application, a perfect sequence, defined as $C_{\mathbf{a}}(\tau) = 0$ for all $\tau : 0 < \tau < N$, i.e., all the out-of-phase of autocorrelation values are zero, has largest SNR, which gives the best performance. For a sequence with (ideal) 2-level autocorrelation defined as $C_{\mathbf{a}}(\tau) = -1$ for all $\tau : 0 < \tau < N$, i.e., all the out-of-phase of autocorrelation values are equal to -1, the SNR is

© Springer International Publishing Switzerland 2014
K.-U. Schmidt and A. Winterhof (Eds.): SETA 2014, LNCS 8865, pp. 109–120, 2014.
DOI: 10.1007/978-3-319-12325-7_10

decreased by 3 dB. The other cases of autocorrelation functions of sequences will induce worse degrading of the performance of a communication system. Thus, in those applications, it is preferable to use either perfect sequences or ideal 2-level autocorrelation sequences.

For practical applications, another factor shaped the way to construct sequences with perfect or ideal 2-level autocorrelation is their easy implementation or not. Therefore, the most of sequences which could be considered in practice are sequences whose elements taken from a finite field or a finite ring, then map to polyphase sequences using additive characters, multiplicative characters or both of them. The research along this line has been attracting researchers since the end of 1950s, see earlier papers [6,12,14,29], just to list a few or a recent survey in [9]. As a result, each term in such a sequence has unit magnitude. Nevertheless, it is hard to construct perfect or ideal 2-level autocorrelation sequences. For perfect sequences with infinite families, the known class is Frank-Zadoff-Chu (FZC) sequences [2,4], and some miscellaneous examples included in [24]. (There exists some other constructions for perfect sequences, for example, see [20,26], which do not belong to the constructions mentioned above.) For ideal 2-level autocorrelation sequences, all known constructions for the binary case are collectively included in [7] and the status remains unchanged until now; for the up-date known non binary cases, it is collected in [9].

However, given a particular application scenario, it is possible that not all the out-of-phase autocorrelation values need to be used. This inspires the studies of almost perfect sequences defined as $C_a(\tau) = 0$ for all $\tau : 0 < \tau < N$ but one. It seems hard as well, because until now, there exist a few constructions for almost perfect sequences for which the most of the known constructions are given for almost perfect ternary sequences, see [13,15,16,19,20,22,27,31,32], to just list a few.

In digital communication, there are three ways to transmit signals: (i) varying amplitude, *called digital pulse amplitude modulation (PAM)* in a baseband case, and *amplitude shift keying (ASK)* in one dimensional signals and *quadrature amplitude modulation (QAM)* in two dimensional signals; (ii) varying phase, called *phase shift keying (PSK)*; (iii) varying frequency, *frequency shift keying (FSK)*. (The reader is referred to [11,28] for the basic concepts and theory about digital communications.) In the laster two methods, each element in a sequence (i.e., signal) has unit magnitude.

In this paper, we will consider ASK sequences and QAM sequences over $\mathbb{F}_p$ with almost perfect autocorrelation. From the sequences over $\mathbb{F}_p$, a finite field with $p > 2$ elements where p is prime, with the two-tuple balance and representing $\mathbb{F}_p$ in a symmetric way, we obtain the autocorrelation functions of those sequences, which are shown in Sect. 3. In Sect. 4, we give a new class of almost perfect quinary ASK sequences, and in Sect. 6, we show a class of almost perfect quinary QAM sequences constructed from almost perfect quinary ASK sequences.

2 Basic Concepts and Definitions

2.1 Notation

We list the following notation and basic properties, which will be used in the paper (for details of the theory of sequences, see [7]).

- $\mathbb{C}$ is the complex field, and $\mathbb{Z}$, an integer ring.
- p is a prime, $n > 1$ a positive integer, $\mathbb{F}_{p^n}$, a finite field with p^n elements, $\mathbb{F}_{p^n}^*$, the multiplicative group of the field, $q = p^n$, $d = \frac{p^n - 1}{p - 1}$, and $N = p^n - 1$.
- $t(x) = x^n - \sum_{i=0}^{n-1} c_i x^i$, $c_i \in \mathbb{F}_p$ is a primitive polynomial over $\mathbb{F}_p$ of degree n and $f(\alpha) = 0, \alpha \in \mathbb{F}_{p^n}$, so α is a primitive element in $\mathbb{F}_{p^n}$. If $\mathbf{a} = \{a_t\}$ satisfies $a_{n+j} = \sum_{i=0}^{n-1} c_i a_{i+j}, j \geq 0$, then $\mathbf{a}$ is an m-sequence of degree n, generated by $t(x)$.
- $Tr(x) = x + x^p + \cdots + x^{p^{n-1}}$ is the trace function from $\mathbb{F}_{p^n}$ to $\mathbb{F}_p$. We have $a_t = Tr(\beta \alpha^t)$, for some $\beta \in \mathbb{F}_{p^n}, t = 0, 1, \ldots$.
- Let $\mathbf{a} = \{a_t\}$ an arbitrary sequence over $\mathbb{F}_p$ of period N. Then we can write $a_t = f(\alpha^t), t = 0, 1, \ldots$ where $f(x)$ is a function from $\mathbb{F}_{p^n}$ to $\mathbb{F}_p$ with $f(0) = 0$, called the trace representation of $\mathbf{a}$, since $f(x)$ can be represented as the sum of monomial trace terms.

2.2 Symmetric Representation of Field Elements in $\mathbb{F}_p$

Throughout the paper, we represent the elements in $\mathbb{F}_p$ as

$$\mathbb{F}_p = \left\{ -\frac{p-1}{2}, \ldots, -2, -1, 0, 1, 2, \ldots, \frac{p-1}{2} \right\}. \tag{1}$$

Thus, $\mathbb{F}_3 = \{-1, 0, 1\}$, and $\mathbb{F}_5 = \{-2, -1, 0, 1, 2\}$.

2.3 Correlation Functions

Let two sequences $\mathbf{a} = \{a_t\}$ and $\mathbf{b} = \{b_t\}$ where $a_t, b_t \in \mathbb{C}$ (note that they may not have unit magnitude) and let their *cross correlation* be defined as

$$C_{\mathbf{a},\mathbf{b}}(\tau) = \sum_{t=0}^{N-1} a_t \bar{b}_{t+\tau} \tag{2}$$

where $\bar{x}$ is the complex conjugate of x and the computation is executed in the complex field. Or equivalently, $C_{\mathbf{a},\mathbf{b}}(\tau) = \langle \mathbf{a}, L^\tau \mathbf{b} \rangle$, the inner product of the vectors $\mathbf{a}$ and τ shift of $\mathbf{b}$, where L is the left-shift operator, i.e., $\mathbf{a} = (a_0, \ldots, a_{N-1})$, then $L\mathbf{a} = (a_1, \ldots, a_{N-1}, a_0)$. If $\mathbf{a} = \mathbf{b}$, then the crosscorrelation function becomes the *autocorrelation function*, denoted as $C_{\mathbf{a}}(\tau)$. When the elements of both $\mathbf{a}$ and $\mathbf{b}$ are real numbers, their correlation, given in (2) becomes

$$C_{\mathbf{a},\mathbf{b}}(\tau) = \sum_{t=0}^{N-1} a_t b_{t+\tau}, \tau = 0, 1, \ldots, . \tag{3}$$

The norm of the sequence $\mathbf{a} = \{a_i\}$ is defined as $||\mathbf{a}|| = \sqrt{\sum_{t=0}^{N-1} |a_t|^2}$ where $|x|^2 = x\overline{x}$.

We define the balance property of $\mathbf{a}$ of period N as follows. Let $\{i_0, \ldots, i_{k-1}\}$ be the subset of $\mathbb{C}$ which consists of all different elements in $\{a_i \mid i = 0, 1, \ldots, N-1\}$. Let $N_j = \{t \mid a_t = i_j, 0 \le t < N\}$. We say that $\mathbf{a}$ is *balanced* if $|N_j - N_i| \le 1$ for all $i \ne j$.

Definition 1. *We call a sequence* perfect *if it is balanced, and $C(\tau) = 0$ for $0 < \tau < N$ and $C(0) = ||\mathbf{a}||^2$; and a sequence* (ideal) two-level *autocorrelation value if it is balanced, and $C(\tau) = -1$, for $0 < \tau < N$ and $C(0) = ||\mathbf{a}||^2$. Furthermore, a sequence of period N is called* almost perfect *if $C(\tau) = 0$ for all $\tau : 0 < \tau < N$ but one.*

Example 1. Let $p = 3$, $n = 3$, $t(x) = x^3 - (x - 1)$, and $\{a_t\}$ an m-sequence generated by $t(x)$ with period 26, i.e.,

$$\{a_t\} = 0\ 0\ -1\ 0\ -1 \quad 1\ -1\ -1 \quad 1\ 0\ -1\ -1\ -1$$
$$0\ 0 \quad 1\ 0 \quad 1\ -1 \quad 1 \quad 1\ -1\ 0 \quad 1 \quad 1 \quad 1$$

The autocorrelation of $\{a_t\}$ is given by

$$C(\tau) = \sum_{t=0}^{25} a_t a_{t+\tau} = \quad 18\ 0\ 0\ 0\ 0\ 0\ 0\ 0\ 0\ 0\ 0\ 0\ 0$$
$$-18\ 0\ 0\ 0\ 0\ 0\ 0\ 0\ 0\ 0\ 0\ 0\ 0$$

Thus, $\{a_t\}$ is a balanced almost perfect ternary sequence. This almost perfect ternary sequence is equal to the one constructed in [25].

Note that many known almost perfect ternary sequences are not balanced.

2.4   Two-Tuple Balance

Let $\mathbf{a} = \{a_i\}$ be a sequence over $\mathbb{F}_p$ with period $N = p^n - 1$ and

$$S = \{(a_i, a_{i+\tau}) \mid i = 0, \ldots, N - 1\}.$$

Recall $d = (p^n - 1)/(p - 1)$.

Definition 2. *With the above notation, for $0 < \tau < p^n - 1$,*

1. *we say that $\mathbf{a}$ is* balanced *if each nonzero element in $\mathbb{F}_p$ occurs p^{n-1} times in one period of $\mathbf{a}$ and zero occurs $p^{n-1} - 1$ times; and*
2. *$\mathbf{a}$ is said to be* two-tuple balanced *when the following two conditions are satisfied:*
   - (a) *If τ is not a multiple of d, then each pair $(0,0) \ne (a, b) \in \mathbb{F}_p \times \mathbb{F}_p$ occurs in S exactly p^{n-2} times and $(0,0)$ occurs $p^{n-2} - 1$ times.*
   - (b) *If $\tau = id$, then $(a, ca), a, c \in \mathbb{F}_p^*$ occurs in S p^{n-1} times and $(0,0)$ occurs $p^{n-1} - 1$ times.*

Let $f(x)$ be the trace representation of $\mathbf{a}$. Sometimes, we interchange to use the term two-tuple balance to either $\mathbf{a}$ or $f(x)$. For $a \in \mathbb{F}_p^*$, we have $f(ax) = a^r f(x)$ where $\gcd(r, p-1) = 1$, then we say that $f(x)$ is $\mathbb{F}_p$ *homomorphic* (called d-form in [17]). Thus, $f(x)$ is $\mathbb{F}_p$ homomorphic, then the second property of the two-tuple balance is satisfied. From [8], we have the following result.

Property 1. If $\mathbf{a}$ a sequence over $\mathbb{F}_p$ with period N is two-tuple balanced, then it is balanced.

Note that until now there are only two known classes of sequences or functions which are two-tuple balanced. The following result on m-sequence is from [33] in 1959, and it can be easily extended to cascaded GMW sequences [18,30] as done in the literature.

Lemma 1. *Any m-sequence or GMW or cascaded GMW sequence over $\mathbb{F}_p$ are two-tuple balanced.*

Remark 1. The ideal distribution of the exponent sequences of m-sequences is investigated in [5], and almost all known almost perfect sequences are constructed using or indirectly using the exponent sequences of m-sequences through the relation to Singer difference sets or divisible difference sets. Especially, all interleaved constructions use the exponent sequences of m-sequences (see [7]). However, the two-tuple balance property is stronger than the ideal distribution of the exponent sequences, which is discussed in [8] using the term, called array structure.

2.5   ASK and QAM Signals and Autocorrelation of QAM

Let $T = \{\phi_i(t) : 0 \le i < h\}$ where $\phi_i(t)$ is integrable in the interval T and the norm of $\phi_i(t)$ is equal to unit. T is called an orthonormal set if $\langle \phi_t(t), \phi_j(t) \rangle = \delta_{ij}$ where

$$\delta_{ij} = \begin{cases} 1 & i = j \\ 0 & i \ne j. \end{cases}$$

Let $x(t) = \sum_{i=0}^{h-1} x_i \phi_i(t), x_i \in \mathbb{Z}$. If we draw the vector $(x_0, \ldots, x_{h-1})$, the coordinates of $x(t)$ in the space spanned by T, then it is referred to as a *signal point*. A diagram of all possible signal points is referred to as a *signal constellation* of the signal set (see [11,28]).

M-ary Digital Pulse Amplitude Modulation (PAM) and ASK Signals. Let

$$\phi_0(t) = \frac{1}{\sqrt{T}} \quad \text{and} \quad \psi_0(t) = \sqrt{\frac{2}{T}} \cos 2\pi f_c t, \ 0 \le t \le T, \tag{4}$$

where f_c is the carrier frequency and T is the symbol interval. With some variation from [28], we define a (symmetric) M-ary PAM (MPAM) or M-ary amplitude shift keying (MASK) sequence of period N as $\mathbf{w} = \{w_i\}$ where

$w_i = x_i\phi(t), x_i \in M_0$ or M_1 where

$$M_0 = \{-(M-1)a, -(M-3)a, \ldots, -3a, -a, a, 3a, \ldots, (M-3)a, (M-1)a\}$$
for M is even
$$M_1 = \{-\tfrac{M-1}{2}a, -(\tfrac{M-1}{2}-1)a, \ldots, -a, 0, a, \ldots, (\tfrac{M-1}{2}-1)a, \tfrac{M-1}{2}a\}$$
for M is odd

(5)

where $a > 0$, a constant and $\phi(t) = \phi_0(t)$ for M-ary PAM and $\phi(t) = \psi_0(t)$ for M-ary ASK (MASK). We note that M-ary PAM and M-ary ASK sequences have the same signal constellation (this means that their error probabilities can be determined in the same way, see [28]). Furthermore, after the RF down converter, the autocorrelation of an MASK sequence becomes the autocorrelation of the sequence $\{x_i\}$, which can be considered as a PAM sequence. Thus, when we talk about the autocorrelation of MASK, we mean the autocorrelation of $\{x_i\}$.

When $\{w_i\}$ is used in CW radar application, the transmitter transmits $\{w_i\}$. At the receiver side, after the RF down conversion if it is of MASK, the receiver computes the correlation, i.e., the autocorrelation in this case, between the incoming signal, which is a shift of $\mathbf{x}$, and a locally generated signal, a shift of $\mathbf{x}$ until it reaches the peak value, which is the autocorrelation function at zero. If the sequence is perfect, the shift corresponding to this peak value gives the round-trip delay time, and thus the range to the target is determined. If the sequence is not perfect, the other autocorrelation values will contribute to a wrong decision for the round-trip delay time. This is referred to as *autocorrelation detection*.

M-ary QAM Sequences and Their Autocorrelation. Let an orthonormal set with two functions be

$$\phi_0(t) = \sqrt{\frac{2}{T}}\cos(2\pi f_c t) \text{ and } \phi_1(t) = \sqrt{\frac{2}{T}}\sin(2\pi f_c t), \ 0 \le t \le T. \quad (6)$$

For a quadrature amplitude modulation (QAM), an M-ary QAM sequence $\mathbf{u} = \{u_i\}$ is defined as

$$u_i = x_i\phi_0(t) + y_i\phi_1(t), 0 \le t \le T, i = 0, 1, \ldots, M-1 \quad (7)$$

where $x_i, y_i \in S$, a finite subset of $\mathbb{Z}$. A receiver of M-ary QAM signal consists of two branches, called in-phase and quadrature branches (see [1]), respectively where the correlation detection is applied to each branch for detecting $\{x_i\}$ and $\{y_i\}$ individually. According to this receiver structure, in this paper, we define the autocorrelation of an M-ary QAM sequence as follows.

Definition 3. *The autocorrelation of $\mathbf{u}$ is defined as*

$$C_{QAM,\mathbf{u}}(\tau) = C_{\mathbf{x}}(\tau) + jC_{\mathbf{y}}(\tau)$$

$j = \sqrt{-1}$, *i.e., it is the linear combination of their respective autocorrelation functions of $\mathbf{x} = \{x_i\}$ and $\mathbf{y} = \{y_i\}$ with respective to the basis $\{1, j\}$.*

In this paper, a sequence we consider here has the elements taken from $\mathbb{F}_p$ represented in the symmetric way in (1) and the computation of the correlation is to treat those elements as integers in the integer ring. Thus, these sequences are not polyphase sequences for which they have unit amplitude for each element in the computation of correlation.

3   Autocorrelation of Sequences over $\mathbb{F}_p$ with Two-Tuple Balance

Theorem 1. *Let* **a** *be a two-tuple balanced sequence over* $\mathbb{F}_p$ *with period* N. *Then the autocorrelation function of* **a** *is given by*

$$C_{\mathbf{a}}(\tau) = \begin{cases} \frac{1}{12}p^n(p^2 - 1), & \tau \equiv 0 \mod N \\ 0, & \tau \not\equiv 0 \mod d, \tau \not\equiv 0 \mod N \\ -p^{n-1}C_{\mathbf{v}}(i), & \tau = id, \ \tau \not\equiv 0 \mod N \end{cases}$$

where $\mathbf{v} = (v_0, \ldots, v_{p-2})$ *is a permutation of* $\mathbb{F}_p^*$.

Proof. **Case 1.** $\tau = 0 \mod N$. From Property 1, we have

$$C_{\mathbf{a}}(0) = \sum_{t=0}^{N-1} a_t^2 = 2 \times p^{n-1}\left(1 + 2 + \cdots + \frac{p-1}{2}\right).$$

The results follows immediately from the sum of the squares of the consecutive integers.

Case 2. $\tau \not\equiv 0 \mod d, \tau \not\equiv 0 \mod N$. Since **a** satisfies the two-tuple balance property, then each $(0,0) \neq (a,b) \in \mathbb{F}_p^2$ occurs p^{n-2} times in $S = \{(a_t, a_{t+\tau}) \mid t = 0, \ldots, N-1\}$ and $(0,0)$ occurs $p^{n-2} - 1$ times. Note that $(a_t, a_{t+\tau}) = (0,0)$ gives zero in $C_{\mathbf{a}}(\tau)$. For each nonzero (a,b), it has another pair $(a, -b)$, which produce their respective products ab and $-ab$ in $C_{\mathbf{a}}(\tau)$. Thus, all terms are cancelled in the sum of $C_{\mathbf{a}}(\tau)$. Thus, the assertion follows.

Case 3. $\tau \equiv 0 \mod d, \tau \not\equiv 0 \mod N$. According to the two-tuple balance, we have

$$C_{\mathbf{a}}(id) = \sum_{t=0}^{N-1} a_t a_{t+\tau} = p^{n-1}\sum_{a \in F_p^*} a(ia) = p^{n-1}C_{\mathbf{v}}(i)$$

where $C_{\mathbf{v}}(i) = \sum_{a \in F_p^*} a(ia)$ where $\mathbf{v} = (v_0, \ldots, v_{p-2})$, a permutation of $\mathbb{F}_p$. $\quad\square$

Example 2. In the following, we use Theorem 1 and compute the other values of $C_{\mathbf{a}}(id)$. Those are almost perfect quinary sequences.

Case 1. Let $p = 5$, $n = 2$, $t(x) = x^2 - (x - 1)$, and $\{a_i\}$ an m-sequence with period 24 generated by $t(x)$, i.e.,

$$2 \quad 1 \quad 2\,0 \quad 1 \quad 1 -1 \quad 2 -1\,0 \quad 2 \quad 2$$
$$-2 -1 -2\,0 -1 -1 \quad 1 -2 \quad 1\,0 -2 -2$$

According to Theorem 1 and the computation of $C_\mathbf{a}(id)$ where $d = 6$, the auto-correlation spectrum is given by

$$50\ 0\ 0\ 0\ 0\ 0\ 0\ 0\ 0\ 0\ 0\ 0\ 0$$
$$-50\ 0\ 0\ 0\ 0\ 0\ 0\ 0\ 0\ 0\ 0\ 0\ 0$$

Case 2. Let $p = 5$, $n = 3$, $t(x) = x^3 - (2x + 2)$, and $\{a_i\}$, an m-sequence of period 124 generated by $t(x)$, and the first 31 elements are given by

$$-2, 0, -1, 1, -2, 0, -2, 1, 1, -2, -1, -2, -1, -1, -1, 1, 1, 0,$$
$$-1, 2, -2, 2, 0, 0, -1, 0, -2, -2, 1, 2, -2.$$

According to Theorem 1 and the computation of $C_\mathbf{a}(id)$ where $d = 31$, the autocorrelation spectrum $C_\mathbf{a}(\tau), \tau = 0, 1, \ldots, 123$ is given by

$$C_\mathbf{a}(\tau) = \begin{cases} 250, & \tau = 0 \\ 0, & \tau \neq 62, 0 < \tau < 124 \\ -250, & \tau = 62 \end{cases}$$

Remark 2. For $p = 3$, the construction in Theorem 1 gives

$$C_\mathbf{a}(\tau) = \begin{cases} 2 \times 3^{n-1}, & \tau = 0 \\ 0, & 0 < \tau < N, \tau \neq d \text{ where } d = \frac{3^n - 1}{2} \\ -2 \times 3^{n-1}, & \tau = d \end{cases}$$

which is almost perfect. However, this is either equal to or is the complement of the almost perfect ternary sequences constructed in [25]. Thus, it does not give new almost perfect ternary sequences.

4   Almost Perfect Quinary ASK Sequences

In the two sequences constructed in Example 2, we have seen the examples of almost perfect quinary ASK sequences. In the following, we present a general construction for a new class of almost perfect quinary ASK sequences.

Theorem 2. *Let* $\mathbf{a}$ *be an* m-*sequence, or a GMW sequence or a cascaded GMW sequence over* $\mathbb{F}_5$, *represented in the symmetric way, of period* $N = 5^n - 1$. *Then* $\mathbf{a}$ *is almost perfect and its autocorrelation function is given by*

$$C_\mathbf{a}(\tau) = \begin{cases} 10 \times 5^{n-1}, & \tau = 0 \\ 0, & 0 < \tau < N, \tau \neq 2d, \text{ where } d = \frac{5^n - 1}{4} \\ -10 \times 5^{n-1}, & \tau = 2d. \end{cases}$$

Proof. In the following, we only show its detailed proof for $\mathbf{a}$ being an m-sequence. The proof for the other cases are similar. From Theorem 1, for $p = 5$, we need to determine $\tau = id$ for $i = 1, 2, 3$ where $d = \frac{5^n - 1}{4}$. We will use the trace representation of the m-sequence $\mathbf{a} = \{a_t\}$, i.e., $a_t = Tr(\beta\alpha^t), t = 0, 1, \ldots$. For $\tau = id, 0 < i < 4$, we have

$$a_{t+\tau} = Tr(\beta\alpha^{t+\tau}) = Tr(\beta\alpha^{id_5}\alpha^t) = \alpha^{id_5}Tr(\beta\alpha^t).$$

Note that $h = \alpha^d$ is a primitive element of $\mathbb{F}_5$. Thus we have $\mathbf{v} = (1, h, h^2, h^3)$ where h^i first reduced by modular 5, then represented in $\mathbb{F}_5$ in the symmetric way. Since there are only two primitive element in $\mathbb{F}_5 = \{-2, -1, 0, 1, 2\}$, which are 2 or -2. Thus, there are only two possibilities for $\mathbf{v}$: $\mathbf{v} = (1, 2, -1, -2)$ or $\mathbf{v} = (1, -2, -1, 2)$. For each case, we have

$$C_{\mathbf{v}}(1) = C_{\mathbf{v}}(3) = 0 \text{ and } C_{\mathbf{v}}(2) = -10.$$

(In fact, those two are decimation equivalent, i.e., one can be obtained from by the decimation operation.) Using Theorem 1, the result is true.

For the case of $\mathbf{a}$ being a GMW or cascaded GMW sequence, let $f(x)$ be its trace representation, then $f(x)$ is $\mathbb{F}_5$ homomorphic. Thus, for $\tau = id$, we have $a_t = f(\alpha^{t+\tau}) = \alpha^{rid}f(\alpha^t)$ where α^{rd} is a primitive element of $\mathbb{F}_5$. Thus the results are established for both GMW or cascaded GMW sequences. $\square$

Remark 3. The result of Theorem 2 is true when $f(x)$ is any function from $\mathbb{F}_{5^n}$ to $\mathbb{F}_5$ which are two-tuple balanced. However, the current known classes of 2-tuple balanced sequences are those considered above. Furthermore, Theorem 2 yields a new class of odd perfect quinary sequences.

5   Almost Perfect Quinary QAM Sequences

In this section, we give a construction for an almost perfect quinary QAM sequence which is constructed through almost perfect quinary sequences in Theorem 2. Let $\mathbf{a} = \{a_i\}$ be an almost perfect quinary sequence in Theorem 2, and $\mathbf{b} = \{b_i\}$ where $b_i = a_i^3$ or $b_i = ca_i^3, c \in \mathbb{F}_5^*$. Let $\mathbf{u} = \{u_i\}$ be a QAM sequence whose elements are given by

$$u_i = a_i\phi_0(t) + b_i\phi_1(t), i = 0, 1, \ldots, N - 1 \tag{8}$$

where $\phi_i(t), i = 0, 1$ are defined in (6) Sect. 2. The following result is directly obtained from the interleaved structure of m-sequences or GMW or cascaded GMW sequences.

Property 2. Let

$$T_0 = \{(0,0), (1,1), (2,-2), (-1,-1), (-2,2)\}$$
$$T_1 = \{(0,0), (1,-2), (2,-1), (-1,2), (-2,1)\}$$
$$T_2 = \{(0,0), (1,-1), (2,2), (-1,1), (-2,-2)\}$$
$$T_3 = \{(0,0), (1,2), (2,1), (-1,-2), (-2,-1)\}.$$

The set consisting of different elements in multi set $\{(a_i, b_i), \,|\, i = 0, \ldots, N-1\}$ of $\mathbf{u}$ is equal to one of $T_i, i = 0, 1, 2, 3$. Thus, $\mathbf{u}$ is a quinary QAM sequence.

Theorem 3. *With the above notation,* **u** *is an almost perfect quinary sequence of period* N *with the following autocorrelation values*

$$C_{QAM,\mathbf{u}}(\tau) = \begin{cases} 20 \times 5^{n-1}, & \tau = 0 \\ 0, & 0 < \tau < N, \tau \neq 2d, \text{ where } d = \frac{5^n - 1}{4} \\ -20 \times 5^{n-1}, & \tau = 2d. \end{cases} \quad (9)$$

Proof. By the definition, the autocorrelation of **u** is given as follows

$$C_{QAM,\mathbf{u}}(\tau) = C_{\mathbf{a}}(\tau) + jC_{\mathbf{b}}(\tau). \quad (10)$$

From the proof of Theorem 2, the quinary sequences **a** and **b** have the identical autocorrelation functions. Thus (9) follows immediately. Hence **u** is almost perfect.   □

The two different signal constellations of **u**, given by T_0 and T_1, are plotted in Fig. 1, and the other two are rotated by 90° from those two. Although they have the same minimum distance, the distance between any two signal points in T_0 is at least or larger than those in T_1.

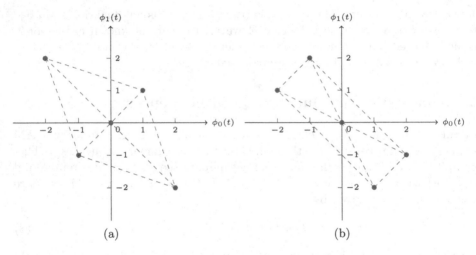

(a)   (b)

Fig. 1. Signal constellation of 5-ary QAM with minimum distance $\sqrt{2}$: (a). T_0 and (b). T_1.

6   Concluding Remarks and Their Applications

In this paper, using the known sequences with two-tuple balance property, we provide a simple construction for almost perfect quinary ASK sequences and almost perfect quinary QAM sequences. These new almost perfect quinary ASK sequences and QAM sequences can be used effectively in (continuous wave) CW radar as well as channel estimation. In the later case, its zero autocorrelation zone is one half of the period of the sequence, which is doubled the length of zero correlation zone of the Golay sequences discussed in [10].

Further investigation is needed for performance comparisons of those sequences in terms their efficiency of implementation as well as their signal-to-noise ratio in both CW radar detection and channel estimation. On the other hand, how to construct almost perfect M-ary ASK sequences for $M = p, p > 5$ or $M = 2^k$ and M'-ary QAM sequences for $M' = M^2$ deserves further study.

Another interesting problem is that the known constructions for almost perfect sequences are strongly related the constructions of odd perfect sequences. In other words, if a sequence is almost perfect with the nonzero autocorrelation value located in the middle point of the shifts, then the sequence is odd perfect. This relation can be seen, say such as [19, 23]. The construction of odd perfect sequences given by Luke in [23] are from a special case of [3], which are not balanced. Those constructed sequences in [19] are not balanced as well according to the ideal symbol distribution. We have seen that the almost perfect quinary sequences constructed in this paper are balanced. It is interested to see how we can construct balanced odd perfect sequences through balanced almost perfect sequences for other values of p where $p \notin \{3, 5\}$.

Acknowledgement. The authors wish to thank Matthew Parker for pointing out that the new almost perfect quinary sequences in Theorem 3 also have odd perfect autocorrelation. The work is supported by NSERC and ORF.

References

1. Barry, J.R., Lee, E.A., Messerschmitt, D.G.: Digital Communication, 3rd edn. Kluwer Academic Publishers, London (2004)
2. Chu, D.C.: Polyphase codes with good periodic correlation properties. IEEE Trans. Inf. Theory **18**, 531–533 (1972)
3. Delsarte, P., Goethals, J.M., Seidel, J.J.: Orthogonal matrices with zero diagonal. II. Can. J. Math. **23**(5), 816–832 (1971)
4. Frank, R., Zadoff, S., Heimiller, R.: Phase shift pulse codes with good periodic correlation properties. IRE Trans. Inf. Theory **8**(6), 381–382 (1962)
5. Games, R.A.: Crosscorrelation of m-sequences and GMW sequences with the same primitive polynomial. Discrete Appl. Math. **12**, 139–146 (1985)
6. Golomb, S.W.: Shift Register Sequences, Holden-Day Inc, San Francisco, 1967, revised edn. Aegean Park Press, Laguna Hills, CA (1982)
7. Golomb, S.W., Gong, G.: Signal Design for Good Correlation for Wireless Communication, Cryptography, and Radar. Cambridge University Press, Cambridge (2005)
8. Gong, G., Song, H.Y.: Two-tuple balance of non-binary sequences with ideal two-level autocorrelation. Discrete Appl. Math. **154**, 2590–2598 (2006)
9. Gong, G.: Character sums and polyphase sequence families with low correlation, DFT and ambiguity. In: Winterhof, A., et al. (eds.) Character Sums and Polynomials, pp. 1–43. De Gruyter, Germany (2013)
10. Gong, G., Huo, F., Yang, Y.: Large zero autocorrelation zone of golay sequences and their applications. IEEE Trans. Commun. **61**(9), 3967–3979 (2013)

11. Haykin, S.: Digital Communications. Wiley, New York (1988)
12. Helleseth, T.: Some results about the cross-correlation functions between two maximal length linear sequences. Discrete Math. **16**(3), 209–232 (1976)
13. Hoholdt, T., Justesen, J.: Ternary sequences with perfect periodic auto-correlation. IEEE Trans. Inf. Theory **29**(4), 597–600 (1983)
14. Helleseth, T., Kumar, P.V.: Sequences with low correlation. In: Pless, V.S., Huffman, W.C. (eds.) Handbook of Coding Theory, vol. 2, pp. 1765–1853. Elsevier, Amsterdam (1998)
15. Ipatov, V.P.: Ternary sequences with ideal autocorrelation properties. Radio Eng. Electron. Phys. **24**, 75–79 (1979)
16. Jungnickel, D., Pott, A.: Perfect and almost perfect sequences. Discrete Appl. Math. **95**(1), 331–359 (1999)
17. Klapper, A.: d-form sequences: families of sequences with low correlation values and large linear spans. IEEE Trans. Inf. Theory **41**(2), 423–431 (1995)
18. Klapper, A., Chan, A.H., Goresky, M.: Cascaded GMW sequences. IEEE Trans. Inf. Theory **39**(1), 177–183 (1993)
19. Krengel, E.I.: Almost-perfect and odd-perfect ternary sequences. In: Helleseth, T., Sarwate, D., Song, H.-Y., Yang, K. (eds.) SETA 2004. LNCS, vol. 3486, pp. 197–207. Springer, Heidelberg (2005)
20. Krengel, E.I.: Some constructions of almost-perfect, odd-perfect and perfect polyphase and almost-polyphase sequences. In: Carlet, C., Pott, A. (eds.) SETA 2010. LNCS, vol. 6338, pp. 387–398. Springer, Heidelberg (2010)
21. Levanon, N.: Radar Principles. Wiley-Interscience, New York (1988)
22. Langevin, P.: Some sequences with good autocorrelation properties. Finite Fields **168**, 175–185 (1994)
23. Lüke, H.D., Schotten, H.D.: Odd-perfect, almost binary correlation sequences. IEEE Trans. Electron. Syst. **31**(1), 495–498 (1995)
24. Lüke, H.D.: Almost-perfect polyphase sequences with small phase alphabet. IEEE Trans. Inf. Theory **43**, 361–363 (1997)
25. Lüke, H.D.: Almost-perfect quadriphase sequences. IEEE Trans. Inf. Theory **47**, 2607–2608 (2001)
26. Milewski, A.: Periodic sequences with optimal properties for channel estimation and fast start-up equalization. IBM J. Res. Dev. **27**(5), 425–431 (1983)
27. Pott, A., Bradley, S.P.: Existence and nonexistence of almost-perfect autocorrelation sequences. IEEE Trans. Inf. Theory **41**, 301–304 (1995)
28. Pursley, M.B.: Introduction to Digital Communications. Prentice Hall, Upper Saddle River (2005)
29. Sarwate, D.V., Pursley, M.B.: Crosscorrelation properties of pseudorandom and related sequences. Proc. IEEE **68**(5), 593–619 (1980)
30. Scholtz, R.A., Welch, L.R.: GMW sequences. IEEE Trans. Inf. Theory **30**(3), 548–553 (1984)
31. Shedd, D.A., Sarwate, D.V.: Construction of sequences with good correlation properties. IEEE Trans. Inf. Theory **25**, 94–97 (1979)
32. Wolfmann, J.: Almost perfect autocorrelation sequences. IEEE Trans. Inf. Theory **38**(4), 1412–1418 (1992)
33. Zierler, N.: Linear recurring sequences. J. Soc. Ind. Appl. Math. **7**(1), 31–48 (1959)

Correlation of Arrays

Correlation of Areas

Inflation of Perfect Arrays Over the Basic Quaternions of Size $mn = (q+1)/2$

Santiago Barrera Acevedo[✉]

Monash University, Melbourne, Australia
Santiago.Barrera.Acevedo@monash.edu

Abstract. Arasu and de Launey showed that every perfect quaternary array (that is perfect arrays over the four roots of unity $\pm 1, \pm i$) of size $m \times n$, can be inflated into another perfect quaternary array of size $mp \times np$, provided $p - mn - 1$ is s prime number. Likewise, they showed that every perfect quaternary array of size $m \times n$, can be inflated into another perfect quaternary array of size $mq \times nq$, provided $q = 2mn - 1$ is a prime number and $q \equiv 3 \pmod 4$. Following from Arasu and de Launey's first construction, Barrera Acevedo and Jolly showed that every perfect array over the basic quaternions, $\{1, -1, i, -i, j, -j, k, -k\}$, of sizes $m \times n$, can be inflated into a new perfect array over the basic quaternions of size $mp \times np$, provided $p = mn - 1$ is s prime number. Combining this construction with the existence of infinitely many modified Lee sequences over $\{1, -1, i, -i, j\}$ (in the sense of Barrera Acevedo and Hall), they showed the existence of infinitely many perfect arrays over the basic quaternions, with appearances of all the basic quaternion elements $1, -1, i, -i, j, -j, k$ and $-k$. In this work, we show that every perfect array over the basic quaternions, of size $m \times n$, can be inflated into a perfect quaternary array of size $mq \times nq$, provided $q = 2mn - 1$ is a prime number and $q \equiv 3 \pmod 4$.

Keywords: Perfect arrays over the basic quaternions · Perfect autocorrelation · Perfect arrays · Quaternions

1 Introduction

Perfect sequences and arrays over the quaternions $\mathbb{H}$ were recently introduced by O. Kuznetsov (2009) and S. Barrera Acevedo (2013), respectively. A proof of the existence of perfect sequences of unbounded lengths over the basic quaternions $\mathbb{H}_8 = \{\pm 1, \pm i, \pm j, \pm k\}$ was recently presented by Barrera Acevedo and Hall (2012). In 2014 Barrera Acevedo and Jolly (2014) generalised an algorithm of Arasu and de Launey (2001) to inflate perfect arrays over four roots of unity, of size $mn = p + 1$, where p is a prime number, to perfect arrays over the basic quaternions. They showed that every array A, of size $m \times n$ over the basic quaternions $\{1, -1, i, -i, j, -j, k, -k\}$, with $p = mn - 1$ a prime number, can be inflated into another perfect array over the basic quaternions of size $mp \times np$.

© Springer International Publishing Switzerland 2014
K.-U. Schmidt and A. Winterhof (Eds.): SETA 2014, LNCS 8865, pp. 123–133, 2014.
DOI: 10.1007/978-3-319-12325-7_11

Following from this, they showed the existence of a family of perfect arrays of unbounded sizes over the basic quaternions (2014).

In this work we generalise another algorithm of Arasu and de Launey (2001) to inflate perfect arrays over four roots of unity, of size $mn = 2q + 1$, where q is a prime number and $q \equiv 3(mod\ 4)$, to perfect arrays over the basic quaternions. We will show that every array A, of size $m \times n$ over the basic quaternions $\{1, -1, i, -i, j, -j, k, -k\}$, with $mn = 2q + 1$, q is a prime number and $q \equiv 3(mod\ 4)$, can be inflated into a perfect array over the basic quaternions of size $mq \times nq$.

2   Preliminaries

A **finite** $m \times n$ **array** $A = (a(r, s))$, where $0 \leq r \leq m$ and $0 \leq s \leq n$, over a set $\mathcal{A}$, is a list of $m \times n$ elements taken from $\mathcal{A} \subset \mathbb{C}$, where repetition is allowed and $\mathbb{C}$ denotes the set of complex numbers. The number $m \times n$ is called the size of the array and $\mathcal{A}$ is called the alphabet. The **shift** of the array A by (t_0, t_1) places is $A^{(t_0,t_1)} = (a(r + t_0, s + t_1))$, where subscripts are calculated modulo m and n, respectively. The **periodic** (t_0, t_1)**-autocorrelation value** of the array A is the inner product of A and $A^{(t_0,t_1)}$, which is given by:

$$AC_A(t_0, t_1) = \sum_{r=0}^{m-1} \sum_{s=0}^{n-1} a(r, s)a^*(r + t_0, s + t_1), \qquad (1)$$

for $0 \leq t_0 \leq m$ and $0 \leq t_1 \leq n$. The indices $r + t_0$ and $s + t_1$ are reduced modulo m and n, respectively, and a^* denotes the conjugate of a. The $m \times n$ array $AC_A = (AC_A(t_0, t_1))$, where $0 \leq t_0 \leq m$ and $0 \leq t_1 \leq n$, of all the autocorrelation values of A, is called the **autocorrelation array** of A. The auto-correlation value $AC_A(0, 0)$ is called the **peak-value** and all the other autocorrelation values are called **off-peak values**. We say that the array A has **constant off-peak autocorrelation**, if all its off-peak autocorrelation values are equal. We call the array A **perfect**, if all its off-peak autocorrelation values are zero.

3   Perfect Arrays Over the Quaternions

Definition 1. *The quaternion algebra* $\mathbb{H}$ *is defined as follows: It is an algebra generated by the elements* i *and* j, *over the real number field* $\mathbb{R}$, *with the following multiplication rules* $i^2 = -1$, $j^2 = -1$ *and* $ij = -ji$. *This last equation makes the algebra non-commutative. For the sake of simplicity the product* ij *is denoted by* k. *Then* $ij = k$, $jk = i$, $ki = j$ *and* $ji = -k$, $kj = -i$, $ik = -j$. *This algebra has conjugation given by* $i^* = -i$, $j^* = -j$ *and* $k^* = -k$. *The quaternion algebra can be regarded as a 4-dimensional* $\mathbb{R}$-*vector space with basis vectors* $1 = (1, 0, 0, 0)$, $i = (0, 1, 0, 0)$, $j = (0, 0, 1, 0)$ *and* $k = (0, 0, 0, 1)$. *The norm of a quaternion* $q = a + ib + jc + kj$, *denoted* $\|q\|$, *is defined by* $\|q\| = qq^* = a^2 + b^2 + c^2 + d^2$.

Since the quaternion algebra $\mathbb{H}$ is non-commutative, left and right cross-correlation and autocorrelation definitions for arrays over the quaternions need to be introduced.

Definition 2. *Let $A = (a(r,s))$ and $B = (b(r,s))$ be two arrays of size $m \times n$ over an arbitrary quaternion alphabet. For any pair of integers (t_0, t_1), the (t_0, t_1)-**right** and **left periodic cross-correlation** values of A and B are*

$$CC_{A,B}^R(t_0, t_1) = \sum_{r=0}^{m-1}\sum_{s=0}^{n-1} a(r,s)b^*(r+t_0, s+t_1) \tag{2}$$

and

$$CC_{A,B}^L(t_0, t_1) = \sum_{r=0}^{m-1}\sum_{s=0}^{n-1} a^*(r,s)b(r+t_0, s+t_1), \tag{3}$$

*respectively. The indices $r+t_0$ and $s+t_1$ are calculated modulo m and n, respectively. When $A = B$, we denote $CC_{A,B}^R(t_0, t_1)$ and $CC_{A,B}^L(t_0, t_1)$ by $AC_A^R(t_0, t_1)$ and $AC_A^L(t_0, t_1)$, respectively, and they are called the (t_0, t_1)-**right** and **left periodic autocorrelation** values of A. Also, as usual, the autocorrelation value of A, for the shift $(0,0)$, is called the **peak value**. The right and left autocorrelation values of A, for all pairs $(t_0, t_1) \neq (0,0)$, are called **right and left off-peak values**.*

The next theorem presents an important property of perfect arrays over the quaternions, namely, right perfection of any array is equivalent to left perfection. This theorem generalises a result for sequences over quaternions (Kuznetsov 2009) to arrays over the quaternions. In preparation for this theorem, we introduce the following lemma.

Lemma 1. *Let $A = (a(r,s))$ be any two-dimensional array of size $m \times n$, with elements in the quaternion algebra $\mathbb{H}$ and let*

$$AC_A^R(u,v) = \sum_{r=0}^{m-1}\sum_{s=0}^{n-1} a_{r,s}a_{r+u,s+v}^* \tag{4}$$

$$C_A^L = \sum_{t=0}^{m-1}\sum_{s=0}^{n-1} a_{r,s}^* a_{r+u,s+v} \tag{5}$$

be the right and left autocorrelation functions of the array A, respectively. Then

$$\sum_{u=0}^{m-1}\sum_{v=0}^{n-1} \|AC_A^L(u,v)\| = \sum_{t_1=0}^{m-1}\sum_{t_2=0}^{m-1}\sum_{s_1=0}^{n-1}\sum_{s_2=0}^{n-1} a_{t_1,s_1}^* (AC_A^R(t_2 - t_1, s_2 - s_1)) a_{t_2,s_2},$$

$$\tag{6}$$

and

$$\sum_{u=0}^{m-1}\sum_{v=0}^{n-1} \|AC_A^R(u,v)\| = \sum_{t_1=0}^{m-1}\sum_{t_2=0}^{m-1}\sum_{s_1=0}^{n-1}\sum_{s_2=0}^{n-1} a_{t_1,s_1} (AC_A^L(t_2 - t_1, s_2 - s_1)) a_{t_2,s_2}^*.$$

$$\tag{7}$$

Proof.

$$\sum_{u=0}^{m-1}\sum_{v=0}^{n-1} \|AC_A^L(u,v)\| =$$

$$\sum_{u=0}^{m-1}\sum_{v=0}^{n-1} \| \sum_{t=0}^{m-1}\sum_{s=0}^{n-1} a_{t,s}^* a_{t+u,s+v}\| =$$

$$\sum_{u=0}^{m-1}\sum_{v=0}^{n-1} \left(\sum_{t_1=0}^{m-1}\sum_{s_1=0}^{n-1} a_{t_1,s_1}^* a_{t_1+u,s_1+v}\right) \left(\sum_{t_2=0}^{m-1}\sum_{s_2=0}^{n-1} a_{t_2,s_2}^* a_{t_2+u,s_2+v}\right)^* = \tag{8}$$

$$\sum_{u=0}^{m-1}\sum_{v=0}^{n-1}\sum_{t_1=0}^{m-1}\sum_{s_1=0}^{n-1}\sum_{t_2=0}^{m-1}\sum_{s_2=0}^{n-1} a_{t_1,s_1}^* a_{t_1+u,s_1+v} a_{t_2+u,s_2+v}^* a_{t_2,s_2} =$$

$$\sum_{t_1=0}^{m-1}\sum_{s_1=0}^{n-1}\sum_{t_2=0}^{m-1}\sum_{s_2=0}^{n-1} a_{t_1,s_1}^* \left(\sum_{u=0}^{m-1}\sum_{v=0}^{n-1} a_{t_1+u,s_1+v} a_{t_2+u,s_2+v}^*\right) a_{t_2,s_2} =$$

$$\sum_{t_1=0}^{m-1}\sum_{s_1=0}^{n-1}\sum_{t_2=0}^{m-1}\sum_{s_2=0}^{n-1} a_{t_1,s_1}^* \left(AC_A^R(t_2-t_1,s_2-s_1)\right) a_{t_2,s_2}.$$

The second equation is proved in a similar way.

Theorem 1. *Let A be any array over an arbitrary quaternion alphabet. Then the array A is right perfect if and only if it is left perfect.*

Proof. Assume that A is a right perfect array. We will show that the sum of the norms of the left off-peak autocorrelation values $\sum_{u=0}^{m-1}\sum_{v=0}^{n-1}\|AC_A^L(u,v)\|$
$(u,v)\neq(0,0)$
is equal to zero, for all $(u,v) \neq (0,0)$. By Lemma (1), Eq. (6) we have

$$\sum_{u=0}^{m-1}\sum_{v=0}^{n-1} \|AC_A^L(u,v)\| = \sum_{t_1=0}^{m-1}\sum_{t_2=0}^{m-1}\sum_{s_1=0}^{n-1}\sum_{s_2=0}^{n-1} a_{t_1,s_1}^* (AC_A^R(t_2-t_1,s_2-s_1)) a_{t_2,s_2}, \tag{9}$$

Since A is right perfect, all right autocorrelation values are equal to zero, for all shifts $(u,v) \neq (0,0)$. Also, it is true that $AC_A^L(0,0) = AC_A^R(0,0)$. Then $AC_A^R(t_2-t_1,s_2-s_1) = 0$, for $t_1 \neq t_2$ or $s_1 \neq s_1 \neq s_2$. In this way, the Eq. (9) above continues to

$$\sum_{u=0}^{m-1}\sum_{v=0}^{n-1} \|AC_A^L(u,v)\| = \sum_{t_1=0}^{m-1}\sum_{s_1=0}^{n-1} a_{t_1,s_1}^* a_{t_1,s_1} AC_A^R(0,0). \tag{10}$$

Thus,

$$\|AC_A^L(0,0)\| + \sum_{\substack{u=0 \\ (u,v)\neq(0,0)}}^{m-1}\sum_{v=0}^{n-1} \|AC_A^L(u,v)\| = AC_A^L(0,0)AC_A^R(0,0). \tag{11}$$

It follows that

$$\sum_{\substack{u=0 \\ (u,v)\neq(0,0)}}^{m-1}\sum_{v=0}^{n-1} \|AC_A^L(u,v)\| = 0. \tag{12}$$

Since the sum of non-negative real numbers is equal to zero, we have that every summand is necessarily equal to zero. Thus, $\|AC_A^L(u,v)\| = 0$, for $(u,v) \neq (0,0)$. So, A is left perfect by definition. The other direction of the statement is proved similarly.

Henceforth, we will say that an array over the quaternions is perfect, if it is right (or left) perfect.

Example 1. The array $\begin{pmatrix} 1,i \\ j,k \end{pmatrix}$ over the quaternions is perfect.

4   Inflation of Perfect Arrays Over the Basic Quaternions

We modify the algorithm of Arasu and de Launey (2001), for inflating perfect quaternary arrays, into an algorithm to inflate perfect arrays over the basic quaternions. The new arrays will have larger size and perfect autocorrelation. We will inflate arrays of size $m \times n$, into arrays of size $mq \times nq$, provided $q = 2mn - 1$ is a prime number and $q \equiv 3(mod\ 4)$.

Definition 3. *For every prime number q, **Legendre sequences**, denoted $L_q = (s_t)$, are defined by*

$$s_t = \begin{cases} 0, & \textit{if } t = 0 \\ 1, & \textit{if } t \textit{ is a quadratic residue mod } q \\ -1, & \textit{if } t \textit{ otherwise} \end{cases} \tag{13}$$

Legendre sequences autocorrelation off-peak values are all equal to -1.

Theorem 2. *If there is a perfect array, over the basic quaternions $\{\pm1, \pm i, \pm j, \pm k\}$, of size $m \times n$, where $q = 2mn - 1$ is a prime number and $q \equiv 3(mod\ 4)$, then there is a perfect array of size $mq \times nq$, over the basic quaternions.*

4.1   Construction

Let A be a perfect array of size $m \times n$ over the basic quaternions, where $q = 2mn - 1$ is a prime number and $q \equiv 3(mod\ 4)$.
(1) Take a Legendre sequence $L_q = (0, s_1, \ldots, s_{q-1})$ of length q and replace the element 0 by $i^{\frac{q+1}{2}}$, to obtain the sequence $S_i = (i^{\frac{q+1}{2}}, s_1, \ldots, s_{q-1})$. The element 0 can also be replaced by $j^{\frac{q+1}{2}}$ or $k^{\frac{q+1}{2}}$, producing the sequence $S_j = (j^{\frac{q+1}{2}}, s_1, \ldots, s_{q-1})$ or $S_k = (k^{\frac{q+1}{2}}, s_1, \ldots, s_{q-1})$, respectively and the following construction is valid for each of these sequences. All three sequences have the same autocorrelation values $(q, -1, \ldots, -1)$. In this construction we use the sequence $S = S_i$.

(2) Produce $q + 1$ arrays, called inflation arrays, from the sequence S and the shifts S^t of S, for $t = 0, 1, \ldots, q - 1$, as follows:

$$B_0 = \begin{pmatrix} S & S & \ldots & S \\ \downarrow & \downarrow & & \downarrow \end{pmatrix}^T, B_1 = \begin{pmatrix} S & S^1 & \ldots & S^{q-1} \\ \downarrow & \downarrow & & \downarrow \end{pmatrix},$$

$$B_2 = \begin{pmatrix} S & S^{(1)2} & \ldots & S^{(q-1)2} \\ \downarrow & \downarrow & & \downarrow \end{pmatrix}, \ldots, B_{q-1} = \begin{pmatrix} S & S^{(1)(q-1)} & \ldots & S^{(q-1)(q-1)} \\ \downarrow & \downarrow & & \downarrow \end{pmatrix},$$

$$B_q = \begin{pmatrix} S & S & \ldots & S \\ \downarrow & \downarrow & & \downarrow \end{pmatrix}$$

(14)

(The following step in this construction introduces the main variation from the construction presented by Barrera Acevedo and Jolly (2014)).

(3) Construct $\frac{q+1}{2}$ inflation arrays of size $q \times q$: for $r = 0, 1, \ldots, \frac{q+1}{2}$, and put

$$C_r = \left(\frac{1+i}{2} \right) (B_{2r} + iB_{2r+1}) \tag{15}$$

All the entries of this matrix are complex fourth roots of unity.

(4) Arrange the arrays $A, C_0, C_1, \ldots, C_{\frac{q-1}{2}}$ into a four-dimensional array D of size $m \times q \times q \times n$ as follows: if $c_{r+ms}(u, v)$ is the (u, v) entry of the inflation array D_{r+ms}, then for $0 \le r \le m - 1$, $0 \le s \le n - 1$ and $0 \le u, v \le q - 1$, we put

$$d(r, u, v, s) = a(r, s)c_{r+ms}(u, v) \tag{16}$$

(5) Reduce the dimensions of the array D from four to two dimensions, obtaining an array E, with same number of entries, as follows. The (r, s) entry of the array E is

$$e(r, s) = d(r (mod\ m), r (mod\ q), s (mod\ q), s (mod\ n)). \tag{17}$$

4.2   Properties of the Inflation Arrays

The following properties of the inflation arrays $C_0, C_1, \ldots, C_{\frac{q+1}{2}}$ ensure that the inflation process produces perfect arrays. These properties are given in (Arasu and de Launey 2001) in polynomial form, and the following matrix form is a simple equivalence.

(1) For $0 \le t_0, t_1 \le q - 1$, the summation of all (t_0, t_1) off-peak autocorrelation values, of the arrays $C_0, \ldots, C_q$, is equal to zero, that is, for $(t_0, t_1) \neq (0, 0)$

$$\sum_{r=0}^{m-1} \sum_{s=0}^{n-1} AC_{C_{r+ns}}(t_0, t_1) = 0 \tag{18}$$

(2) For $0 \le r, s \le q$, with $r \neq s$, the cross-correlation values of C_r and C_s are always one, that is, for all $0 \le t_0, t_1 \le q - 1$

$$CC_{C_r, C_s}(t_0, t_1) = 1 \tag{19}$$

(3) The autocorrelation values of the arrays $C_0, \ldots, C_q$ are either q^2 or $-q$.

The proofs in polynomial form of the above properties of the inflation arrays are far simpler than what they would be in matrix form.

Lemma 2. *The array D in Eq. (16) has perfect autocorrelation.*

Proof. We need to prove that all off-peak autocorrelation values of the array D are zero. First, we write the autocorrelation function of the array D in terms of the arrays $A, B_0, \ldots, B_q$. The right (t_0, t_1, t_3, t_2)-autocorrelation value of D is given by the equation

$$AC_D^R(t_0, t_1, t_2, t_3) = \sum_{r=0}^{m-1} \sum_{u=0}^{q-1} \sum_{v=0}^{q-1} \sum_{s=0}^{n-1} d(r, u, v, s) d^*(r+t_0, u+t_1, v+t_2, s+t_3) \quad (20)$$

where

$$d(r, u, v, s) = a(r, s) c_{r+ms}(u, v) \quad (21)$$

and $c_{r+ms}(u, v)$ is the (u, v) entry of the inflation array C_{r+ms}. So, we can write the right (t_0, t_1, t_2, t_3)-autocorrelation value of C as follows

$$AC_D^R(t_0, t_1, t_2, t_3) =$$

$$\sum_{r=0}^{m-1} \sum_{u=0}^{q-1} \sum_{v=0}^{q-1} \sum_{s=0}^{n-1} a(r, s) c_{r+ms}(u, v) \left(a(r + t_0, s + t_3) c_{r+t_0+n(s+t_3)}(u + t_1, v + t_2) \right)^* =$$

$$\sum_{r=0}^{m-1} \sum_{s=0}^{n-1} a(r, s) \left(\sum_{u=0}^{q-1} \sum_{v=0}^{q-1} c_{r+ms}(u, v) c^*_{r+t_0+m(s+t_3)}(u + t_1, v + t_2) \right) a^*(r + t_0, s + t_3)$$

$$(22)$$

Since the expression $\sum_{u=0}^{q-1} \sum_{v=0}^{q-1} c_{r+ms}(u, v) c^*_{r+t_0+n(s+t_3)}(u + t_1, v + t_2)$, in Eq. (22), is the cross-correlation of the arrays C_{r+ms} and $B_{r+t_0+m(s+t_3)}$, we can write Eq. (22) as

$$AC_D^R(t_0, t_1, t_2, t_3) =$$
$$\sum_{r=0}^{m-1} \sum_{s=0}^{n-1} a(r, s) \left(CC_{C_{r+ms}, C_{r+t_0+m(s+t_3)}}(t_1, t_2) \right) a^*(r + t_0, s + t_3) \quad (23)$$

In order to prove that D is perfect, we consider the following four cases: $t_0 \neq 0, t_1 \neq 0, t_2 \neq 0$ and $t_3 \neq 0$.

Case (1) $t_0 \neq 0$. Then C_{r+ms} and $C_{r+t_0+m(s+t_3)}$ are different arrays. So, by Eq. (19), the (t_1, t_2) cross-correlation value of B_{r+ms} and $B_{r+t_0+m(s+t_3)}$ is 1. So Eq. (23) becomes

$$AC_D^R(t_0, t_1, t_2, t_3) = \sum_{r=0}^{m-1} \sum_{s=0}^{n-1} a(r, s) (1) a^*(r + t_0, s + t_3) \quad (24)$$

Now, since A is a perfect array, from Eq. (24), we have

$$AC_D^R(t_0, t_1, t_2, t_3) = 0 \quad (25)$$

Case (2) $t_2 \neq 0$. Similar to Case 1.

Case (3) $t_1 \neq 0$. We divide this case into three sub-cases.

Case (a) $t_0 = t_3 = 0$. Since $t_0 = t_3 = 0$, we have $C_{r+t_0+m(s+t_e)} = C_{r+ms}$ and so, the (t_1, t_2) cross-correlation value of $C_{r+t_0+m(s+t_3)}$ and C_{r+ms} becomes the (t_1, t_2) autocorrelation value of C_{r+ms}. Equation (23) can be written as

$$AC_D^R(t_0, t_1, t_2, t_3) = \sum_{r=0}^{m-1} \sum_{s=0}^{n-1} a(r,s) \left(AC_{C_{r+ms}}(t_1, t_2) \right) a^*(r,s) \qquad (26)$$

From Property (3) in Section (4.2), $AC_{C_{r+ms}}(t_1, t_2)$ is either q^2 or $-q$, which are integers and commute with quaternions. Equation (26) becomes

$$AC_C^R(t_0, t_1, t_2, t_3) =$$

$$\sum_{r=0}^{m-1} \sum_{s=0}^{n-1} a(r,s) a^*(r,s) \left(AC_{C_{r+ms}}(t_1, t_2) \right) = \qquad (27)$$

$$\sum_{r=0}^{m-1} \sum_{s=0}^{n-1} 1 \left(AC_{C_{r+ms}}(t_1, t_2) \right)$$

From Eq. (18), we have that

$$\sum_{r=0}^{m-1} \sum_{s=0}^{n-1} AC_{C_{r+ms}}(t_1, t_2) = 0 \qquad (28)$$

Thus, $AC_C^R(t_0, t_1, t_2, t_3) = 0$.

Case (b) $t_0 \neq 0$. See Case 1.

Case (c) $t_2 \neq 0$. See Case 2.

Case (4) $t_3 \neq 0$. Similar to Case 3.

This completes the proof of Lemma (2)

Theorem 3. *The array E in Eq. (17) has perfect autocorrelation.*

Proof. We will show that each off-peak autocorrelation value of the array E is equal to an off-peak autocorrelation value of the array D, which is perfect. We will do this by showing that, if the shift (t_0, t_1) of E is non trivial, then the shift of D associated with (t_0, t_1) is also non trivial.

For $(t_0, t_1) \in \mathbb{Z}_{mq} \times \mathbb{Z}_{nq} \setminus \{(0,0)\}$, we use the equation

$$e(r,s) = d(r(mod\ m), r(mod\ q), s(mod\ q), s(mod\ n)) \qquad (29)$$

to produce the right (t_0, t_1)-autocorrelation value of E as follows

$$AC_E^R(t_0, t_1) = \sum_{r=0}^{mq-1} \sum_{s=0}^{nq-1} e(r,s) e^*(r+t_0, s+t_1) =$$

$$\sum_{r=0}^{mq-1} \sum_{s=0}^{nq-1} d(r(mod\ m), r(mod\ q), s(mod\ q) s(mod\ n)) \qquad (30)$$

$$d(r(mod\ m) + t_0(mod\ m), r(mod\ q) + t_0(mod\ q), s(mod\ q) + t_1(mod\ q),$$
$$s(mod\ n) + t_1(mod\ n))$$

So Eq. (30) above continues

$$\sum_{u=0}^{m-1} \sum_{v=0}^{q-1} \sum_{x=0}^{q-1} \sum_{y=0}^{n-1} d(u, v, x, y)$$

$$d^*(u + t_0(mod\ m), v + t_0(mod\ q), x + t_1(mod\ q), y + t_1(mod\ n)) = \qquad (31)$$

$$AC_D^R(t_0(mod\ m), t_0(mod\ q), t_1(mod\ q), t_1(mod\ n))$$

Case (1) $t_0 \neq 0$. Since $GCD(m, q) = 1$, we have that no number less than mq is divisible by m and q. Therefore, for $0 < t_0 \leq mq - 1$, if $t_0 \equiv 0(mod\ m)$, then $t_0 \not\equiv 0(mod\ q)$, and similarly if $t_0 \equiv 0(mod\ q)$, then $t_0 \not\equiv 0(mod\ m)$. Thus, the shift $(t_0(mod\ m), t_0(mod\ q), t_1(mod\ q), t_1(mod\ n))$ of D is not equivalent to the shift $(0, 0, 0, 0)\ mod\ (m, q, q, n)$.
Case (2) $t_1 \neq 0$. Similar to Case (1). Thus E is perfect.

Example 2. The perfect two-dimensional array

$$\begin{pmatrix} 1 & i \\ j & -k \end{pmatrix} \qquad (32)$$

of size 2×2 is inflated into a perfect two-dimensional array of size 14×14. Since $7 \equiv 3(mod\ 4)$, we construct 8 binary inflation arrays of size 7×7 as follows:

$$B_0 = \begin{pmatrix} 1\ 1\ 1 -1\ 1 -1 -1 \\ 1\ 1\ 1 -1\ 1 -1 -1 \\ 1\ 1\ 1 -1\ 1 -1 -1 \\ 1\ 1\ 1 -1\ 1 -1 -1 \\ 1\ 1\ 1 -1\ 1 -1 -1 \\ 1\ 1\ 1 -1\ 1 -1 -1 \\ 1\ 1\ 1 -1\ 1 -1 -1 \end{pmatrix} \qquad B_1 = \begin{pmatrix} 1 -1 -1\ \ 1 -1\ \ 1\ \ 1 \\ 1\ \ 1 -1 -1\ \ 1 -1\ \ 1 \\ 1\ \ 1\ \ 1 -1 -1\ \ 1 -1 \\ -1\ \ 1\ \ 1\ \ 1 -1 -1\ \ 1 \\ 1 -1\ \ 1\ \ 1\ \ 1 -1 -1 \\ -1\ \ 1 -1\ \ 1\ \ 1\ \ 1 -1 \\ -1 -1\ \ 1 -1\ \ 1\ \ 1\ \ 1 \end{pmatrix}$$

$$\qquad (33)$$

$$B_2 = \begin{pmatrix} 1 -1 -1\ \ 1 -1\ \ 1\ \ 1 \\ 1 -1\ \ 1\ \ 1\ \ 1 -1 -1 \\ 1\ \ 1 -1 -1\ \ 1 -1\ \ 1 \\ -1\ \ 1 -1\ \ 1\ \ 1\ \ 1 -1 \\ 1\ \ 1\ \ 1 -1 -1\ \ 1 -1 \\ -1 -1\ \ 1 -1\ \ 1\ \ 1\ \ 1 \\ -1\ \ 1\ \ 1\ \ 1 -1 -1\ \ 1 \end{pmatrix} \qquad B_3 = \begin{pmatrix} 1\ \ 1\ \ 1 -1\ \ 1 -1 -1 \\ 1 -1\ \ 1 -1 -1\ \ 1\ \ 1 \\ 1 -1 -1\ \ 1\ \ 1\ \ 1 -1 \\ -1\ \ 1\ \ 1\ \ 1 -1\ \ 1 -1 \\ 1\ \ 1 -1\ \ 1 -1 -1\ \ 1 \\ -1\ \ 1 -1 -1\ \ 1\ \ 1\ \ 1 \\ -1 -1\ \ 1\ \ 1\ \ 1 -1\ \ 1 \end{pmatrix}$$

$$
B_4 = \begin{pmatrix}
1 & -1 & -1 & 1 & -1 & 1 & 1 \\
1 & 1 & 1 & -1 & -1 & 1 & -1 \\
1 & -1 & 1 & 1 & 1 & -1 & -1 \\
-1 & -1 & 1 & -1 & 1 & 1 & 1 \\
1 & 1 & -1 & -1 & 1 & -1 & 1 \\
-1 & 1 & 1 & 1 & -1 & -1 & 1 \\
-1 & 1 & -1 & 1 & 1 & 1 & -1
\end{pmatrix}
\quad
B_5 = \begin{pmatrix}
1 & 1 & 1 & -1 & 1 & -1 & -1 \\
1 & -1 & -1 & 1 & 1 & 1 & -1 \\
1 & 1 & -1 & 1 & -1 & -1 & 1 \\
-1 & -1 & 1 & 1 & 1 & -1 & 1 \\
1 & -1 & 1 & -1 & -1 & 1 & 1 \\
-1 & 1 & 1 & 1 & -1 & 1 & -1 \\
-1 & 1 & -1 & -1 & 1 & 1 & 1
\end{pmatrix}
$$

$$
B_6 = \begin{pmatrix}
1 & 1 & 1 & -1 & 1 & -1 & -1 \\
1 & 1 & -1 & 1 & -1 & -1 & 1 \\
1 & -1 & 1 & -1 & -1 & 1 & 1 \\
-1 & 1 & -1 & -1 & 1 & 1 & 1 \\
1 & -1 & -1 & 1 & 1 & 1 & -1 \\
-1 & -1 & 1 & 1 & 1 & -1 & 1 \\
-1 & 1 & 1 & 1 & -1 & 1 & -1
\end{pmatrix}
\quad
B_7 = \begin{pmatrix}
1 & 1 & 1 & 1 & 1 & 1 & 1 \\
1 & 1 & 1 & 1 & 1 & 1 & 1 \\
1 & 1 & 1 & 1 & 1 & 1 & 1 \\
-1 & -1 & -1 & -1 & -1 & -1 & -1 \\
1 & 1 & 1 & 1 & 1 & 1 & 1 \\
-1 & -1 & -1 & -1 & -1 & -1 & -1 \\
-1 & -1 & -1 & -1 & -1 & -1 & -1
\end{pmatrix}
\tag{34}
$$

We use the above arrays and Eq. (15) to produce 4 inflation arrays of size 7×7

$$
C_0 = \begin{pmatrix}
i & 1 & 1 & -1 & 1 & -1 & -1 \\
i & i & 1 & -i & i & -i & -1 \\
i & i & i & -i & 1 & -1 & -i \\
1 & i & i & -1 & 1 & -i & -1 \\
i & 1 & i & -1 & i & -i & -i \\
1 & i & 1 & -1 & i & -1 & -i \\
1 & 1 & i & -i & i & -1 & -1
\end{pmatrix}
\quad
C_1 = \begin{pmatrix}
i & -1 & -1 & 1 & -1 & 1 & 1 \\
i & -i & i & 1 & 1 & -1 & -1 \\
i & 1 & -i & -1 & i & -1 & 1 \\
-i & i & -1 & i & 1 & i & -i \\
i & i & 1 & -1 & -i & 1 & -1 \\
-i & -1 & 1 & -i & i & i & i \\
-i & 1 & i & i & -1 & -i & i
\end{pmatrix}
$$

$$
C_2 = \begin{pmatrix}
i & -1 & -1 & 1 & -1 & 1 & 1 \\
i & 1 & 1 & -1 & -1 & i & -i \\
i & -1 & 1 & i & 1 & -i & -1 \\
-i & -i & i & -1 & i & 1 & i \\
i & 1 & -1 & -i & 1 & -1 & i \\
-i & i & i & i & -i & -1 & 1 \\
-i & i & -i & 1 & i & i & -1
\end{pmatrix}
\quad
C_3 = \begin{pmatrix}
i & i & i & -1 & i & -1 & -1 \\
i & i & -1 & i & -1 & -1 & i \\
i & -1 & i & -1 & -1 & i & i \\
-i & 1 & -i & -i & 1 & 1 & 1 \\
i & -1 & -1 & i & i & i & -1 \\
-i & -i & 1 & 1 & 1 & -i & 1 \\
-i & 1 & 1 & 1 & -i & 1 & -i
\end{pmatrix}
\tag{35}
$$

We use the inflation arrays C_0, C_1, C_2 and C_3, to produce, from the array $\begin{pmatrix} 1 & i \\ j & -k \end{pmatrix}$, the perfect array

$$\begin{pmatrix}
i & -i & 1 & i & 1 & i & -1 & -1 & 1 & -i & -1 & -i & -1 & i \\
-k & j & -k & j & j & k & j & j & k & k & j & k & j & j \\
i & -i & i & -1 & 1 & 1 & -i & -1 & i & i & -i & i & -1 & -i \\
k & -k & j & j & j & -k & k & j & -k & j & -k & -k & -k & -k \\
i & i & i & 1 & i & -i & -i & -1 & 1 & -i & -1 & i & -i & -1 \\
k & j & j & -k & -k & j & -k & j & j & -k & k & -k & -k & -k \\
1 & -1 & i & i & i & -1 & -1 & 1 & 1 & 1 & -i & -1 & -1 & -i \\
-k & j & j & k & j & k & j & j & j & j & j & j & j & k \\
i & i & 1 & -i & i & -1 & -1 & -1 & i & i & -i & -i & -i & 1 \\
-k & k & k & k & -k & j & j & j & j & j & j & k & j & j \\
1 & 1 & i & -i & 1 & i & -1 & 1 & i & -1 & -1 & -1 & -i & -1 \\
-k & k & j & j & k & j & j & j & -k & k & j & j & j & k \\
1 & -1 & 1 & -1 & i & -i & -i & 1 & i & -1 & -1 & 1 & -1 & i \\
k & -k & -k & -k & j & -k & -k & j & j & -k & -k & j & k & j
\end{pmatrix} \qquad (36)$$

of size 14×14, with appearances of all the basic quaternion elements 1, -1, i, $-i$, j, $-j$, k and $-k$.

5   Conclusion

In this paper we showed that every array A, of size $m \times n$ over the basic quaternions $\{1, -1, i, -i, j, -j, k, -k\}$, with $q = 2mn - 1$ a prime number and $q \equiv 3 (mod\ 4)$, can be inflated into another perfect array over the basic quaternions of size $mq \times nq$. This construction paves the way for constructing new families of perfect arrays over the basic quaternions with different sizes to those showed in Barrera Acevedo and Jolly's work (2014).

References

Arasu, K.T., de Launey, W.: Two-dimensional perfect quaternary arrays. IEEE Trans. Inf. Theory **47**, 1482–1493 (2001)

Barrera Acevedo, S.: Perfect sequences and arrays of unbounded lengths and sizes over the basic quaternions. Ph.D. Thesis, Monash University (2013)

Barrera Acevedo, S., Jolly, N.: Perfect arrays of unbounded sizes over the basic quaternions. Crypt. Commun. **6**, 47–57 (2014)

Acevedo, S.B., Hall, T.E.: Perfect sequences of unbounded lengths over the basic quaternions. In: Helleseth, T., Jedwab, J. (eds.) SETA 2012. LNCS, vol. 7280, pp. 159–167. Springer, Heidelberg (2012)

Kuznetsov, O.: Perfect sequences over the real quaternions. In: WSDA '09: Fourth International Workshop on Signal Design and its Applications in Communications, vol, 1, pp. 17–20 (2009)

Kuznetsov, O., Hall, T.E.: Perfect sequences over the real quaternions of longer length. Online J. Math. Stat. **1**, 8–11 (2009) (The 2010 World Congress on Mathematics and Statistics, WCMS 10)

Lee, C.E.: Perfect q-ary sequences from multiplicative characters over $GF(p)$. Electron. Lett. **28**, 833–835 (1992)

Luke, H.D.: BTP transform and perfect sequences with small phase alphabet. IEEE Trans. Aerosp. Electro. Syst. **32**, 497–499 (1996)

Families of 3D Arrays for Video Watermarking

Sam Blake[1], Oscar Moreno[2], and Andrew Z. Tirkel[3]([✉])

[1] School of Mathematical Sciences, Monash University, Melbourne, Australia
[2] Gauss Research Foundation, San Juan, Puerto Rico
[3] Scientific Technology, Melbourne, Australia
atirkel@bigpond.net.au

Abstract. This paper presents new constructions of families of binary and ternary arrays with low off-peak (periodic) autocorrelation and low cross-correlation for application to video watermarking. The constructions are based on the composition method which uses a shift sequence to cyclically shift a commensurate "column" sequence/array. The shift sequence/array has auto and cross-hit values constrained to 1 or 2, while the column sequence/array is pseudonoise. The shift sequences are new, while the column sequence is a Sidelnikov sequence, and the column array is a multi-dimensional Legendre array. The shift sequence constructions involve mapping the elements of a finite field onto its associated multiplicative group, the field plus infinity and other such variations. The constructions yield families of arrays suitable for embedding into video as watermarks. Examples of such watermarks are presented. The watermarks survive H264 compression, and are being considered for a video security standard.

Keywords: Array · Correlation · Periodic · Costas · Legendre · Multi dimensional · Shift sequence

1 Introduction

Since 1993, the area of digital watermarking has undergone an explosion in activity. Digital watermarks have been applied to still images, audio, video, text, sheet music, etc. Watermarking techniques have been used to provide copyright protection, access control, audit trail, traitor tracing, provide certificates of authenticity, etc. Watermark embedding and recovery techniques have been studied extensively and have been tailored to use the masking effect of the human visual system and human auditory system. Almost all of these advances have occurred in the applications domain. Major advances have occurred in protecting watermarks against unintentional distortions (compression, cropping, geometrical effects etc.) and against deliberate cryptographic attack. New forms of attack have emerged as a result of these advances. By contrast, the generators or sequences used to carry the message have not changed significantly. As a consequence, watermarks can benefit significantly by using families of sequences or arrays with good auto and cross-correlation. This is because multiple sets

© Springer International Publishing Switzerland 2014
K.-U. Schmidt and A. Winterhof (Eds.): SETA 2014, LNCS 8865, pp. 134–145, 2014.
DOI: 10.1007/978-3-319-12325-7_12

of such sequences or arrays can be embedded as composite watermarks. Such composite watermarks have three significant advantages: they are more secure against cryptographic attack, they can carry more information, and where the watermarks are used as fingerprints, composite watermarks can have immunity to collusion attack.

One popular watermarking technique that has been developed uses a statistical method to generate the watermark patterns, employing a random number generator or a noisy physical process. It is simple and effective, easy to implement, and can be made resistant to standard compression methods. Its weakness is that it cannot specify a probability that the watermarks generated by this process are "unique", or at least sufficiently dissimilar, so as never to be confused. This is not a problem for proof of ownership or copyright applications, where there are few watermarks needed, and many recipients of the media receive the same watermark. This is not true for video surveillance cameras, nor for audit trail applications, where a large number of watermarks are required. It should be noted that the statistical method can be adapted, so that any similar watermarks are "filtered out". However, this only applies to a single node of watermarking, and is difficult or impossible to implement in a distributed watermarking system, such as a network of surveillance cameras.

By contrast, the watermark method developed by our group is based on an algebraic construction [1]. Originally, it used m-sequences to embed watermark information line by line in an image. It was primitive, difficult to implement, and to make resistant to compression and attack. It also suffered from visibility problems, due to the fact that each watermark was embedded in a small portion of the image: a line. However, it was free from the weakness of other methods, in that the probability of missed or mistaken detection could be specified for a set of watermarks generated using this method.

While many video watermarking solutions have been proposed, few of them are appropriate for hardware implementation. In addition, most are implemented as post-processing steps after the initial video was obtained. This means that an unwatermarked version of the image or data already exists, and that constitutes a security vulnerability. The paper is organised as follows: Sect. 2 outlines the method of array construction. Section 3 introduces the multi-dimensional grid which is used in the construction as well as the generation of the multi-dimensional Legendre array, which is used as a "column array" in our 3D construction. Section 4 describes various 3D constructions in detail, whilst Sects. 5 and 6 demonstrate how the arrays have been applied to video watermarking.

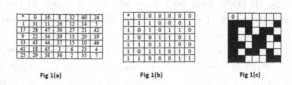

Fig 1(a)   Fig 1(b)   Fig 1(c)

Fig. 1. Two dimensional Legendre array

2   Construction

Our constructions are based on a method developed in [2] and described briefly in [3] and applied to 2D constructions for wireless communications in [4]. The essential ingredients are a column sequence with good autocorrelation, and a shift sequence which is applied as a cyclic shift to the column sequences to form a watermarking array. The sequence below is adapted from one developed for frequency hopping [5].

$$s_i = log_\alpha(A\alpha^{2i} + B\alpha^i + C) \tag{1}$$

α is a primitive element of a finite field GF(q) where q is the number of elements and is a prime power $q = p^n$ where p is prime and n is any positive integer, including 1. i is an index taking on the values $0, 1, 2, ..., q - 2$. s_i takes on the values $0, 1, 2, ..., q - 2, \infty$, where ∞ results from the argument of the log function being equal to 0. A, B, C are suitably chosen entries from GF(q).
Here $GF(q) = 0, \alpha^1, \alpha^2, ..., \alpha^{q-1}$.

In this context log refers to $log_\alpha x = j$ implies that $x = \alpha^j$. Note that the log mapping is 1:1 i.e. there is a single value of s_i for each i.

The arrays from (1) have the following property: for any non-zero doubly periodic shift of such an array, its auto correlation is 0 or 1. Some of the arrays generated are shifts of each other, and hence have bad correlation. There is an equivalence relation which makes $(q - 1)^2$ choices of A, B or C redundant, and hence there are approximately q inequivalent arrays in the family. It can be shown that all inequivalent quadratics can be represented by q choices of C in **S** where

$$s_i = log_\alpha(x^2 + x + C). \tag{2}$$

Each of these arrays from (1) can be assigned to a different user. A doubly periodic cross-correlation between any pair of such arrays is also 0 or 1. These arrays may also find application in modulating radar signals for multi-target recognition and in OCDMA (Optical Code Division Multiple Access) [6].

The watermarking array construction relies on replacing any column i of array **S** with a 1 in it (notice that each column has either 0's and a 1 or an ∞) by a known column over roots of unity, with good correlation in a cyclic shift equal to s_i for that column in **S**. Note that columns commensurate with this construction are: Sidelnikov sequences [7], Legendre sequences, m-sequences and Hall sequences and others. This is the first time Sidelnikov sequences have been used in such a construction.

Columns with ∞ in them can be replaced by a column of 0's. This reduces the peak autocorrelation by $q - 1$, but has almost no effect on the off-peak autocorrelation, or the cross-correlation. Where there is only one column with an ∞, the column can be replaced by a column of constant values, including +1 or −1. The autocorrelation is even better than when the constant is 0 whilst the cross-correlation can increase by $q - 1$. When there are two or more entries with ∞, the best option is to replace them by a string of 0's. This reduces the peak autocorrelation even further, and makes such arrays less desirable.

3 Multi-dimensional Grid

We write the elements of $GF(p^2)$ as doubletons (a, b) based on α, a primitive element, following [8].

$$\alpha^1 = (1,0) \quad \alpha^2 = (.,.), \ \alpha^3 = (.,.), \ldots, \alpha^{p^2-1} = (0,1) \tag{3}$$

where each doubleton has entries from Z_p. Consequently, the doubletons define an integer grid in two dimensions. This is shown in Fig. 1(a) where the field elements are written in exponent notation.

The map of Fig. 1(a) can itself be converted into a two-dimensional array over $(0, 1, \infty)$ as in Fig. 1(b), or over the symbols $+1, -1, 0$ as in Fig. 1(c). The latter can be achieved by reducing the numbered entries in Fig. 1 modulo 2 and mapping 0 onto $+1$, and 1 onto -1. The ∞ is mapped onto 0. We call this new array a two-dimensional Legendre array. A two dimensional 7×7 Legendre array derived from Fig. 1(a) is shown in Fig. 1(b). The two dimensional periodic autocorrelation of this array is -1 for all non-trivial shifts. This method of alphabet reduction can be applied modulo k, as long as k divides m. It can be performed on arrays of this type in dimensions higher than 2. This array will be employed in the construction of families of higher dimensional arrays in constructions C. Its existence is vital.

4 Three Dimensional Constructions

In the following discussion the construction numbers are chosen to be consistent with [9–11]. New three dimensional constructions can be obtained by using a partition of the finite field $GF(p^2)$ to generate a unique (Costas) grid.

As in Sect. 3, we write the elements of $GF(p^2)$ as doubletons based on α, a primitive element.

$$\alpha^1 = (1,0) \quad \alpha^2 = (.,.), \ \alpha^3 = (.,.), \ldots, \alpha^{p^2-1} = (0,1) \tag{4}$$

where each doubleton has entries from Z_p. Consequently, the doubletons define an integer grid in two dimensions, which can be used as a basis for a three dimensional periodic Costas array. The construction can be generalized to m dimensions.

This grid can be used to construct a single array in three dimensions with array correlation 1.

Consider $s = log_\alpha X$ where $X \in GF(p^2)$.

Specifically for $X = \alpha^i$, $s_i = log_\alpha \alpha^i = i$.

The sequence scheme is to use the doubleton representation of α^i to determine the coordinates (location) on the two dimensional integer grid defined above. s_i is a periodic sequence with period $p^2 - 1$. An example of this method of mapping is shown in Fig. 1(a). It displays the map for $GF(7^2)$. For our Costas type construction we take the grid point location belonging to s_i, and place a *1* at position i in a column of length $p^2 - 1$ located below the grid point, and zeros in all other entries in that column. This array has an off-peak autocorrelation of at most 1.

The array can be converted into a binary or higher alphabet array by substituting a pseudonoise sequence in place of the sequence of 0's and 1's. This process is illustrated in Fig. 2. Figure 2(a)(i) shows the powers of a primitive element in $GF(3^2)$ raised to all its powers in the grid format. Figure 2(a)(ii) shows a logarithmic mapping of Fig. 2(a)(i). Figure 2(a)(iii) shows a column sequence of length 8, (a Sidelnikov sequence) which is used to generate a three dimensional array. The array is generated by first placing a column of length 8 below every entry in the grid. The column contains all zeros if the entry is * and otherwise contains a solitary entry of one in the position determined by the entry in the corresponding grid location. This is illustrated in Fig. 2(b). The columns of Fig. 2(b) are then substituted by corresponding cyclic shifts of the Sidelnikov sequence of Fig. 2(a)(iii). The all zeros column is not substituted. The resulting array is shown in Fig. 2(c).

The array in Fig. 2(c) is solitary, so by itself it does not address the requirement of delivering large families of arrays with low off-peak autocorrelation and low cross-correlation. However, the modifications described below do deliver such families. In the next two constructions we use the grid just like the one in Fig. 1, which is for arrays of size 7*7, or the one of Fig. 2, which is for arrays of size 3*3. We show how to construct two sets of families of arrays whose auto and cross-correlation of 0 or 1.

4.1   Column Based Constructions

Construction C1

Take A, B, C $\in GF(p^2)$, $A \neq 0$. Let:

$$s_{kl} = log_\alpha(AX^2 + BX + C) \tag{5}$$

Here $X = \alpha^i$ with α being a primitive element of $GF(p^2)$ and k and l refer to the grid coordinates of α^i. In this family, two shift arrays s_{kl} and s'_{kl} are equivalent if the watermark arrays they generate are multi-periodical shifts of each other. The number of non-equivalent classes is approximately p^2.

Construction C2

Take A, B, C, D $\in GF(p^2)$, $AD - BC \neq 0$

$$s_{kl} = log_\alpha \frac{AX + B}{CX + D} \tag{6}$$

The equivalence classes are defined similarly to Construction C1, and the number of non-equivalent arrays is also similar to Construction C1.

Construction C1 can be generalized by using polynomials with degree greater than 2 and Construction C2 with polynomials of degree greater than 1.

A watermarking array is constructed by using s_i belonging to each coordinate on the grid to cyclically shift a binary Sidelnikov sequence of length $p^2 - 1$.

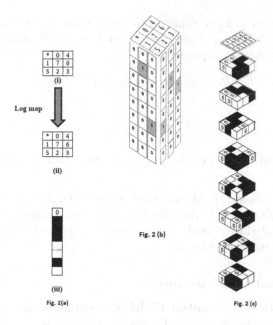

Log map

(i)

(ii)

(iii)

Fig. 2(a)

Fig. 2 (b)

Fig. 2 (c)

Fig. 2. Construction C1

Our construction guarantees that no more than 2 such Sidelnikov columns can match and therefore the worst case autocorrelation and cross-correlation is of the order of $2p^2$. The peak autocorrelation is of order p^4.

Constructions C1 and C2 can be generalized to produce an $m+1$ dimensional watermarking array by using the grid mapping method to map $GF(p^m)$ onto a $p \times p \times p \times p$ grid by representing each power of a primitive element as an m-tuple. The resultant watermarking array is of size $p \times p \times p \times p \times (p^m - 1)$.

The method of connecting elements of $GF(p^m) - \{0\}$ with Z_{p^m-1} using a logarithmic function has an inverse. In one dimension this has led to logarithmic and exponential Costas array constructions. Here it leads to even more new multi-periodic multidimensional arrays.

Observe than $Z_p^m - \{0\}$ and Z_{p^m-1} have the same cardinality. Consequently, there exists an inverse function to the one that gives a generic Costas Array, since it is a 1-1 onto function. Consider now the inverse function

$$g : Z_{p^m-1} \longrightarrow Z_p^m - \{0\},$$

for the case of the generic Welch Costas Array. We take α, the corresponding primitive element of the finite field. As in coding theory, we write the elements of GF(p^m) as m-tuples based on α, a primitive element.

$$\alpha^1 = (0,0,0,\dots,1,0), \alpha^2 = (.,.,.,\dots,.,.)), ,\dots, \alpha^{p^m-1} = (0,0,0,\dots,0,1) \quad (7)$$

where each m-tuple has entries from Z_p. Consequently, the m-tuples define an integer grid in m dimensions, which can be used as a basis for a $(m+1)$-periodic Costas arrays.

α^i can be written as an m-tuple on the grid. Consequently we can take g to be $g = \alpha^i$ for any $i \in Z_{p^m-1}$. Note that g is multi-periodic, and we now see that it has the distinct difference property:

$$\forall h \neq 0, \alpha^{i+h} - \alpha^i = \alpha^{j+h} - \alpha^j \implies i \equiv j \tag{8}$$

This is true since

$$\alpha^{i+h} - \alpha^i = \alpha^i \left(\alpha^h - 1\right) = \alpha^j \left(\alpha^h - 1\right) = \alpha^{j+h} - \alpha^j \tag{9}$$

Therefore divide by $\left(\alpha^h - 1\right)$ since $h \neq$ and we obtain $\alpha^i = \alpha^j$. This implies that $i = j$.

Definition: An elementary Abelian Costas array $f : Z_{p^m-1} \longrightarrow Z_p^m - \{0\}$ is a 1-1 onto function which is periodic and with the distinctness of differences property. Note that since $+$ and $-$ are the operations of the Abelian Group, we also say this is elementary Abelian.

4.2   Plane Based Constructions

Construction D1 (Exponential Welch Generalization)
$f(i) = \alpha^i$ is an elementary Abelian Costas Array. $f : Z_{p^m-1} \longrightarrow Z_p^m - \{0\}$. This is a solitary array.

Construction D1 is illustrated by the example in Fig. 3. Figure 3(a)(i) shows the starting 3×3 grid, as described before. Figure 3(a)(ii) shows the inverse mapping, which produces an array of cells containing the two dimensional shifts associated with each entry. Figure 3(a)(iii) shows the 3×3 Legendre array constructed by the method described above, and illustrated for a 7×7 array in Fig. 1. Figure 3(b) shows how each of the two dimensional shifts of Fig. 3(a)(ii) are used to shift the Legendre array of Fig. 3(a)(iii) to produce 8 layers of a three-dimensional array.

Construction D2 (Quadratic Generalization – Family of Arrays)

$$f_{A,B,C}(i) = A(\alpha^i)^2 + B\alpha^i + C, \quad f : Z_{p^m-1} \longrightarrow Z_p^m - \{0\} \tag{10}$$

where A, B, C are elements of the finite field $GF(p^m)$, gives a family of arrays. Any two arrays which are multi-dimensional cyclic shifts of one another are called equivalent. The autocorrelation of our arrays and the cross-correlation between any non-equivalent arrays is bounded by two.

Construction D3
With the same assumptions and conclusions as in the previous constructions, consider now:

$$f_{A,B,C,D}(i) = \frac{A\alpha^i + B}{C\alpha^i + D}, \quad f : Z_{p^m-1} \longrightarrow Z_p^m - \{0\} \tag{11}$$

where A, B, C, D are elements of the finite field $GF(p^m)$. The autocorrelation of our arrays and the cross-correlation between any non-equivalent arrays is bounded by two.

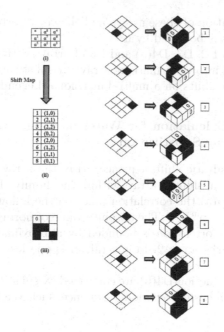

Fig. 3. Construction D2

Construction D4(a)

The elements of a finite field together with ∞ can be written in an order determined by Moreno-Maric. Each of these elements, except ∞ can also be expressed as m-tuple commensurate with the grid described in the preamble to Construction D1. The fractional function of Construction 3 can then be used to map the $q + 1$ entries of $\{GF(q^m) \cup \infty\}$ onto $GF(q^m)$. Whenever the result is ∞, that entry is left blank. The autocorrelation of these arrays and the cross-correlation between any non-equivalent arrays is bounded by two. We obtain a watermarking array by substituting the coordinates of a grid entry with corresponding cyclic shifts of a commensurate Legendre array.

Construction D4(b)

In a manner similar to Construction 4(a), the entries in $\{GF(q^m) \cup \infty\}$ resulting from the fractional function map of Construction D3 can be mapped onto the multiplicative group $GF(q^m)/0$ by using the log function described before. The autocorrelation of these arrays and the cross-correlation between any non-equivalent arrays is bounded by two. We obtain a watermarking array by substituting the entries on the grid with cyclic shifts of a commensurate Sidelnikov or m-sequence.

Construction D5

In a way similar to previous generalizations, the quadratic of construction D2 can be generalized to a degree n polynomial with coefficients from the finite field. The autocorrelation of these arrays and the cross-correlation between any non-equivalent arrays is bounded by n.

The constructions D2, D3, D4(a) and D5 of arrays with constrained correlation can be converted into watermarking arrays by substituting the grids of shift m'tuples into periodic shifts on a multi-dimensional Legendre array.

4.3   Correlation Calculation for Watermarking Array Using Legendre Array Substitution

For a non-trivial quadratic shift sequence/array the array can match 0, 1, or 2 columns/planes of a shifted array within the family. Hence, for the constructions D2, D3, D4(a) the correlation takes on the following possible values: $(p^m - 1)^2, p^m + 1, +1, -p^m + 1$. The absolute value of normalized off-peak autocorrelation and cross-correlation is bounded by approximately p^{-m}. For three dimensional watermarks, m = 2. In general, m is one less than the number of dimensions.

Arrays from Construction D4(b) are converted by substituting the integers 0, 1, ..., p^{m-2} by a commensurate periodic sequence, such as a Sidelnikov sequence.

4.4   Higher Dimensions

Another application of the grid is a four dimensional construction E1, which uses a map from a finite field onto itself. We use the same grid as in Constructions C and Constructions D. Now we apply the following mapping to the powers of α

$$f(x) \rightarrow Ax^2 + Bx + C \tag{12}$$

where x, A, B, C are elements of $GF(q)$ and x is a variable.

$f(x)$ is a mapping from GF(q) to GF(q) where q = p^n.

Both x and f(x) can be seen as n-tuples using the grid. We obtain a watermarking array by substituting the coordinates of a grid entry with corresponding cyclic shifts of a commensurate multi-dimensional Legendre array.

Figure 4 illustrates the method of constructing a 4 dimensional array. Figure 4(a)(i) shows the starting grid. Figure 4(a)(ii) shows a simple quadratic map $x \rightarrow x^2$. Figure 4(a)(iii) shows the array of shifts obtained by looking up Fig. 4(a)(i). This array of shifts is then used to cyclically shift a 3×3 Legendre array shown in its (0, 0) shift in Fig. 4(a)(iv). The resulting four dimensional array is shown schematically in Fig. 4(b), where only some of the entries are shown. There are 8 shift inequivalent arrays corresponding to 8 quadratics of the type $x \rightarrow Ax^2$ where in this case A is any non-zero element of the finite field GF(3^2). The autocorrelation of each array is bounded by one and the cross-correlation between any pair of the eight arrays is bounded by two.

The constructions in two, three and higher dimensions discussed above are designed for image, video and multimedia watermarking, but may be applied to or adapted to multiple target recognition in radar and optical communications.

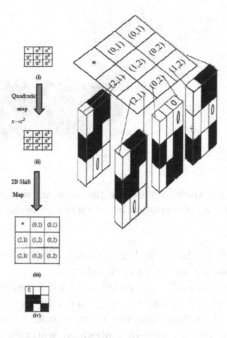

Fig. 4. Four dimensional construction

5   Examples of Arrays

Arrays derived from Construction C1 and D2 were generated using Wolfram Mathematica. An example of C1 for p = 17 is available on http://youtu.be/CU6-UEz8_UY.

6   Application to Video Watermarking

Arrays produced by Construction C1 and D2 were embedded in various videos and extracted by filtered correlation. In order to comply with current industry data transmission standards, the video was subjected to H264 compression prior to watermark extraction. A histogram of the cross-correlation of the array with the compressed video is shown in Fig. 5 (left). This demonstrates that the video and the watermark are uncorrelated, since the bulk of histogram displays Gaussian characteristics.

The autocorrelation peak of the watermark is an outlier as indicated by the arrow and provides unambiguous proof of the existence of the watermark in the video. This figure justifies the use of Gaussian statistics in watermark analysis. The autocorrelation peak is more than 5 standard deviations from the mean. The probability of error is 1/390,682,215,445. This is below the industry standard of 10^{-12}. The watermark patterns were undetectable according to several viewers, or were indistinguishable from compression artefacts. A correlation plot as a

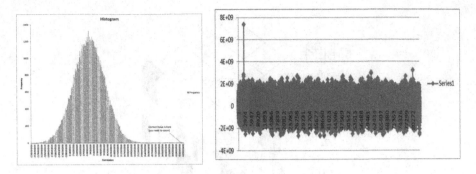

Fig. 5. left = Histogram of the correlation of the reference array with a watermarked copy of "The Fringe", right = Correlation as a function of cyclic shift

function of cyclic shift is shown in Fig. 5(right). The watermark is capable of carrying 16 bits of information in its cyclic shift.

Since the arrays discussed in this paper have low cross-correlation, it is possible to embed and extract more than one array in the same video. Figure 6 shows the extraction of multiple concurrent watermarks. Figure 6(a) shows the baseline correlation in the absence of embedded watermarks. The probability of false detection is low. Figure 6(b) shows the correlation in the presence of 3 watermarks, clearly demonstrating the existence of 3 correlation peaks. Such multiplexing increases the data payload from 16 bits to 48 bits. In the case of construction D2, the arrays have a common frame structure. It is possible to extract two correlation peaks from a single frame, as shown in Fig. 6(c). Such integration of 2D and 3D techniques is being investigated.

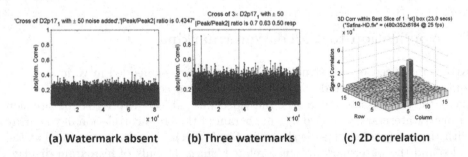

(a) Watermark absent   (b) Three watermarks   (c) 2D correlation

Fig. 6. Cross-correlation of reference watermark with an unwatermarked video and with a video in which multiple watermarks were embedded

7   Conclusions

Our results demonstrate that our families of 3D arrays with good correlation can be used as superior watermarks for video. We show several methods of construction of the critical shift sequence/array. The examples presented use a quadratic polynomial, but in principle, this can be extended to higher degrees, resulting in virtually inexhaustible supplies of 3D video watermarks with slightly worse correlation. The watermarks use Sidelnikov sequences or multi-dimensional Legendre arrays as column sequences. Our arrays have extremely high linear complexity, which makes them cryptographically secure. Research into quantifying multi-dimensional complexity is ongoing.

References

1. Tirkel, A.Z., Rankin, G.A., van Schyndel, R.M., Ho, W.J., Mee, N.R.A., Osborne, C.F.: Electronic water mark. In: DICTA 93, pp. 666–673, Macquarie University (1993)
2. Tirkel, A.Z., Osborne, C.F., Hall, T.E.: Steganography - applications of coding theory. In: IEEE - IT Workshop, pp. 57–59, Svalbard, Norway (1997)
3. Tirkel, A.Z., Hall, T.E.: Array construction using cyclic shifts of a column. In: ISIT'05, pp. 2050–2054 (2005)
4. Moreno, O., Tirkel, A.: New optimal low correlation sequences for wireless communications. In: Helleseth, T., Jedwab, J. (eds.) SETA 2012. LNCS, vol. 7280, pp. 212–223. Springer, Heidelberg (2012)
5. Moreno, O., Maric, S.: A class of frequency hop codes with nearly ideal characteristics for multiple-target recognition. In: ISIT, p. 126 (1997)
6. Moreno, O., Ortiz-Ubarri, J.: Double periodic arrays with optimal correlation for applications in watermarking. In: Golomb, S.W., Gong, G., Helleseth, T., Song, H.-Y. (eds.) SSC 2007. LNCS, vol. 4893, pp. 82–94. Springer, Heidelberg (2007)
7. Sidelnikov, V.M.: Some k-valued pseudo-random sequences and nearly equidistant codes. Probl. Inf. Transm. 5(1), 12–16 (1969)
8. Bomer, L., Antweiler, M.: Optimizing the aperiodic merit factor of binary arrays. Signal Process. (Elsevier) 30, 1–13 (1993)
9. Digital Watermarking PCT/AU2010/000990, WO2011/050390, US 2012213402, EP2494516, AU2010312302
10. Digital Communications AU 2011904698, PCT/AU2012/001377
11. Algebraic Generators of Sequences for Communication Signals AU2011905002, PCT/AU2012/001473

Relative Difference Sets

The Nonexistence of $(18, 3, 18, 6)$ Relative Difference Sets

David Clark and Vladimir Tonchev$^{(\boxtimes)}$

Department of Mathematical Sciences, Michigan Technological University,
Houghton, MI 49931, USA
{dcclark,tonchev}@mtu.edu

Abstract. It is known that relative difference sets with parameters $(18, 3, 18, 6)$ in a group of order 54 with normal subgroup N of order 3 do not exist in any abelian group of order 54. In this paper, using the recent classification of all generalized Hadamard matrices of order 18 over a group of order 3 [5], we show that such relative difference sets do not exist in any non-abelian group of order 54 as well. Our results are validated computationally using the computer algebra system Magma.

1 Introduction

Suppose G is a finite group of order mn. Let N be a normal subgroup of G, $|N| = n$, called the "forbidden" subgroup. Let $R \subset G$, and let

$$D = \{xy^{-1} : x, y \in R, x \neq y\}.$$

The set R is called a (m, n, k, λ) Relative Difference Set (RDS) in G relative to N, if

- $|R| = k$,
- D contains every element of $G \setminus N$ exactly λ times, and
- D contains no element of N.

A *difference set* is an RDS with a forbidden group of order $n = 1$. An RDS is called *abelian* if G is abelian, and *non-abelian* otherwise.

  Relative difference sets are closely related to difference sets, group-divisible designs, generalized Hadamard matrices, symmetric nets, and finite geometry [2,6,9]. Similarly to ordinary difference sets, relative difference sets yield sequences with interesting correlation properties [7]. A comprehensive survey on RDS is the paper by Pott [8]. The existence problem of (p^a, p^b, p^a, p^{a-b}) RDSs is considered to be one of the most important questions concerning RDSs [8]. The existence of abelian (p^a, p^b, p^a, p^{a-b}) RDS was studied by Schmidt [10].

  In this paper, we prove the nonexistence of $(18, 3, 18, 6)$ RDSs. The nonexistence of abelian RDSs with parameters $(18, 3, 18, 6)$ is known and follows from [1, Theorem 2.1], by taking $m = 18$, $n = 3$, $k = 18$, $\lambda_1 = 0$, $\lambda_2 = 6$, $|U| = 2$, $t = 1$, and $p = 2$ (we note that this argument does not apply to the non-abelian case). We extend this result to show that $(18, 3, 18, 6)$ RDSs do not exist

© Springer International Publishing Switzerland 2014
K.-U. Schmidt and A. Winterhof (Eds.): SETA 2014, LNCS 8865, pp. 149–153, 2014.
DOI: 10.1007/978-3-319-12325-7_13

in non-abelian groups either. We use the computer algebra package Magma [3] for our computations. Our method is based on the close relation between RDS with forbidden normal subgroup N and generalized Hadamard matrices over the group N. The authors used this method to enumerate all RDS with parameters $(16, 4, 16, 4)$ [4], by utilizing the classification of all generalized Hadamard matrices Hadamard matrices of order 16 over a group of order 4 [6]. In this paper, we use the recent classification of all generalized Hadamard matrices of order 18 over a group of order 3 [5], to prove the non-existence of non-abelian $(18, 3, 18, 6)$ relative difference sets.

2   RDSs and Symmetric Nets

We simplify our search for $(18, 3, 18, 6)$ RDSs by taking advantage of their connection to symmetric nets, generalized Hadamard matrices and combinatorial designs of certain type.

A t-(v, k, λ) design is a pair $D = (\mathcal{P}, \mathcal{B})$ of a set $\mathcal{P}$ of v points and a collection $\mathcal{B}$ of blocks, such that each block $B \in \mathcal{B}$ is a k-subset of $\mathcal{P}$, and every t-subset of points appears in exactly λ blocks. The *dual design* D^* is obtained by interchanging the roles of points an blocks of D. A design is *symmetric* if it has the same number of blocks and points. Further background concerning designs may be found in [2].

A *parallel class* in a design is a collection of blocks which partition the point set $\mathcal{P}$. A *resolution* is a partition of the blocks $\mathcal{B}$ into parallel classes. A design is *resolvable* if it contains a resolution. A design is *affine resolvable* (or just *affine*) if there exists a constant $m \neq 0$ such that, for every pair of blocks B_1, B_2 from distinct parallel classes, $|B_1 \cap B_2| = m$.

An *automorphism* of a design is a permutation of $\mathcal{P}$ which preserves $\mathcal{B}$. The set of all automorphisms forms a group, called the (full) automorphism group of D. Subgroups of $Aut(D)$ are called automorphism groups. Two designs $D_1 = (\mathcal{P}_1, \mathcal{B}_1)$ and $D_2 = (\mathcal{P}_2, \mathcal{B}_2)$ are called *isomorphic* if there exists a bijection from $\mathcal{P}_1$ to $\mathcal{P}_2$ which takes $\mathcal{B}_1$ to $\mathcal{B}_2$.

A *symmetric (m, n) net* is a symmetric 1-(mn^2, mn, mn) design D such that both D and D^* are affine, namely, both the points and the blocks can be partitioned uniquely into parallel classes of size n. If a symmetric (m, n)-net admits a group of automorphisms G of order n which acts transitively and regularly on every point and block parallel class, then the net is called *class regular* and G is called the *group of bitranslations* [2].

A *generalized Hadamard matrix* $H(m, n)$ over a multiplicative group G of order n is a $mn \times mn$ array with entries from G with the property that for every $i, j, 1 \leq i < j \leq mn$, the multi-set

$$\{h_{is} h_{js}^{-1} \mid 1 \leq s \leq nm\}$$

contains every element of G exactly m times. Every generalized Hadamard matrix $H(m, n)$ determines a class-regular symmetric (m, n)-net with a group of bitranslations G, and conversely, every class-regular (m, n)-net with a group of bitranslations G gives rise to a generalized Hadamard matrix $H(m, n)$ [2].

In this paper, we consider symmetric $(6, 3)$ nets, which correspond to generalized Hadamard matrices $H(6, 3)$. All such matrices were recently enumerated in [5]. Any generalized Hadamard matrices $H(6, 3)$ gives rise to a symmetric $(6, 3)$ net, which is also a symmetric 1-$(54, 18, 18)$ affine resolvable design, and its dual design is also affine resolvable, with all parallel classes of size 3, on which the group of order 3 acts semi-regularly, and every two blocks from different parallel classes sharing exactly 6 points.

Consider a $(18, 3, 18, 6)$ RDS R in a group G of order 54 relative to a normal subgroup $N \leq G$ of order 3. Then G can be associated with a class-regular $(6, 3)$ net D. The points of D are the elements of G, and blocks are the subsets $B_g \subseteq G$ of the form

$$B_g = \{Rg : g \in G\}.$$

The point partition of the net is given by the partition of G into cosets of N. As a result, G is an automorphism group of D, and $N \equiv Z_3$ acts transitively on each point group and on each parallel class.

Thus, every $(18, 3, 18, 6)$ RDS corresponds to a class regular symmetric $(6, 3)$ net which admits a regular automorphism group. All nonisomorphic class regular symmetric $(6, 3)$ nets were enumerated by Harada, Lam, Munemasa, and Tonchev in [5]. Up to isomorphism, there are 53 such nets, having a group of bitranslations $N = Z_3$.

Consequently, the enumeration of $(18, 3, 18, 3)$ RDSs is reduced to the following problem: for each class regular $(6, 3)$ symmetric net D which admits an automorphism group i $Aut(D)$ acting transitively on its points and on its blocks, find all regular subgroups G of $Aut(D)$ such that Z_3 is a normal subgroup of G, and G is transitive on the blocks of D. Clearly, every $(18, 3, 18, 3)$ RDS corresponds to a symmetric $(6, 3)$ net having an automorphism group with the described properties.

We note that our method essentially uses the fact that N is a normal subgroup. We do not consider RDSs for which N is not normal.

There are 15 nonisomorphic groups of order 54: three abelian, 12 are nonabelian. Up to monomial equivalence, there are 85 generalized Hadamard matrices $(6, 3)$ over a group of order 3 [5]. Among the 53 pairwise nonisomorphic symmetric $(6, 3)$-nets corresponding to these matrices, 17 admit an automorphism group which is transitive on the points, with exactly one of these groups is also transitive on the blocks. Within this single automorphism group (which corresponds to net #4 in [5]), Magma finds 12 conjugacy classes of order 54 subgroups which are regular on the 54 points of the net. However, none of these subgroups are also transitive on the blocks of the net. Therefore, there cannot exist any RDSs within the automorphism groups of these nets. Thus, we have the following.

Theorem 1. *There are no $(18, 3, 18, 6)$ RDSs.*

It is known [8] that there can be no abelian RDSs with parameters $(18, 3, 18, 6)$, which is confirmed by this enumeration.

3   Validation via Backtrack Search

To validate these results, we also performed an exhaustive search for RDSs in all 15 groups of order 54. Using Magma, we identified all normal subgroups of order 3 within each such group. For each group of order 54 and each normal subgroup of order 3, we attempted to construct all possible $(18, 3, 18, 6)$ RDSs via a backtrack search. To simplify this search, we made several assumptions, outlined below.

Lemma 1. *Let R be any RDS in a group G relative to $N \leq G$. Then for any $g \in G$, Rg is also an RDS.*

Proof. We calculate the multiset

$$\{xy^{-1} : x, y \in Rg, \ x \neq y\} = \{ag(bg)^{-1} : a, b \in R, \ a \neq b\}$$
$$= \{ab^{-1} : a, b \in R, \ a \neq b\}.$$

This is exactly the multiset obtained from the RDS R.

Let G be a group of order 54, and $N \leq G$ be a normal subgroup of order 3. Let e be the identity element of G. Let R be a $(18, 3, 18, 6)$ RDS contained in G, relative to N. By Lemma 1, we may assume that $e \in R$. Indeed, for any $r \in R$, Rr^{-1} is also such an RDS which must contain e.

Our next result applies specifically to RDSs with $m = k$:

Lemma 2. *Let R be an (m, n, m, λ) RDS in a group G relative to $N \leq G$. For any $g \in G$, R contains exactly one element of the coset Ng, that is, $|Ng \cap R| = 1$.*

Proof. Without loss of generality, assume that $e \in R$. No other element n of N can be contained in R, or else $ne^{-1} = n \in N$ appears in the multiset of differences. Suppose $r_1, r_2 \in R$ where $r_1 = n_1 g$ and $r_2 = n_2 g$ for some $n_1, n_2 \in N$. Then

$$r_1 r_2^{-1} = n_1 g(n_2 g)^{-1} = n_1 g g^{-1} n_2^{-1} = n_1 n_2^{-1},$$

which must be in N. Thus, $n_1 = n_2$, because only $e \in N$ is also in R, and so at most one element from Ng is contained in R. Because $|G| = mn$ and $|N| = n$, there are exactly m cosets of N in G. Thus, each such coset of N must contribute exactly one element to R.

For each group G of order 54 and each normal subgroup N of order 3, we arbitrarily ordered the cosets of N in G. We then used a backtrack search to construct a partial RDS R'. Each level of the backtrack search corresponded to one of the cosets of N within G. At each level of the search, we added a single element from the corresponding coset. By Lemma 1, we chose e from the coset N, and by Lemma 2 we needed exactly one element from each coset. This significantly reduced the computational cost.

At each level of the backtrack search, we calculated the multiset

$$D = \{xy^{-1} : x, y \in R', \ x \neq y\}.$$

If D contained any element more than 6 times, the partial RDS was rejected, and the most recently added element was removed. The backtrack search then appended the next possible element from the same coset of N. If at any point every element within a given coset had been attempted, then the element added from the previous coset was then rejected, and so forth. In this way, all possible valid combinations of one element from each coset of N were tested.

This search took approximately 24 h (running on a single desktop computer), and confirmed Theorem 1: There are no $(18, 3, 18, 6)$ RDSs within any groups of order 54.

Acknowledgments. The authors wish to thank the unknown referees for their constructive remarks, and Bernhard Schmidt and Alexander Pott for the helpful discussion on the nonexistence of abelian $(18, 3, 18, 6)$ RDSs.

References

1. Arasu, K.T., Jungnickel, D., Pott, A.: The mann test for divisible difference sets. Graphs Combin. **7**, 209–217 (1991)
2. Beth, T., Jungnickel, D., Lenz, H.: Design Theory, 2nd edn. Cambridge University Press, Cambridge (1999)
3. Bosma, W., Cannon, J.: Handbook of Magma Functions. Department of Mathematics, University of Sydney (1994)
4. Clark, D., Tonchev, V.D.: Enumeration of (16, 4, 16, 4) relative difference sets. Electron. J. Comb. **20**(1), P72 (2013)
5. Harada, M., Lam, C., Munemasa, A., Tonchev, V.D.: Classification of generalized Hadamard matrices $H(6, 3)$ and quaternary Hermitian self-dual codes of length 18. Electron. J. Comb. **17**, 1–14 (2010)
6. Harada, M., Lam, C., Tonchev, V.D.: Symmetric (4, 4)-nets and generalized Hadamard matrices over groups of order 4. Des. Codes Crypt. **34**, 71–87 (2005)
7. Kumar, P.V.: On the existence of square dot-matrix patterns having a specified three-valued periodic correlation function. IEEE Trans. Inform. Theory **34**, 271–277 (1988)
8. Pott, A.: A survey on relative difference sets. In: Arasu, K.T., Dillon, J.F., Harada, K., Seghal, S.K., Solomon, R.I. (eds.) Groups, Difference Sets, and the Monster, pp. 195–233. DeGruyter Verlag, Berlin (1996)
9. Röder, M.: The quasiregular projective planes of order 16. Glas. Mat. **43**, 231–242 (2008)
10. Schmidt, B.: On (p^a, p^b, p^a, p^{a-b})-relative difference sets. J. Algebraic. Combin. **6**, 279–297 (1997)

Aperiodic Correlation

Exhaustive Search for Optimal Minimum Peak Sidelobe Binary Sequences up to Length 80

Anatolii N. Leukhin[(✉)] and Egor N. Potekhin[(✉)]

Mari State University, Yoskar-Ola 424000, Russia
leukhinan@list.ru, potegor@yandex.ru

Abstract. A brief survey of the problem to find binary sequences with minimum peak sidelobe of aperiodic autocorrelation is given. Results of an exhaustive search for minimum peak sidelobe level sequences are presented. Several techniques for efficiency implementation of search algorithm are described. A table of number of non-equivalent optimal binary sequences with minimum peak sidelobe level up to length 80 is given. Such sequence families are important in low probability of intercept radar. Examples of optimal binary minimum peak sidelobe sequences having high merit factor for each length $N \in [2, 80]$ are shown.

Keywords: Exhaustive search · Minimum peak sidelobe · Aperiodic autocorrelation function · Binary sequences · Merit factor

1 Introduction

Binary sequences with low autocorrelation sidelobe levels have many applications in communication, cryptography and radar engineering. The length of a binary sequence represents the pulse compression ratio achieved via its use. In pulse radar detection thresholds must be set higher than peak sidelobe (PSL) from the compression of the largest target return expected in a scene. So there are two problems: the problem of estimating the optimal PSL for a binary sequence of length N and the problem of design such binary sequences.

Let $A = (a_0, a_1, ..., a_{N-1})$ be a binary sequence of length N, where $a_n = 1$ or -1 for each $n = 0, 1, ..., N - 1$.

The aperiodic autocorrelation function (AACF) of A at shift τ defined as

$$C_\tau = \sum_{n=0}^{N-1-\tau} a_n \cdot a_{n+\tau} \tag{1}$$

There are two principal measures of level of sidelobe level. The primary measure is the peak sidelobe level (PSL):

$$PSL(A) = \max_{1 \leq \tau \leq N-1} |C_\tau| \tag{2}$$

© Springer International Publishing Switzerland 2014
K.-U. Schmidt and A. Winterhof (Eds.): SETA 2014, LNCS 8865, pp. 157–169, 2014.
DOI: 10.1007/978-3-319-12325-7_14

For optimal binary sequences by PSL criteria the peak sidelobe has to be minimum

$$MPS(N) = \min_{A} PSL \tag{3}$$

taken over all binary sequences A of length N. Such sequences are called minimum peak sidelobe (MPS) sequences.

A secondary measure, is the merit factor (MF)

$$MF(C) = \frac{N^2}{2\sum_{\tau=1}^{N-1}[C_\tau]^2} \tag{4}$$

PSL affects the maximum of self interference of the sequence and MF determines average interference. There are three transformations that preserve PSL in binary codes: (1) reversal, which is $R(a_n) = a_{N-1-n}$, (2) negation, which is $N(a_n) = -a_n$, and (3) alternating sign $S(a_n) = (-1)^n a_n$. Such sequences form an equivalence class.

The periodic autocorrelation function (PACF) of A at shift τ defined as

$$R_\tau = \sum_{n=0}^{N-1} a_n \cdot a_{n+\tau \ (mod \ N)} \tag{5}$$

First let us mention several relevant results to estimation of typical PSL. Moon and Moser [1] proved that for almost all binary sequences

$$k(N) \leq PSL(A) \leq (2+\varepsilon)\sqrt{N \ln N},$$

for any $k(N) = o\left(\sqrt{N}\right)$.

Mercer [2] improved this result and showed that

$$MPS(N) \leq \left(\sqrt{2}+\varepsilon\right)\sqrt{N \ln N}.$$

Dmitriev and Jedwab [3] conjectured and provided an experimental evidence that the typical PSL of random sequences

$$\sqrt{N} \leq PSL(A) \leq \sqrt{N \ln N}.$$

Recently [34] the result of Moon and Moser has been strengthened to

$$(\sqrt{2}-e)\sqrt{N \cdot \log(N)} < PSL(A) < (\sqrt{2}+e)\sqrt{N \cdot \log(N)},$$

for every $e > 0$ and almost all binary sequences A of sufficiently large length N. Equivalently (and more precisely), if A is a random binary sequence of length N, then

$$PSL(A)/\sqrt{N \cdot \log(N)} \to \sqrt{2} \text{ in probability.}$$

This also proves the conjecture attributed to Dmitriev and Jedwav [3]. Also in the paper [3] evidence was proved that

$$PSL(A) = O(\sqrt{N} \cdot \log \log N)$$

for all m-sequences A of length N (not just for typical m-sequence).

Also in the prior paper [35] by Jedwab and Yoshida it was proved that

$$PSL(A) = \Omega(N)$$

for all m-sequence.

Schmidt [4] gave a construction for another family of binary sequences of length N with proved value that the PSL is at most

$$\sqrt{2 \cdot N \cdot \log(2 \cdot N)}$$

There are some interesting constructions based on difference sets [5] and almost difference sets [6–8] that allowed get binary sequences for a wide range of different lengths N with PSL level even less than $\sqrt{N}$. It should be noted that this result can be experimentally estimated and there is no theoretical proven. At the same time it is conjectured the result of Ein-Dor et al. [9] based on a heuristic argument that

$$MPS(N)/\sqrt{N} \to d, \text{ where } d = 0,435\cdots, N \to \infty.$$

There is no known analytical technique to construct sequences with lowest aperiodic autocorrelation, and exhaustive searches have to be made in order to find the lowest autocorrelation binary sequence (LABS) for a give length. Many authors have put considerable computational effort in finding binary sequences with lowest or small peak sidelobe level.

Let us mention known results to computer search of such binary sequences. There are two search strategies of finding binary sequences of desired length and optimal aperiodic autocorrelations: global and local methods. The main idea [10] to use an exhaustive search for optimal MF sequences is based on a branch and bound (BB) algorithm. Symmetry breaking procedures for identifying equivalent sequences allow the search space to be reduced to approximately one-eighth with the runtime complexity $O\left(1,85^N\right)$. A similar approach has been applied to exhaustive search for optimal MPS sequences. In addition to branch and bound strategy using preserving operations and peak sidelobe level breaking, sidelobe invariant transform, symmetry breaking, partitioning and parallelizing it is further possible to reduce the runtime complexity. Lindner [11] in 1975 did an exhaustive search for binary MPS sequences up to $N = 40$. Cohen et al. [12] in 1990 continued up to $N = 48$. Coxson and Russo [13] in 2005 performed an exhaustive search of binary MPS sequences for $N = 64$. The authors of these works presented tables with identical numbers of the obtained MPS sequences. Elders-Boll et al. [14] in 1997 found binary MPS sequences for the lengths up to 61 but they did not present the numbers of such sequences, but just gave sample codes with lowest peak sidelobe for each lengths from $N \in [49, 61]$. Authors of this paper did an exhaustive search for binary MPS sequences up to length $N = 74$ in 2013 [15] and size of non-equivalent MPS sequences for each length from $N \in [2, 74]$ were reported. Also authors published in [16] the result on exhaustive search of binary MPS sequence and the number of non-equivalent MPS sequences for the length $N = 76$.

The authors of [12], p. 633 make the statement about "some improvement to the theory, which results in a computational growth rate of the problem of $1, 4^N$ rather than 2^N". This value of runtime complexity repeatedly mentioned in difference papers (for example [4]). However the runtime complexity was not proved. Furthermore, the only way today to determine the runtime complexity both for global and for local search algorithms is experimental computing. This observation was not presented in [12]. Finally the runtime complexity of exhaustive search algorithm for binary MPS sequence should depend on peak sidelobe level due to procedure of peak sidelobe level breaking. Experimentally the runtime complexity of algorithm for exhaustive search of binary MPS sequences was determined in [15, 16] by the authors of this paper. Just only for the level $PSL = 2$ the runtime complexity is approximately equal to $O\left(1, 42^N\right)$. The other results are the next: for $PSL = 3$ is $O\left(1, 57^N\right)$, for $PSL = 4$ is $O\left(1, 7^N\right)$ and for $PSL = 5$ is $O\left(1, 79^N\right)$. The only disadvantage of an exhaustive search method is that it takes exponentially long time and can not be used today for the lengths more than $N = 100$.

Apart from known results of global exhaustive search of binary MPS sequences, there are some useful results of local search of binary sequences with low aperiodic autocorrelation. Kerdock et al. [17] in 1986 found binary sequences for lengths $N = 51, 69, 88$ with $PSL = 3, 4, 5$ respectively. Coxson and Russo [13] in 2005 continued the list of best known binary sequences.PSL sequences with $PSL = 4$ up to length $N = 70$. In [18], binary sequences with sidelobe level $PSL = 4$ were found up to length $N = 82$, while those with sidelobe level $PSL = 4$ were found for lengths $N \in [83, 105]$. There are some different stochastic local search methods for binary sequences. Evolutionary algorithm (EA) algorithm [19] developed by Militzer and et.al in 1998 has a runtime complexity "better than BB algorithm". In 2001, Prestwich [20] proposed a hybrid branch and bound algorithm and local search, called constrained local search (CLS). The algorithm was estimated to run in time $O\left(1, 68^N\right)$. Later in 2007, Prestwich [21] modified local search algorithm and constructed local search relaxation (LSR) algorithm with runtime complexity $O\left(1, 51^N\right)$. In 2003, Brgles et al. [22] proposed EAs algorithm for escaping local minima with two new termination criterion using Kernighan-Lin (KL) solver and evolutionary strategy (ES) solver. Runtime complexity of ES algorithm $O\left(1, 4^N\right)$ and runtime complexity of KL algorithm is $O\left(1, 46^N\right)$. In 2005, Borwein et al. [23] developed direct stochastic search (DSC) algorithm with runtime complexity $O\left(1, 5^N\right)$. In 2006, Dotu and van Henteryck [24] proposed tabu search (TS) algorithm with runtime complexity $O\left(1, 49^N\right)$. In 2007, Gallarado et al. [25] proposed memetic algorithm (MA) with runtime complexity $O\left(1, 32^N\right)$. There are some other stochastic local search algorithms: ants colony optimization [26], simulated annealing [27], genetic [28], iterative local search [29], scatter search [30], variable neighborhood search [31] etc. Using some of these local search methods the binary sequences with low level aperiodic autocorrelation were founded and published in [32] for the lengths $N \in [100, 300]$, for some lengths between $N = 353$ and $N = 1019$, for some lengths between $N = 1024$ and $N = 4096$. Local search methods require relatively short time but have shortcomings. Although they find "reasonable" answers, they can-not guarantee optimality. Also local search methods usually based on initial codes, then determine next codes by using an intermediate criterion. So, if all optimum sequences are required, the only solution is a global search.

Let us present our main results.

1. In this paper a new modification of exhaustive search algorithm for optimal MPS sequences is developed. Our algorithm takes into account analytical dependences between sidelobes of a periodic and an aperiodic autocorrelation and the distribution of sidelobes of aperiodic autocorrelation. The runtime complexity of proposed global search method is approximately equal to the runtime complexity of some known local search methods and it is the best known result for exhaustive search methods today. Also we have some new improvements of global algorithm for exhaustive search due to programming techniques.

2. This paper adds to available knowledge for record length of binary MPS sequences and provides numbers of non-equivalent sequences for all lengths from $N \in [75, 80]$, this result improves our previous results for an exhaustive search of binary sequences from the lengths $N = 62, 63$ and $N \in [65, 74]$. Moreover the results published in [13] and in this paper are expanded the list of known numbers of non-equivalent MPS sequences from the range $N \in [2, 48]$ and $N = 64$ to the range $N \in [2, 80]$ which amount 40 percents to previous known results. Also it proved that founded binary sequences in [12, 16, 17] for the lengths up to $N = 80$ are optimal minimum peak sidelobe sequences.

3. Also sample binary MPS sequences with highest value of merit factor (MF) are shown for lengths 2 to 80. Most of these samples are new non-equivalent to previous known binary MPS sequences and firstly published in this paper. Such samples are very interesting especially for the lengths where number of non-equivalent MPS sequence is a few units.

It should be noted that an exhaustive search of binary MPS sequences was performed off and on during 12 months using 1 supercomputer Flagman RX240 on the base of 8 NVIDIA TESLA C2059 with 3584 parallel graphical processors and on the base of 2 processors Intel Xeon X5670 (up to Six-Core) and using CUDA compilation. For example an exhaustive search for the length $N = 80$ was performed in the background for 1 month.

2   Algorithm of Exhaustive Search

First of all we use all ideas of modification of BB algorithm for exhaustive search of minimum peal sidelobe sequences presented in [12,13]. Also we add some new improvements.

1. Our modification uses next analytical dependences between periodic and aperiodic autocorrelations

$$R_\tau = R_{N-1-\tau} = C_\tau + C_{N-1-\tau}, \tau = 1, 2, ..., \lfloor (N-1)/2 \rfloor \qquad (6)$$

where $\lfloor x \rfloor$ - ceil part of x, $-PSL \leq C_\tau \leq PSL$.

It is easy to show that number of levels of periodic autocorrelation function is $\leq (PSL + 1)$. Sidelobes of periodic autocorrelation (6) can be determined from next conditions:

$$R_\tau = [\min, \min +4, ..., \max], \tag{7}$$

where

$\min = -2 \cdot PSL + 2 \cdot (PSL \bmod 2), \max = 2 \cdot PSL - 2 \cdot (PSL \bmod 2),$
if N (mod 4) = 0;

$\min = -2 \cdot PSL + 1 + 2 \cdot (PSL \bmod 2), \max = 2 \cdot PSL - 3 + 2 \cdot (PSL \bmod 2),$
if N (mod 4) = 1;

$\min = -2 \cdot PSL + 2 \cdot (PSL + 1 \bmod 2), \max = 2 \cdot PSL - 2 \cdot (PSL + 1 \bmod 2),$
if N (mod 4) = 2;

$\min = -2 \cdot PSL + 3 - 2 \cdot (PSL \bmod 2), \max = 2 \cdot PSL - 1 - 2 \cdot (PSL \bmod 2),$
if N (mod 4) = 3.

So, we can determine last sidelobes of aperiodic autocorrelation $C_\tau, \tau = \lceil (N-1)/2 \rceil, \lceil (N-1)/2 \rceil + 1, ..., N - 1$, by first known sidelobes $C_\tau, \tau = 1, 2, ..., \lfloor (N-1)/2 \rfloor$. Let us note

$$Left_\tau = C_\tau, Right_\tau = C_{N-1-\tau}, \tag{8}$$

Now we can find

$$Left_\tau = R_\tau - Right_\tau. \tag{9}$$

R_τ is the same for all cyclically shifted copies of binary sequence A. So it is once necessary to determine PACF R_τ for initial sequence using both $Left$ and $Right$ parts of AACF (6). For all others $(N - 1)$ cycle shifted copies it is sufficient calculate only a $Right$ part of AACF and determine $Left$ part by Eq. (9).

2. Our next idea is about distribution of aperiodic autocorrelation sidelobes. We assume that a number of excluding sequences at the first shifts more than a number of excluding sequences at the central shifts, due to numbers of additions. After calculating the last sidelobes for shifts $\tau = N - 1, N - 2, ..., \lfloor (N - 1)/2 \rfloor$ we jump to $\tau = 1$ sidelobe, than to $\tau = 2$ sidelobe and so on, i.e. we change direction of calculation of aperiodic autocorrelation after central index $\lfloor N/2 \rfloor$. This is a way to calculate AACF and PACF for initial sequence. This technique allows to reduce runtime complexity approximately to 25 %.

3. Partitioning and parallelizing give the way to reduce the number of initial codes. For example for the length $N = 50$ we get $P = 2122026$ "initial" codes with 13 bits $(a_0, a_1, ..., a_{12})$ and $(a_{N-12}, a_{N-11}, ..., a_{N-1})$ from each sides of sequence. For the length $N = 80$ there are $P = 1907802$ "initial" codes with 13 bits from each sides of sequence.

4. Also we used "package" regime to find some binary PSL sequences with the lengths $N, N + 2, N + 4, ...$, because cross correlation functions for left and right parts of the sequences with lengths $N, N + 2, N + 4, ...$, are the same, so they are computing just only once for minimum N.

Let us experimentally estimate the runtime complexity of offered algorithm. The runtime complexity of the full search algorithm is determined as

$$O\left((N - 1) \cdot 2^N\right) \tag{10}$$

Table 1. Runtime complexity of modified BB algorithm for exhaustive search of binary MPS sequences

$PSL = 2$	$PSL = 3$	$PSL = 4$	$PSL = 5$
$O\left(20.7 \cdot 1.42^N\right)$	$O\left(18.3 \cdot 1.57^N\right)$	$O\left(9.9 \cdot 1.7^N\right)$	$O\left(6.9 \cdot 1.79^N\right)$

where 2^N is the number of possible binary sequences of length N, $(N - 1)$ is the number of frames of AACF sidelobes. The experimental results from calculating the computational complexity are approximated according to the law

$$O\left(c \cdot b^N\right) \tag{11}$$

The runtime complexity of the algorithm for finding binary peak sidelobe sequences with $PSL = 2, 3, 4, 5$ is shown in the Table 1. The results from calculating the experimental and theoretical runtime complexity of the new algorithm for finding MPS sequences in the range of lengths $N \in [10, 50]$ are shown in the Fig. 1. The number of calculations for AACF frames is shown on the vertical axis in logarithmic scale, while the length of a sequence is shown on the horizontal axis.

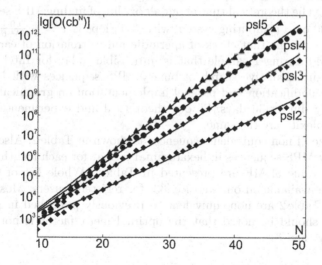

Fig. 1. Computational complexity of the new algorithm for global search for MPS sequences

3   Program Implementation

1. Our main idea is to use new assembler instructions for computing autocorrelation function of binary sequences. We can find side lobes of aperiodic autocorrelation using XOR operation. To determine the level of sidelobe we have to calculate the numbers of zeros and units for each shift of sequences. New Intel

processors have microarchitecture Intel Core of version SSE4.2 which operating with the set of command on low level. For example C/C++ Microsoft compiler has function_popcnt64 of intrin library and also compilersGCC and G++ has function_mm_popcnt_u64 of smmintrin library for calculation the number of units in binary sequences by 1 cycle.

2. We used recursive implementation of our algorithm using inline options for all external operations.

3. For excluding equivalent sequences we used reverse transformation for two bytes at the time instead of each bit. All possible reverses are stored in static massive with 65536 different bit variations.

4. We realized parallel computing in multiprocessor system for all set of non-equivalent sequences separately each from other. We implemented our algorithm on CUDA SDK using function_popcll() for calculating number of unit bits.

4   Results of Exhaustive Search for Binary MPS Sequences

Using our modification of BB algorithm we are able to find all non-equivalent classes of binary MPS sequences for each length N up to $N = 80$. Today it is the record length. The theoretical time of our algorithm of finding MPS sequences for length $N = 80$ on a computing system with 3 TFlops is $T = \frac{9.9 \cdot 1.7^{80}}{3 \cdot 10^{12} \cdot 60 \cdot 60 \cdot 24} \approx 31$ days. Here we assume that check of aperiodic autocorrelation of each sequence is implemented during 1 sample that is impossible today for Intel processors. We realized an exhaustive search of binary MPS sequences with length $N = 80$ using our modification and parallel implementation on graphical processors approximately during 30 days. So our theoretical and experimental results of runtime complexity are the same.

Family size of non-equivalent sequences is shown in Table 2. Also the examples of binary MPS sequences in hexadecimal format for each length $N \in [2, 80]$ with highest value of MF are presented in Table 2. Whole list of synthesized sequences are available on our website [33] for registered users. Most sample of codes in the Table 2 are non-equivalent to previously published in papers [11–14,17,18]. It should be noted that the optimal merit factor is not known for lengths larger than 61.

Table 2. Results of an exhaustive search of binary MPS sequences

Length	PSL	MF	Optimal or best known by MF?	Sequence	Size of set
2	1	2	yes	0	1
3	1	4.5	yes	3	1
4	1	4	yes	2	1

(*Continued*)

Table 2. (*Continued*)

Length	PSL	MF	Optimal or best known by MF?	Sequence	Size of set
5	1	6.25	yes	02	1
6	2	2.571	yes	0B	4
7	1	8.167	yes	0D	1
8	2	4	yes	16	8
9	2	3.375	yes	029	10
10	2	3.846	yes	076	5
11	1	12.1	yes	0ED	1
12	2	7.2	yes	0A6	16
13	1	14.083	yes	00CA	1
14	2	5.158	yes	019A	9
15	2	4.891	no	0329	13
16	2	4.571	no	1DDA	10
17	2	4.516	yes	0192B	4
18	2	6.48	yes	0168C	2
19	2	4.878	no	07112	1
20	2	5.263	no	04D4E	3
21	2	6.485	no	005D39	3
22	3	6.205	yes	013538	378
23	3	5.628	yes	084BA3	515
24	3	8	yes	31FAB6	858
25	2	7.102	no	031FAB6	1
26	3	7.511	no	07015B2	242
27	3	9.851	yes	0F1112D	388
28	2	7.84	yes	1E2225B	2
29	3	6.782	yes	031FD5B2	284
30	3	7.627	yes	03F6D5CE	86
31	3	7.172	yes	00E326A5	251
32	3	7.111	no	01E5AACC	422
33	3	8.508	yes	003CB5599	139
34	3	8.892	yes	0CC01E5AA	51
35	3	7.562	no	0CC01E5AA	111
36	3	6.894	no	3314A083E	161
37	3	6.985	no	006C94A8E7	55
38	3	8.299	yes	003C34AA66	17

(*Continued*)

Table 2. (*Continued*)

Length	PSL	MF	Optimal or best known by MF?	Sequence	Size of set
39	3	6.391	no	13350BEF3C	30
40	3	7.407	yes	2223DC3A5A	57
41	3	7.504	no	038EA520364	15
42	3	8.733	yes	04447B874B4	4
43	3	6.748	no	005B2ACCE1C	12
44	3	6.286	no	202E2714B96	15
45	3	6.575	no	02AF0CC6DBF6	4
46	3	6.491	no	03C0CF7B6556	1
47	3	7.126	no	069A7E851988	1
48	3	6.128	no	24AC8847B87C	4
49	4	8.827	yes	05E859E984451	49088
50	4	8.17	yes	07837FB996B2A	25169
51	3	7.517	no	0E3F88C89524B	1
52	4	8.145	yes	50AE3808C8DB6	33058
53	4	7.89	no	07C0CFBDB4CD56	23673
54	4	7.327	no	116E1DF7D2C6E6	10808
55	4	7.451	no	1658A2BC0A133B	11987
56	4	8.167	yes	0C790164F6752A	15289
57	4	7.963	no	01B4DE3455B93BF	9476
58	4	8.538	yes	008D89574E1349E	4026
59	4	8.328	no	1CAD63EFF126A2E	4624
60	4	8.108	no	119D01522ED3C34	5542
61	4	7.563	no	0024BA568EB83731	3246
62	4	8.179	yes	000C67247C59568B	1212
63	4	9.587	yes	1B3412F0501539CE	1422
64	4	9.846	yes	26C9FD5F5A1D798C	1859
65	4	8.252	no	04015762C784EC369	1003
66	4	7.751	no	03FEF2CCB0B8CAC54	324
67	4	7.766	no	073C2FADC44255264	381
68	4	8.438	no	562B8CA48E0C9027E	489
69	4	7.988	no	0292582AC6A767CC03	248
70	4	7.313	no	01C2FFD4AF33356596	72
71	4	8.105	no	12493BE76A5EE2A3F1	115
72	4	7.2	no	27C8D6E165A71577FE	107

(*Continued*)

Table 2. (*Continued*)

Length	PSL	MF	Optimal or best known by MF?	Sequence	Size of set
73	4	8.327	no	012DE781C9167577AB7	46
74	4	7.039	no	00ABFA66C560E3094C2	18
75	4	7.878	no	0E0038AEB50B59C99B6	16
76	4	7.113	no	2CD864E4AA90B8073DE	17
77	4	6.959	no	066B7BDB752AA6F80E3C	10
78	4	7.548	no	0C4852361E77C0574BAC	1
79	4	7.308	no	0028AE35C3A59AC4ED89	7
80	4	6.349	no	01A4F07798EA85AE6C48	8

5   Conclusion

This paper presents some improvements to previously known exhaustive search algorithm for minimum peak sidelobe sequences. Such optimal sequences are highly sought after in radar. Also there are some improvements due to programming techniques and parallel implementation on graphical processors. It is experimentally proven that the runtime complexity of offered global algorithm of exhaustive search for binary minimum peak sidelobe sequences depends on level of PSL and it the same as some of known local algorithms.

The list of size of non-equivalent optimal binary MPS sequences set is extended approximately to 40 percent and it is included each lengths from 2 to 80. Examples of optimal binary MPS sequences having high merit factor up to length 80 are shown.

We conjecture that it is possible to find the maximum length of MPS binary sequence with the minimum achievable sidelobe level $PSL = 4$ with the length more than 84 (today the known maximum length is 82).

Acknowledgments. This work is supported by the grants: grant of Russian Foundation of Basic Research 12-07-00552 and project of Russian Ministry of Education and Science 1856.

References

1. Moon, J.W., Moser, L.: On the correlation function of random binary sequences. SLAM J. Appl. Math. **16**(12), 340–343 (1968)
2. Mercer, I.D.: Autocorrelations of random binary sequences. Comb. Probab. Comput. **15**(5), 663–671 (2006)
3. Dmitriev, D., Jedwab, J.: Bounds on the growth rate of the peak sidelobe level of binary sequences. Adv. Math. Commun. **1**(4), 461–475 (2007)
4. Schmidt, K.-U.: Binary sequences with small peak sidelobe level. IEEE Trans. Inf. Theor. **58**(4), 2512–2515 (2012)

5. Golomb, S.W., Gong, G.: Signal Design for Good Correlation: For Wireless Communication, Cryptography, and Radar. Cambridge University Press, Cambridge (2005)
6. Davis, J.A.: Almost difference sets and reversible divisible difference sets. Arch. Math. (Basel) **59**(6), 595–602 (1992)
7. Arasu, K.T., Ding, C., Helleseth, T., Kumar, P.V., Martisen, H.: Almost difference sets and their sequences with optimal autocorrelation. IEEE Trans. Inf. Theor. **47**, 2934–2943 (2001)
8. Ding, C., Pott, A., Wang, Q.: Constructions of almost difference sets from finite fields. Des. Codes Crypt. **72**(3), 581–592 (2013)
9. Ein-Dor, L., Kanter, I., Kinzel, W.: Low autocorrelated multiphase sequences. Phys. Rev. (E) **65**(2), 020102-1–020102-4 (2002)
10. Mertens, S.: On the ground state of the Bernasconi model. J. Phys. A: Math. Gen. **41**, 3731–3749 (1998)
11. Lindner, J.: Binary sequences up to length 40 with best possible autocorrelation function. Electron. Lett. **11**(21), 507 (1975)
12. Cohen, M.N., Fox, M.R., Baden, J.M.: Minimum peak sidelobes pulse compression codes. In: Proceedings of the IEEE International Radar Conference, Arlington, VA, pp. 633–638, May 1990
13. Coxson, G.E., Russo, J.: Efficient exhaustive search for optimal-peak-sidelobe binary codes. IEEE Trans. Aerosp. Electron. Syst. **41**, 302–308 (2005)
14. Elders-Boll, H., Schotten, H., Busboom, A.: A comparative study of optimization methods for the synthesis of binary sequences with good correlation properties. In: 5th IEEE Symposium on Communication and Vehicular Technology in the Benelux, pp. 24–31. IEEE (1997)
15. Leukhin, A.N., Potekhin, E.N.: Optimal peak sidelobe level sequences up to length 74. In: IEEE Conference Publications: Conference Proceedings "European Microwave Conference, EuMC'2013", Nuremberg, Germany, pp. 1807–1810, 7–10 October 2013
16. Leukhin, A.N., Shuvalov, A.S., Potekhin, E.N.: A Bernascony model for constructing ground-state spin systems. Bull. Russ. Acad. Sci. Phys. **78**(3), 207–209 (2014)
17. Kerdock, A.M., Mayer, R., Bass, D.: Longest binary pulse compression codes with given peak sidelobe levels. Proc. IEEE **74**(2), 366 (1986)
18. Nunn, C.J., Coxson, G.E.: Best-known autocorrelation peak sidelobe levels for binary codes of length 71 to 105. IEEE Trans. Aerosp. Electron. Syst. **44**(1), 392–395 (2008)
19. Militzer, B., Zamparelli, M., Beule, D.: Evalutionary search for low autocorrelated binary sequences. IEEE Trans. Evol. Comput. **2**(1), 34039 (1998)
20. Prestwich, S.: A hybrid local search algorithm for low-autocorrelation binary sequences, Technical report, Department of computer science, National University of Ireland at Cork (2001)
21. Prestwich, S.: Exploiting relaxation in local search for LABS. Ann. Oper. Res. **1**, 129–141 (2007)
22. Brglez, F., Viao, Y., Stallmann, M., Militzer, B.: Reliable cost predictions for finding optimal solutions to LABS problem: evolutionary and alternative algorithms. In: International Workshop on Frontiers in Evolutionary Algorithms (2003)
23. Borwein, P.; Ferguson, R.; Knauer, J.: The merit factor problem
24. Dotú, I., Van Hentenryck, P.: A note on low autocorrelation binary sequences. In: Benhamou, F. (ed.) CP 2006. LNCS, vol. 4204, pp. 685–689. Springer, Heidelberg (2006)

25. Gallarado, J., Cotta, C., Fernandez, A.: A memetic algorithm for the low auto-correlation binary sequence problem. In: Genetic and Evolutionary Computation Conference, pp. 1226–1233. ACM (2007)
26. Dorigo, M., Stutzle, T.: Ants Colony Optimization. MIT Press, Cambridge (2004)
27. Kirkpatrick, S., Gelatt, D., Veechi, M.: Optimization by simulated annealing. Science **220**, 671–680 (1983)
28. Holland, J.: Adaptation in Natural and Artificial Ecosystems, 2nd edn. MIT Press, Cambridge (1992)
29. Stutzle, T.; Hoos, H.; Analyzing the run-time behavior of iterated local search for the TSP. In: 3rd Metaheuristics International Conference, pp. 449–453 (1999)
30. Rego, C., Alidaee, B.: Tabu Search and Scatter Search. Kluwer Academic Publishers, Norwell (2005)
31. Hansen, P., Mladenovic, N.: A tutorial on variable neighborhood search, TR G-2003-16, Gerad (2003)
32. Du, K.L., Wu, W.H., Mow, W.H.: Determination of long binary sequences having low autocorrelation functions. United States patent, no. US 8,493,245 B2, 23 July 2013
33. Signalslab. http://signalslab.volgatech.net
34. Schmidt, K.-U.: The peak sidelobe level of random binary sequences. Bull. Lond. Math. Soc. **46**(3), 643–652 (2014)
35. Jedwab, J., Yoshida, K.: The peak sidelobe level of families of binary sequences. IEEE Tran. Inform. Theor. **52**, 2247–2254 (2014)

Invited Paper

The Inverse of the Star-Discrepancy Problem and the Generation of Pseudo-Random Numbers

Josef Dick[1]($\boxtimes$) and Friedrich Pillichshammer[2]

[1] School of Mathematics and Statistics,
The University of New South Wales, Sydney, NSW, Australia
josef.dick@unsw.edu.au
[2] Department of Financial Mathematics,
Johannes Kepler University, Linz, Austria
friedrich.pillichshammer@jku.at

Abstract. The inverse of the star-discrepancy problem asks for point sets $P_{N,s}$ of size N in the s-dimensional unit cube $[0,1]^s$ whose star-discrepancy $D_N^*(P_{N,s})$ satisfies

$$D_N^*(P_{N,s}) \le C\sqrt{s/N},$$

where $C > 0$ is a constant independent of N and s. The first existence results in this direction were shown by Heinrich, Novak, Wasilkowski, and Woźniakowski in 2001, and a number of improvements have been shown since then. Until now only proofs that such point sets exist are known. Since such point sets would be useful in applications, the big open problem is to find explicit constructions of suitable point sets $P_{N,s}$. We review the current state of the art on this problem and point out some connections to pseudo-random number generators.

1 Introduction

The star-discrepancy is a quantitative measure for the irregularity of distribution of a point set $P_{N,s} = \{\boldsymbol{x}_0, \boldsymbol{x}_1, \ldots, \boldsymbol{x}_{N-1}\}$ in the s-dimensional unit cube $[0,1)^s$. It is defined as the L_∞-norm of the local discrepancy

$$\Delta(\boldsymbol{t}) := \frac{\#\{n \in \{0,1,\ldots,N-1\} \ : \ \boldsymbol{x}_n \in [\boldsymbol{0},\boldsymbol{t})\}}{N} - \lambda_s([\boldsymbol{0},\boldsymbol{t})),$$

for $\boldsymbol{t} = (t_1, t_2, \ldots, t_s) \in [0,1]^s$, where $[\boldsymbol{0},\boldsymbol{t}) = \prod_{j=1}^s [0,t_j)$ and λ_s denotes the s-dimensional Lebesgue measure. In other words, the star-discrepancy (or L_∞-discrepancy) of $P_{N,s}$ is

$$D_N^*(P_{N,s}) = \sup_{\boldsymbol{t} \in [0,1]^s} |\Delta(\boldsymbol{t})|.$$

Its significance arises from the classical Koksma-Hlawka inequality [23,25], which states that

$$\left| \int_{[0,1]^s} f(\boldsymbol{x}) \, \mathrm{d}\boldsymbol{x} - \frac{1}{N} \sum_{n=0}^{N-1} f(\boldsymbol{x}_n) \right| \le V(f) D^*(P_{N,s}), \tag{1}$$

© Springer International Publishing Switzerland 2014
K.-U. Schmidt and A. Winterhof (Eds.): SETA 2014, LNCS 8865, pp. 173–184, 2014.
DOI: 10.1007/978-3-319-12325-7_15

where $V(f)$ denotes the variation of f in the sense of Hardy and Krause, see, e.g., [11,28,31]. This is the fundamental error estimate for quasi-Monte Carlo rules $Q(f) = (1/N) \sum_{n=0}^{N-1} f(\boldsymbol{x}_n)$.

To provide some insight into this inequality, we prove a simple version of it. Let $f : [0, 1] \to \mathbb{R}$ be absolutely continuous, then for $x \in [0, 1]$ we have

$$f(x) = f(1) - \int_0^1 \mathbf{1}_{[x,1]}(t) f'(t) \, dt, \tag{2}$$

where $\mathbf{1}$ denotes the indicator function. Using (2) we obtain

$$\int_{[0,1]^s} f(x) \, dx - \frac{1}{N} \sum_{n=0}^{N-1} f(x_n) = \int_0^1 f'(t) \left[\frac{1}{N} \sum_{n=0}^{N-1} \mathbf{1}_{[x_n,1]}(t) - \int_0^1 \mathbf{1}_{[x,1]}(t) \, dx \right] dt$$

$$= \int_0^1 f'(t) \left[\frac{1}{N} \sum_{n=0}^{N-1} \mathbf{1}_{[0,t]}(x_n) - t \right] dt.$$

This implies that

$$\left| \int_{[0,1]^s} f(x) \, dx - \frac{1}{N} \sum_{n=0}^{N-1} f(x_n) \right| \le \int_0^1 |f'(t)| \, dt \sup_{0 \le t \le 1} \left| \frac{1}{N} \sum_{n=0}^{N-1} \mathbf{1}_{[0,t]}(x_n) - t \right|.$$

The right-most expression in the above inequality is simply the star-discrepancy of the point set $\{x_0, x_1, \ldots, x_{N-1}\}$ and for absolutely continuous functions f, the term $\int_0^1 |f'(t)| \, dt$ coincides with the Hardy-Krause variation. This approach can be generalized to the s-dimensional unit cube $[0, 1]^s$, yielding a version of the Koksma-Hlawka inequality (1). To obtain quasi-Monte Carlo rules with small quadrature error, it is therefore of importance to design point sets with small star-discrepancy.

In many papers the star-discrepancy is studied from the viewpoint of its asymptotic behavior in N (for a fixed dimension s). Define the Nth *minimal star-discrepancy* in $[0, 1)^s$ as

$$\mathrm{disc}(N, s) := \inf_{P_{N,s}} D_N^*(P_{N,s}),$$

where the infimum is extended over all N-element point sets in $[0, 1)^s$. It is well known that $\mathrm{disc}(N, s)$ behaves like

$$\frac{(\log N)^{(s-1)/2+\delta_s}}{N} \ll_s \mathrm{disc}(N, s) \ll_s \frac{(\log N)^{s-1}}{N}, \tag{3}$$

where $\delta_s \in (0, 1/2)$ is an unknown quantity depending only on s. Here $A(N, s) \ll_s B(N, s)$ means that there is a constant $c_s > 0$ depending only on s but not on N such that $A(N, s) \le c_s B(N, s)$ for all large enough N. The lower bound was shown by Bilyk, Lacey and Vagharshakyan [5] improving a famous result of Roth [33]. For the upper bound several explicit constructions of point sets are known whose star-discrepancy achieves such a bound. See, e.g., [11, 31]. Thus the upper bound on the Nth minimal star-discrepancy is of order of magnitude $O(N^{-1+\varepsilon})$ for every $\varepsilon > 0$. The problem however is that the function $N \mapsto (\log N)^{s-1}/N$ does not decrease to zero until $N > \exp(s - 1)$. For $N \leq \exp(s - 1)$ this function is increasing which means that for N in this range our discrepancy bound is useless.

In the following we mention two applications which illustrate that there is a need for point sets with small star-discrepancy where $N \leq \exp(s - 1)$, thus motivating the need for different types of discrepancy bounds (which we discuss in the next section). One such application was studied in [29], which deals with partial differential equations with random coefficients. Without going into any details, in this paper one always has $N = s^\kappa$ for some $0 < \kappa \leq 1$ (the interested reader may consult [29, Theorem 8]). Another case arises when the dimension s is very large. For instance, in some applications from financial mathematics, the dimension s can be several hundreds, see for instance [32]. If $s = 100$, then $\exp(s - 1) \approx 10^{43}$. Due to the limitations of the current technology, the number of points N we can use is much smaller than $\exp(s - 1)$ in this case.

In the next section we discuss bounds on the star-discrepancy which are of interest for high-dimensional applications.

2   The Inverse of the Star-Discrepancy Problem

We review the current literature on the inverse of the star-discrepancy problem as first studied in [20]. To analyze the problem systematically the so-called *inverse of the star-discrepancy* is defined as

$$N(s, \varepsilon) = \min\{N \in \mathbb{N} : \operatorname{disc}(N, s) \leq \varepsilon\} \quad \text{for} \quad s \in \mathbb{N} \text{ and } \varepsilon \in (0, 1].$$

This is the minimal number of points which is required to achieve a star-discrepancy less than ε in dimension s. The following theorem is the first classic result in this direction. By $A(N, s) \ll B(N, s)$ we mean that there is a constant $c > 0$ which is independent of N and s such that $A(N, s) \leq cB(N, s)$.

Theorem 1 (Heinrich, Novak, Wasilkowski, Woźniakowski [20]). *We have*

$$\operatorname{disc}(N, s) \ll \sqrt{\frac{s}{N}} \quad \textit{for all} \quad N, s \in \mathbb{N}. \tag{4}$$

Hence

$$N(s, \varepsilon) \ll s\varepsilon^{-2} \quad \textit{for all} \quad s \in \mathbb{N} \ \textit{and} \ \varepsilon > 0.$$

The bound (4) does not achieve the optimal rate of convergence for fixed dimension s as the number of points N goes to ∞. However, the dependence on the dimension s is much weaker than in (3). Thus such point sets are more suited for integration problems where the dimension s is large.

The proof of Theorem 1 is based on the probabilistic method. It is shown that the probability that the absolute local discrepancy $|\Delta(t)|$ of a randomly chosen point set is larger than a certain quantity δ is extremely small. Then one applies a union bound over all $t \in [0,1]^s$ and chooses δ such that this union bound is strictly less than one, which then implies the result. In this particular instance the authors of [20] used a large deviation inequality for empirical processes on Vapnik-Červonenkis classes due to Talagrand and Haussler. Details can be found in [8,20]. A simplified proof which leads in addition to explicit constants was given recently by Aistleitner [1].

It is also known that the dependence on the dimension s of the inverse of the star-discrepancy cannot be improved. Hinrichs [21] proved the existence of constants $c, \varepsilon_0 > 0$ such that

$$N(s, \varepsilon) \geq cs\varepsilon^{-1} \quad \text{for all } \varepsilon \in (0, \varepsilon_0) \text{ and } s \in \mathbb{N}$$

and $\mathrm{disc}(N, s) \geq \min(\varepsilon_0, cs/N)$. The exact dependence of $N(s, \varepsilon)$ on ε^{-1} is still an open question which seems to be very difficult.

Doerr [13] on the other hand showed that, with very high probability, the star-discrepancy of a random point set is at least of order $\sqrt{s/N}$, which shows in some sense that the upper bound of [20] is asymptotically sharp.

A similar but slightly weaker result compared to Theorem 1 is the following:

Theorem 2 (Heinrich, Novak, Wasilkowski, Woźniakowski [20]). *We have*

$$\mathrm{disc}(N, s) \ll \sqrt{\frac{s}{N}} \sqrt{\log s + \log N} \quad \textit{for all } N, s \in \mathbb{N}, \tag{5}$$

and

$$N(s, \varepsilon) \ll s\varepsilon^{-2} \log(s/\varepsilon) \quad \textit{for all } s \in \mathbb{N} \text{ and } \varepsilon > 0.$$

The proof of this result is based on similar ideas as used in the proof of the previous theorem, but instead of the result of Talagrand and Haussler, here the authors of [20] used Hoeffding's inequality, which is an estimate for the deviation from the mean for sums of independent random variables. We give a short sketch of the proof which offers some insights. More details can be found in [11,20,30].

Sketch of the proof of Theorem 2. Hoeffding's inequality (in the form required here) states that if $X_0, \ldots, X_{N-1}$ are independent real valued random variables with mean zero and $|X_i| \leq 1$ for $i = 0, \ldots, N-1$ almost surely, then for all $t > 0$ we have

$$\mathrm{Prob}\left(\left|\sum_{i=0}^{N-1} X_i\right| > t\right) \leq 2\exp\left(-\frac{t^2}{2N}\right).$$

Now let $P_{N,s} = \{\boldsymbol{x}_0, \ldots, \boldsymbol{x}_{N-1}\}$ where $\boldsymbol{x}_0, \ldots, \boldsymbol{x}_{N-1}$ are independent and uniformly distributed in $[0, 1)^s$. We want to show that

$$\mathrm{Prob}\left(D_N^*(P_{N,s}) \leq 2\varepsilon\right) > 0$$

where 2ε is the right hand side in (5). That amounts to the task to show that the event

$$|\Delta(\boldsymbol{x})| > 2\varepsilon \quad \text{at least for one } \boldsymbol{x} \in [0, 1)^s$$

has a probability smaller than 1. These are infinitely many constraints, but it can be shown that $|\Delta(\boldsymbol{x})| > 2\varepsilon$ implies $|\Delta(\boldsymbol{y})| > \varepsilon$ for one of the points in a rectangular equidistant grid $\Gamma_{m,s}$ of mesh size $1/m$ with $m = \lceil s/\varepsilon \rceil$. Actually, this holds either for the grid point directly below left or up right from $\boldsymbol{x}$. Since this grid $\Gamma_{m,s}$ has cardinality $(m+1)^s$, a union bound shows that it is enough to prove

$$\mathrm{Prob}\left(|\Delta(\boldsymbol{y})| > \varepsilon\right) < (m+1)^{-s}$$

for every $\boldsymbol{y} \in \Gamma_{m,s}$. But now

$$N\Delta(\boldsymbol{y}) = \sum_{i=0}^{N-1} \left(\mathbf{1}_{[0,\boldsymbol{y})}(\boldsymbol{x}_i) - \lambda_s([\mathbf{0},\boldsymbol{y}))\right)$$

is the sum of the N random variables $X_i = \mathbf{1}_{[0,\boldsymbol{y})}(\boldsymbol{x}_i) - \lambda_s([\mathbf{0},\boldsymbol{y}))$, which have mean 0 and obviously satisfy $|X_i| \leq 1$. So we can apply Hoeffding's inequality and obtain

$$\mathrm{Prob}\left(|\Delta(\boldsymbol{y})| > \varepsilon\right) = \mathrm{Prob}\left(\left|\sum_{i=0}^{N-1} X_i\right| > N\varepsilon\right) \leq 2\exp\left(\frac{-N\varepsilon^2}{2}\right) < (m+1)^{-s},$$

where the last inequality is satisfied for the chosen values of the parameters. $\square$

The results in Theorems 1 and 2 are only existence results. Until now no *explicit* constructions of N-element point sets $P_{N,s}$ in $[0,1)^s$ for which $D_N^*(P_{N,s})$ satisfy (4) or (5) are known. A first constructive approach was given by Doerr, Gnewuch and Srivastav [15], which was further improved by Doerr and Gnewuch [14], Doerr, Gnewuch, and Wahlström [17] and Gnewuch, Wahlström and Winzen [18]. There, a deterministic algorithm is presented that constructs point sets $P_{N,s}$ in $[0,1)^s$ satisfying

$$D_N^*(P_{N,s}) \ll \sqrt{\frac{s}{N}}\ \sqrt{\log(N+1)}$$

in run-time $O(s\log(sN)(\sigma N)^s)$, where $\sigma = \sigma(s) = O((\log s)^2/(s\log\log s)) \to 0$ as $s \to \infty$ and where the implied constants in the O-notations are independent of s and N. However, this is by far too expensive to obtain point sets for high dimensional applications. A slight improvement for the run time is presented in Doerr, Gnewuch, Kritzer and Pillichshammer [16], but this improvement has to be payed with by a worse dependence of the bound on the star-discrepancy on the dimension.

3   The Weighted Star-Discrepancy

In the paper [34], Sloan and Woźniakowski introduced the notion of weighted star-discrepancy and proved a "weighted" Koksma-Hlawka inequality. The idea is that in many applications some projections are more important than others and that this should also be reflected in the quality measure of the point set.

We start with some basic notation: let $[s] = \{1, 2, \ldots, s\}$ denote the set of coordinate indices. Let $\boldsymbol{\gamma} = (\gamma_j)_{j \geq 1}$ be a sequence of nonnegative reals. For $\mathfrak{u} \subseteq [s]$ we write $\gamma_{\mathfrak{u}} = \prod_{j \in \mathfrak{u}} \gamma_j$, where the empty product is one by definition. The real number $\gamma_{\mathfrak{u}}$ is the "weight" corresponding to the group of variables given by $\mathfrak{u}$. Let $|\mathfrak{u}|$ be the cardinality of $\mathfrak{u}$. For a vector $\boldsymbol{z} \in [0, 1]^s$ let $\boldsymbol{z}_{\mathfrak{u}}$ denote the vector from $[0, 1]^{|\mathfrak{u}|}$ containing the components of $\boldsymbol{z}$ whose indices are in $\mathfrak{u}$. By $(\boldsymbol{z}_{\mathfrak{u}}, 1)$ we mean the vector $\boldsymbol{z}$ from $[0, 1]^s$ with all components whose indices are not in $\mathfrak{u}$ replaced by 1.

For an N-element point set $P_{N,s}$ in $[0, 1)^s$ and given weights $\boldsymbol{\gamma} = (\gamma_j)_{j \geq 1}$, the *weighted star-discrepancy* $D^*_{N,\gamma}$ is given by

$$D^*_{N,\gamma}(P_{N,s}) = \sup_{\boldsymbol{z} \in [0,1]^s} \max_{\emptyset \neq \mathfrak{u} \subseteq [s]} \gamma_{\mathfrak{u}} |\Delta(\boldsymbol{z}_{\mathfrak{u}}, 1)|.$$

If $\gamma_j = 1$ for all $j \geq 1$, then the weighted star-discrepancy coincides with the classical star-discrepancy.

Quite similar to the classical case, we define the *Nth minimal weighted star-discrepancy*

$$\mathrm{disc}_\gamma(N, s) = \inf_{P_{N,s}} D^*_{N,\gamma}(P_{N,s})$$

and the *inverse of the weighted star-discrepancy*

$$N_\gamma(s, \varepsilon) = \min\{N \in \mathbb{N} : \mathrm{disc}_\gamma(N, s) \leq \varepsilon\}.$$

Now we recall two notions of tractability. Tractability means that we control the dependence of the inverse of the weighted star-discrepancy on s and ε^{-1} and rule out the cases for which $N_\gamma(s, \varepsilon)$ depends exponentially on s or on ε^{-1}.

- We say that the weighted star-discrepancy is *polynomially tractable*, if there exist nonnegative real numbers α and β such that

$$N_\gamma(s, \varepsilon) \ll s^\beta \varepsilon^{-\alpha} \quad \text{for all } s \in \mathbb{N} \text{ and } \varepsilon \in (0, 1). \tag{6}$$

The infima over all $\alpha, \beta > 0$ such that (6) holds are called the ε-exponent and the s-exponent, respectively, of polynomial tractability.
- We say that the weighted star-discrepancy is *strongly polynomially tractable*, if there exists a nonnegative real number α such that

$$N_\gamma(s, \varepsilon) \ll \varepsilon^{-\alpha} \quad \text{for all } s \in \mathbb{N} \text{ and } \varepsilon \in (0, 1). \tag{7}$$

The infimum over all $\alpha > 0$ such that (7) holds is called the ε-exponent of strong polynomial tractability.

In both cases the implied constant in the $\ll$ notation is independent of s and ε.

We collect some known results for the weighted star-discrepancy. The first result is an extension of Theorem 2 to the weighted star-discrepancy.

Theorem 3 (Hinrichs, Pillichshammer, Schmid [22]). *We have*

$$\mathrm{disc}_\gamma(N, s) \ll \frac{\sqrt{\log s}}{\sqrt{N}} \max_{\emptyset \neq \mathfrak{u} \subseteq [s]} \gamma_{\mathfrak{u}} \sqrt{|\mathfrak{u}|}.$$

Note that the result holds for every choice of weights. It is a pure existence result. Under very mild conditions on the weights, Theorem 3 implies polynomial tractability with s-exponent zero. See [22] for details. A slightly improved and numerically explicit version of Theorem 3 can be found in the recent paper of Aistleitner [2].

Theorem 4 (Dick, Leobacher, Pillichshammer [9]). *For every prime number p, every $m \in \mathbb{N}$ and for given weights $\gamma = (\gamma_j)_{j \geq 1}$ with $\sum_j \gamma_j < \infty$ one can construct (component-by-component) a p^m-element point set $P_{p^m,s}$ in $[0,1)^s$ such that for every $\delta > 0$ we have*

$$D^*_{p^m,\gamma}(P_{p^m,s}) \ll_{\gamma,\delta} \frac{1}{p^{m(1-\delta)}}.$$

Note that the point set $P_{p^m,s}$ from Theorem 4 depends on the choice of weights. The result implies that the weighted star-discrepancy is strongly polynomially tractable with ε-exponent equal to one, as long as the weights γ_j are summable. See [9–11, 22] for more details.

The next result (which follows implicitly from [37]) is about Niederreiter sequences in prime-power base q. For the definition of Niederreiter sequences we refer to [11, 31].

Theorem 5 (Wang [37]). *For the weighted star-discrepancy of the first N elements $P_{N,s}$ of an s-dimensional Niederreiter sequence in prime-power base q we have*

$$D^*_{N,\gamma}(P_{N,s}) \leq \frac{1}{N} \max_{\emptyset \neq \mathfrak{u} \subseteq [s]} \prod_{j \in \mathfrak{u}} [\gamma_j(C\, j \log(j+q) \log(qN))],$$

with a suitable constant $C > 0$.

One can easily deduce from Theorem 5 that the weighted star-discrepancy of the Niederreiter sequence can be bounded independently of the dimension whenever the weights satisfy $\sum_j \gamma_j j \log j < \infty$. This implies strong polynomial tractability with ε-exponent equal to one. A similar result can be shown for Sobol' sequences and for the Halton sequence (see [36, 37]).

We also have the following recent existence result:

Theorem 6 (Aistleitner [2]). *For product weights satisfying $\sum_j e^{-c\gamma_j^{-2}} < \infty$, for some $c > 0$, we have*

$$\mathrm{disc}_\gamma(N,s) \ll_\gamma \frac{1}{\sqrt{N}} \quad \textit{for all} \ s, N \in \mathbb{N}.$$

Consequently, the weighted star-discrepancy for such weights is strongly polynomially tractable, with ε-exponent at most 2.

All results described so far have either been existence results of point sets with small star-discrepancy, or results for point sets with small star-discrepancy which can be obtained via computer search. The Ansatz via computer search remains difficult and is limited to a rather small number of points N and dimensions s (in fact, it is known that the computation of the star-discrepancy is NP-hard as shown by Gnewuch, Srivastav, and Winzen [19], which makes it difficult to obtain good point sets via computer search). To make the random constructions useful in applications, Aistleitner and Hofer [3] show that with probability δ one can expect point sets with discrepancy of order $c(\delta)\sqrt{s/N}$. Another Ansatz for obtaining explicit constructions is contained in [35].

In the following section we discuss results for explicit constructions of point sets with low weighted star-discrepancy. Since the existence proofs above are based on randomly selected point sets, it may not be so surprising that the explicit constructions below are related to pseudo-random number generators.

4   The Weighted Star-Discrepancy of Korobov's p-sets

Let p be a prime number. For a nonnegative real number x let $\{x\} = x - \lfloor x \rfloor$ denote the fractional part of x. For vectors we use this operation component-wise.

We consider the so-called p-sets in $[0,1)^s$, a term which goes back to Hua and Wang [24]:

- Let $P_{p,s} = \{\boldsymbol{x}_0, \ldots, \boldsymbol{x}_{p-1}\}$ with

$$\boldsymbol{x}_n = \left(\left\{\frac{n}{p}\right\}, \left\{\frac{n^2}{p}\right\}, \ldots, \left\{\frac{n^s}{p}\right\}\right) \quad \text{for } n = 0, 1, \ldots, p-1.$$

The point set $P_{p,s}$ was introduced by Korobov [27].
- Let $Q_{p^2,s} = \{\boldsymbol{x}_0, \ldots, \boldsymbol{x}_{p^2-1}\}$ with

$$\boldsymbol{x}_n = \left(\left\{\frac{n}{p^2}\right\}, \left\{\frac{n^2}{p^2}\right\}, \ldots, \left\{\frac{n^s}{p^2}\right\}\right) \quad \text{for } n = 0, 1, \ldots, p^2-1.$$

The point set $Q_{p,s}$ was introduced by Korobov [26].
- Let $R_{p^2,s} = \{\boldsymbol{x}_{a,k} \ : \ a, k \in \{0, \ldots, p-1\}\}$ with

$$\boldsymbol{x}_{a,k} = \left(\left\{\frac{k}{p}\right\}, \left\{\frac{ak}{p}\right\}, \ldots, \left\{\frac{a^{s-1}k}{p}\right\}\right) \quad \text{for } a, k = 0, 1, \ldots, p-1.$$

Note that $R_{p^2,s}$ is the multi-set union of all Korobov lattice point sets[1] with modulus p. The point set $R_{p^2,s}$ was introduced by Hua and Wang (see [24, Sect. 4.3]).

We present some weighted star-discrepancy estimates for the p-sets.

Theorem 7 (Dick, Pillichshammer [12]). *Assume that the weights γ_j are non-increasing.*

---

[1] A Korobov lattice point set with modulus p and generator $a \in \{0, 1, \ldots, p-1\}$ consists of the points $\boldsymbol{x}_n = (\{\frac{n}{p}\}, \{\frac{an}{p}\}, \ldots, \{\frac{a^{s-1}n}{p}\})$ for $n = 0, 1, \ldots, p-1$.

1. *If $\sum_j \gamma_j < \infty$, then for all $\delta > 0$ we have*

$$D^*_{p,\gamma}(P_{p,s}) \ll_{\gamma,\delta} \frac{1}{p^{1/2-\delta}}, \quad D^*_{p^2,\gamma}(Q_{p^2,s}) \ll_{\gamma,\delta} \frac{1}{p^{1-\delta}}, \quad and$$

$$D^*_{p^2,\gamma}(R_{p^2,s}) \ll_{\gamma,\delta} \frac{1}{p^{1-\delta}},$$

where in all cases the implied constant is independent of p and s. This implies strong polynomial tractability.

2. *If there exists a real $\tau > 0$ such that $\sum_j \gamma_j^\tau < \infty$, then for all $\delta > 0$ we have*

$$D^*_{p,\gamma}(P_{p,s}) \ll_{\gamma,\delta} \frac{s}{p^{1/2-\delta}}, \quad D^*_{p^2,\gamma}(Q_{p^2,s}) \ll_{\gamma,\delta} \frac{s}{p^{1-\delta}}, \quad and$$

$$D^*_{p^2,\gamma}(R_{p^2,s}) \ll_{\gamma,\delta} \frac{s}{p^{1-\delta}},$$

where in all cases the implied constant is independent of p and s. This implies polynomial tractability.

The proof of Theorem 7 is based on an Erdős–Turan–Koksma-type inequality for the weighted star-discrepancy and the following estimates for exponential sums. For details we refer to [12].

Lemma 1. *Let p be a prime number and let $s \in \mathbb{N}$. Then for all $h_1, \ldots, h_s \in \mathbb{Z}$ such that $p \nmid h_j$ for at least one $j \in [s]$ we have*

$$\left| \sum_{n=0}^{p-1} \exp(2\pi i(h_1 n + h_2 n^2 + \cdots + h_s n^s)/p) \right| \leq (s-1)\sqrt{p}, \tag{8}$$

$$\left| \sum_{n=0}^{p^2-1} \exp(2\pi i(h_1 n + h_2 n^2 + \cdots + h_s n^s)/p^2) \right| \leq (s-1)p, \quad and$$

$$\left| \sum_{a=0}^{p-1} \sum_{k=0}^{p-1} \exp(2\pi i k(h_1 + h_2 a + \cdots + h_s a^{s-1})/p) \right| \leq (s-1)p.$$

Inequality (8) is known as the Weil bound [38] and is often used in the area of pseudo-random number generation. Constructions related to the p-sets have also been considered in [7]. All of these constructions are related to the generation of (streams of) pseudo-random numbers (rather than low-discrepancy point sets and sequences). We discuss pseudo-random number generators in the next section more generally.

5   Complete Uniform Distribution and Pseudo-random Number Generators

Pseudo-random number generators are commonly used in computer simulations to replace real random numbers for various reasons. These point sets are based on deterministic constructions with the aim to mimic randomness. A number of quality criteria are applied to such pseudo-random number generators to assess their quality. One such criterion is complete uniform distribution.

Let $u_1, u_2, \ldots \in [0, 1]$ be a sequence of real numbers. For $s, n \in \mathbb{N}$ we define

$$\boldsymbol{u}_n^{(s)} = (u_{(n-1)s+1}, \ldots, u_{ns}) \in [0, 1]^s.$$

Then the sequence $(u_n)_{n \geq 1}$ is completely uniformly distributed if for every $s \geq 1$

$$\lim_{N \to \infty} D_N^*(\{\boldsymbol{u}_1^{(s)}, \ldots, \boldsymbol{u}_N^{(s)}\}) = 0.$$

The concept of complete uniform distribution measures independence between successive numbers $u_i, u_{i+s}, u_{i+2s}, \ldots$. For simulation purposes it is often assumed that the random numbers are i.i.d. uniform, thus their discrepancy goes to 0 (in probability), and so one wants pseudo-random numbers with the same property.[2]

Markov chain algorithms are a staple tool in statistics and the applied sciences for generating samples from distributions for which only partial information is available. As such they are an important class of algorithms which use pseudo-random number generators. In [6] it was shown that if the random numbers which drive the Markov chain are completely uniformly distributed, then the Markov chain consistently samples the target distribution (i.e. yields the correct result). For instance, [6, Theorem 4] requires pseudo-random numbers $(u_n)_{n \geq 1}$ such that for every sequence of natural numbers $(s_N)_{N \geq 1}$ with $s_N = \mathcal{O}(\log N)$, we have

$$\lim_{N \to \infty} D_N^*(\{\boldsymbol{u}_1^{(s_N)}, \ldots, \boldsymbol{u}_N^{(s_N)}\}) = 0.$$

In this case, bounds like (3) are not strong enough due to their dependence on the dimension. Even a bound of the form $C^s N^{-\delta}$, with $C > 1$ and some $\delta > 0$ which does not depend on the dimension s, is not strong enough, since for $s = c \log N$ with $c > \frac{\delta}{\log C}$, we have $C^s N^{-\delta} = C^{c \log N} N^{-\delta} = N^{-\delta + c \log C} \geq 1$ for all $N \in \mathbb{N}$ and so we do not get any convergence.

Thus it would be interesting for applications to explicitly construct a deterministic sequence $(u_n)_{n \geq 1}$ such that, say

$$D_N^*(\{\boldsymbol{u}_1^{(s)}, \ldots, \boldsymbol{u}_N^{(s)}\}) \leq C \frac{\sqrt{s \log N}}{\sqrt{N}} \quad \text{for all } N, s \in \mathbb{N}.$$

The existence of such a sequence has already been shown in [6, p. 684] and an improvement has been shown in [4]. Such a sequence has good properties when viewed as a pseudo-random sequences but is also useful as a deterministic sequence in quasi-Monte Carlo integration.

---

[2] For instance, the classic construction by van der Corput $(\phi(n))_{n \geq 0}$ in base 2, given by

$$\phi(n) = \frac{n_0}{2} + \frac{n_1}{2^2} + \cdots + \frac{n_m}{2^{m+1}},$$

where n has dyadic expansion $n = n_0 + n_1 2 + \cdots + n_m 2^m$, is not completely uniformly distributed, since $\phi(2n)$ lies in the interval $[0, 1/2)$, whereas $\phi(2n - 1)$ lies in the interval $[1/2, 1)$.

Acknowledgments. J. Dick is the recipient of an Australian Research Council Queen Elizabeth II Fellowship (project number DP1097023). F. Pillichshammer is supported by the Austrian Science Fund (FWF): Project F5509-N26, which is a part of the Special Research Program "Quasi-Monte Carlo Methods: Theory and Applications". The authors are grateful to Ch. Aistleitner, P. Kritzer and A. Winterhof for helpful comments.

References

1. Aistleitner, C.: Covering numbers, dyadic chaining and discrepancy. J. Complex. **27**, 531–540 (2011)
2. Aistleitner, C.: Tractability results for the weighted star-discrepancy. J. Complex. **30**, 381–391 (2014)
3. Aistleitner, C., Hofer, M.: Probabilistic discrepancy bound for Monte Carlo point sets. Math. Comput. **83**, 1373–1381 (2014)
4. Aistleitner, C., Weimar, M.: Probabilistic star discrepancy bounds for double infinite random matrices. In: Dick, J., Kuo, F.Y., Peters, G.W., Sloan, I.H. (eds.) Monte Carlo and quasi-Monte Carlo methods 2012. Springer, Berlin (2013)
5. Bilyk, D., Lacey, M.T., Vagharshakyan, A.: On the small ball inequality in all dimensions. J. Funct. Anal. **254**, 2470–2502 (2008)
6. Chen, S., Dick, J., Owen, A.B.: Consistency of Markov chain quasi-Monte Carlo on continuous state spaces. Ann. Stat. **39**, 673–701 (2011)
7. Dick, J.: Numerical integration of Hölder continuous, absolutely convergent Fourier, Fourier cosine, and Walsh series. J. Approx. Theory **183**, 14–30 (2014)
8. Dick, J., Hinrichs, A., Pillichshammer, F.: Proof techniques in quasi-Monte Carlo theory. J. Complexity (2014) (Submitted)
9. Dick, J., Leobacher, G., Pillichshammer, F.: Construction algorithms for digital nets with low weighted star-discrepancy. SIAM J. Numer. Anal. **43**, 76–95 (2005)
10. Dick, J., Niederreiter, H., Pillichshammer, F.: Weighted star-discrepancy of digital nets in prime bases. In: Talay, D., Niederreiter, H. (eds.) Monte Carlo and Quasi-Monte Carlo Methods 2004. Springer, Berlin (2006)
11. Dick, J., Pillichshammer, F.: Digital Nets and Sequences - Discrepancy Theory and Quasi-Monte Carlo Integration. Cambridge University Press, Cambridge (2010)
12. Dick, J., Pillichshammer, F.: The weighted star discrepancy of Korobov's p-sets. Proc. Amer. Math. Soc. (2014) (Submitted)
13. Doerr, B.: A lower bound for the discrepancy of a random point set. J. Complex. **30**, 16–20 (2014)
14. Doerr, B., Gnewuch, M.: Construction of low-discrepancy point sets of small size by bracketing covers and dependent randomized rounding. In: Keller, A., Heinrich, S., Niederreiter, H. (eds.) Monte Carlo and Quasi-Monte Carlo Methods 2006, pp. 299–312. Springer, Berlin (2007)
15. Doerr, B., Gnewuch, M., Srivastav, A.: Bounds and constructions for the star-discrepancy via δ-covers. J. Complex. **21**, 691–709 (2005)
16. Doerr, B., Gnewuch, M., Kritzer, P., Pillichshammer, F.: Component-by-component construction of low-discrepancy point sets of small size. Monte Carlo Methods Appl. **14**, 129–149 (2008)
17. Doerr, B., Gnewuch, M., Wahlström, M.: Algorithmic construction of low-discrepancy point sets via dependent randomized rounding. J. Complex. **26**, 490–507 (2010)

18. Gnewuch, M., Wahlström, M., Winzen, C.: A new randomized algorithm to approximate the star discrepancy based on threshold accepting. SIAM J. Numer. Anal. **50**, 781–807 (2012)
19. Gnewuch, M., Srivastav, A., Winzen, C.: Finding optimal volume subintervals with k-points and calculating the star discrepancy are NP-hard problems. J. Complex. **25**, 115–127 (2009)
20. Heinrich, S., Novak, E., Wasilkowski, G.W., Woźniakowski, H.: The inverse of the star-discrepancy depends linearly on the dimension. Acta Arith. **96**, 279–302 (2001)
21. Hinrichs, A.: Covering numbers, Vapnik-Červonenkis classes and bounds for the star-discrepancy. J. Complex. **20**, 477–483 (2004)
22. Hinrichs, A., Pillichshammer, F., Schmid, W.C.: Tractability properties of the weighted star-discrepancy. J. Complex. **24**, 134–143 (2008)
23. Hlawka, E.: Funktionen von beschränkter Variation in der Theorie der Gleichverteilung. Ann. Mat. Pura Appl. **54**, 325–333 (1961)
24. Hua, L.K., Wang, Y.: Applications of Number Theory to Numerical Analysis. Springer, Berlin (1981)
25. Koksma, J.F.: Een algemeene stelling uit de theorie der gelijkmatige verdeeling modulo 1. Math. B (Zutphen) **11**, 7–11 (1942–1943)
26. Korobov, N.M.: Approximate calculation of repeated integrals by number-theoretical methods. (Russ.) Dokl. Akad. Nauk SSSR (N.S.) **115**, 1062–1065 (1957)
27. Korobov, N.M.: Number-theoretic methods in approximate analysis. (Russian) Gosudarstv. Izdat. Fiz.-Mat. Lit., Moscow (1963)
28. Kuipers, L., Niederreiter, H.: Uniform Distribution of Sequences. Wiley, New York, 1974. Reprint, Dover Publications, Mineola, NY (1974) (2006)
29. Kuo, F.Y., Schwab, C., Sloan, I.H.: Quasi-Monte Carlo finite element methods for a class of elliptic partial differential equations with random coefficients. SIAM J. Numer. Anal. **50**, 3351–3374 (2012)
30. Leobacher, G., Pillichshammer, F.: Introduction to Quasi-Monte Carlo Integration and Applications. Compact Textbooks in Mathematics. Birkhäuser, Basel (2014)
31. Niederreiter, H.: Random Number Generation and Quasi-Monte Carlo Methods. SIAM, Philadelphia (1992)
32. Paskov, S.H., Traub, J.: Faster evaluation of financial derivatives. J. Portfolio Manag. **22**, 113–120 (1995)
33. Roth, K.F.: On irregularities of distribution. Mathematika **1**, 73–79 (1954)
34. Sloan, I.H., Woźniakowski, H.: When are quasi-Monte Carlo algorithms efficient for high dimensional integrals? J. Complex. **14**, 1–33 (1998)
35. Temlyakov, V.N.: Greedy-type approximation in Banach spaces and applications. Constr. Approx. **21**, 257–292 (2005)
36. Wang, X.: A constructive approach to strong tractability using quasi-Monte Carlo algorithms. J. Complex. **18**, 683–701 (2002)
37. Wang, X.: Strong tractability of multivariate integration using quasi-Monte Carlo algorithms. Math. Comput. **72**, 823–838 (2003)
38. Weil, A.: On some exponential sums. Proc. Natl. Acad. Sci. U.S.A. **34**, 204–207 (1948)

Pseudorandom Sequences
and Stream Ciphers

An Equivalence-Preserving Transformation of Shift Registers

Elena Dubrova[✉]

Royal Institute of Technology (KTH), Electrum 229, 164 40 Kista, Sweden
dubrova@kth.se

Abstract. The Fibonacci-to-Galois transformation is useful for reducing the propagation delay of feedback shift register-based stream ciphers and hash functions. In this paper, we extend it to handle Galois-to-Galois case as well as feedforward connections. This makes possible transforming Trivium stream cipher and increasing its keystream data rate by 27 % without any penalty in area. The presented transformation might open new possibilities for cryptanalysis of Trivium, since it induces a class of stream ciphers which generate the same set of keystreams as Trivium, but have a different structure.

1 Introduction

Shift register-based cryptographic systems are the fastest and the most power-efficient cryptographic systems for hardware implementations [1]. The speed and the power are two crucial factors for future cryptographic systems, since they are expected to support very high data rates in 5G ultra-low power products and applications. The 5G is envisioned to have 1000 times higher traffic volume compared to current LTE deployments while providing a better quality of service [2]. Consumer data rates of hundreds of Mbps are expected to be available in a general scenario. In special scenarios, such as office spaces or dense urban outdoor environments, reliably achievable data rates of multi-Gbps are foreseen.

An n-bit shift register implements an n-variate mapping $\{0,1\}^n \rightarrow \{0,1\}^n$ of type

$$\begin{pmatrix} x_0 \\ \cdots \\ x_{n-1} \end{pmatrix} \rightarrow \begin{pmatrix} f_0(x_0,\ldots,x_{n-1}) \\ \cdots \\ f_{n-1}(x_0,\ldots,x_{n-1}) \end{pmatrix} \tag{1}$$

where each Boolean function f_i, $i \in \{0,1,\ldots,n-1\}$, is of type:

$$f_i = x_{i+1} \oplus g_i(x_0,\ldots,x_i,x_{i+2},\ldots,x_{n-1}) \tag{2}$$

where "$\oplus$" is addition modulo 2 and "$+$" is addition modulo n.

Note that the function g_i in (2) does not depend on x_{i+1}. This is a necessary condition for invertibility of mappings implemented by a shift register [3]. A mapping $x \rightarrow f(x)$ on a finite set is called *invertible* if $f(x) = f(y)$ if and only if $x = y$. Stream ciphers usually use invertible mappings to prevent incremental reduction of the entropy of the state [4].

© Springer International Publishing Switzerland 2014
K.-U. Schmidt and A. Winterhof (Eds.): SETA 2014, LNCS 8865, pp. 187–199, 2014.
DOI: 10.1007/978-3-319-12325-7_16

Another desirable property is long period. The *period* of a mapping is the length of the longest cycle in its state transition graph. Obviously, if we iterate a mapping a large number of times, we do not want the sequence of generated states to be trapped in a short cycle. Furthermore, as demonstrated by the cryptanalysis of A5, short cycles can be exploited to greatly reduce the complexity of the attack [5].

In this paper, we present a transformation which preserves both, invertibility and period, of a mapping. It makes possible to construct classes of shift registers which have structurally isomorphic state transition graphs and generate equivalent sets of output sequences. This is useful for optimizing the hardware performance of shift register-based stream ciphers [6–10] and hash functions [11]. We apply the presented transformation to Trivium [7] and show that it increases its keystream data rate by 27 % without any penalty in area. The transformation can also be potentially useful for cryptanalysis since, within the class of shift registers generating equivalent sets of output sequences, some might be easier to cryptanalysize than others.

The presented transformation extends Fibonacci-to-Galois transformation of Non-Linear Feedback Shift Registers (NLFSR) [12] to the more general case of shift registers. Two main differences are:

1. The presented transformation can be applied to shift registers with both, feedback and feedforward connections (e.g. Trivium).
2. The presented transformation can be applied to any Galois NLFSR. The transformation [12] is applicable to uniform NLFSRs only[1].

The paper is organized as follows. Section 2 summarises basic notations used in the sequel. Section 3 describes previous work. Section 4 gives an informal description of the presented transformation. Section 5 formalizes the main result. Section 6 shows how the presented transformation can be applied to Trivium. Section 7 concludes the paper and discusses open problems.

2　Preliminaries

Throughout the paper, we use "$\oplus$" and "$\cdot$" to denote the *GF(2)* addition and multiplication, respectively.

The Algebraic Normal Form (ANF) [13] of a Boolean function $f : \{0,1\}^n \to \{0,1\}$ is a polynomial in *GF(2)* of type

$$f(x_0, x_1, \ldots, x_{n-1}) = \sum_{i=0}^{2^n-1} c_i \cdot x_0^{i_0} \cdot x_1^{i_1} \cdot \ldots \cdot x_{n-1}^{i_{n-1}},$$

where $c_i \in \{0,1\}$ and $(i_0 i_1 \ldots i_{n-1})$ is the binary expansion of i.

---

[1] An n-bit NLFSR is *uniform* if, for all $i \in \{\tau, \tau+1, \ldots, n-1\}$, the largest index of variables of function g_i in (2) is smaller than or equal to τ, where τ is the maximal index such that, for all $j \in \{0, 1, \ldots, \tau-1\}$, $g_j = 0$.

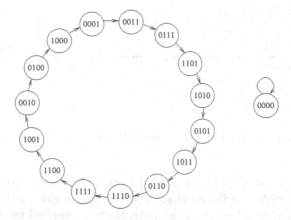

Fig. 1. The state transition graph of the mapping (3). Each 4-tuple represents a state (x_0, x_1, x_2, x_3).

The *dependence set* [14] of a Boolean function is defined by

$$dep(f) = \{j \mid f(x_j = 0) \neq f(x_j = 1)\},$$

where $f(x_j = k) = f(x_0, \ldots, x_{j-1}, k, x_{j+1}, \ldots, x_{n-1})$ for $k \in \{0, 1\}$.

Throughout the paper we also use the expression "dependence set of a monomial of the ANF". It should not create any ambiguity since each monomial of the ANF represents a Boolean function. For example, for $m = x_1 x_3$, $dep(m) = \{1, 3\}$.

The *state* of an n-variate mapping $\{0, 1\}^n \to \{0, 1\}^n$ is any specific assignment of $\{x_0, x_1, \ldots, x_{n-1}\}$. The *State Transition Graph* (STG) is a directed graph in which the nodes represent the states and the edges show possible transitions between the states.

For example, the STG of the 4-variate mapping $\{0, 1\}^4 \to \{0, 1\}^4$:

$$\begin{pmatrix} x_0 \\ x_1 \\ x_2 \\ x_3 \end{pmatrix} \rightarrow \begin{pmatrix} x_1 \\ x_2 \\ x_3 \oplus x_1 x_2 \\ x_0 \oplus x_3 \end{pmatrix}. \tag{3}$$

is shown in Fig. 1. This mapping is invertible. It has period 15.

Any n-variate mapping $\{0, 1\}^n \to \{0, 1\}^n$ can be implemented by an n-bit *shift register* shown in Fig. 2. It consists of n binary storage elements, called *stages*, and n *updating functions* $f_i : \{0, 1\}^n \to \{0, 1\}$ which determine how the values of stages are updated [3]. At every clock cycle, the next state is computed from the current state by updating the values of all stages simultaneously to the values of the corresponding updating functions.

The *degree of parallelization* of a shift register is the number of bits of output which are produced at each clock cycle.

A shift register can be implemented either in the *Fibonacci* or in the *Galois* configuration [12]. In the former, all updating functions except f_{n-1} are of type

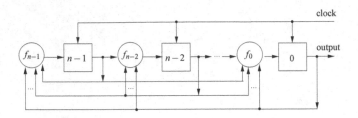

Fig. 2. The general structure of an n-bit shift register with updating functions.

$f_i(x) = x_{i+1}$, for $i \in \{0, 1, \ldots, n-2\}$. In other words, feedback/feedforward connections are applied to the input stage of the shift register only. In the latter, feedback/feedforward connections can potentially be applied to every stage.

Two shift registers are *equivalent* if their sets of output sequences are equal.

3   Previous Work

For LFSRs, there exist a one-to-one mapping between the Fibonacci and the Galois configurations. The Galois LFSR generating the same sets of output sequences as a given Fibonacci LFSR can be obtained by reversing the order of the feedback taps and adjusting the initial state. Several transformations aiming to optimize the traditional LFSRs with respect to different parameters were presented, including [15–19].

For NLFSRs, however, the Galois configuration is not unique. Usually, there are many n-bit Galois NLFSRs which are equivalent to a given n-bit Fibonacci NLFSR. On the other hand, not every n-bit Galois NLFSR has an equivalent n-bit Fibonacci NLFSR. The latter is because, while an output sequence of every n-bit Fibonacci NLFSR can be described by a nonlinear recurrence of order n [3], for an n-bit Galois NLFSR such a recurrence may not exist. It was shown in [12] that a nonlinear recurrence of order n always exists for uniform n-bit NLFSRs. An algorithm for constructing a best uniform Galois NLFSR which is equivalent to a given Fibonacci NLFSR was presented in [20].

A interesting type of NLFSRs was introduced by Massey and Liu in [21]. Similarly to the Fibonacci NLFSRs, these NLFSRs have a single feedback function Boolean function, f, of the state variables $x_0, x_1, \ldots, x_{n-1}$. However, the output of $f(x_0, x_1, \ldots, x_{n-1})$ is fed not only to the bit $n-1$ but also to other bits, namely

$$f_i(x_0, x_1, \ldots, x_{n-1}) = f(x_0, x_1, \ldots, x_{n-1}), \text{ for } i = n-1$$
$$f_i(x_0, x_1, \ldots, x_{n-1}) = x_{i+1} \oplus w_i \cdot f(x_0, x_1, \ldots, x_{n-1}), \text{ for } i = \{0, 1, \ldots, n-2\}$$

where $w_i \in \{0, 1\}$. It was shown in [21] that every NLFSR of this type has an equivalent NLFSR in the Fibonacci configuration. Moreover, the mapping between the two configurations is one-to-one.

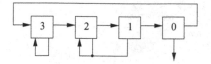

Fig. 3. The 4-bit ring with connections corresponding to the monomials of ANFs of Boolean functions induced by the mapping (3).

4   Intuitive Description

We start with an intuitive description of the presented transformation.

Consider an n-variate mapping of type (1). It can be represented by an n-bit ring with connections corresponding to the monomials of ANFs of functions f_i induced by the mapping[2]. Each connection has a single sink and one or more sources. The sources originate in the stages corresponding to the state variables of the monomial. The sink points to the stage i with the index of the updating function f_i represented by the ANF, $i \in \{0, 1, \ldots, n-1\}$. The output is represented by an outgoing edge from the corresponding stage.

For example, if we assume that the output is taken from the stage 0, then the 4-variate mapping (3) is represented by the 4-bit ring shown in Fig. 3. The connection with sources 1,2 and sink 2 corresponds to the monomial $x_1 x_2$ of f_2.

The transformation presented in the paper moves a connection either left or right in the ring, without changing its length or shape, i.e. the sink and all sources are moved by the same number of stages. For example, if the monomial $x_1 x_2$ of f_2 in the mapping (3) is moved one stage right, we get the mapping

$$\begin{pmatrix} x_0 \\ x_1 \\ x_2 \\ x_3 \end{pmatrix} \rightarrow \begin{pmatrix} x_1 \\ x_2 \oplus x_0 x_1 \\ x_3 \\ x_0 \oplus x_3 \end{pmatrix}. \tag{4}$$

Its STG is shown in Fig. 4.

Indexes crossing the 0 to $n-1$ border of the ring are updated modulo n. So, if we move the monomial x_3 of f_3 in the mapping (3) one stage left, we get

$$\begin{pmatrix} x_0 \\ x_1 \\ x_2 \\ x_3 \end{pmatrix} \rightarrow \begin{pmatrix} x_1 \oplus x_0 \\ x_2 \\ x_3 \oplus x_1 x_2 \\ x_0 \end{pmatrix}. \tag{5}$$

Its STG is shown in Fig. 5.

---

[2] We use an n-bit ring as a simplification of an n-bit shift register which shows the structure of its feedback/feedforward connections. The gates implementing $GF(2)$ addition (XORs) are omitted and the gates implementing $GF(2)$ multiplication (ANDs) are represented by a dot. Everything unnecessary for structural analysis is removed.

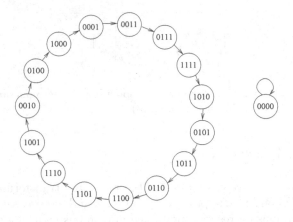

Fig. 4. The state transition graph of the mapping (4).

Three conditions should hold for the transformation to preserve the cycle structure of the STG. First, only the connections corresponding to the monomials of functions g_i in the Eq. (2), $i \in \{0, 1, \ldots, n-1\}$, can be moved. The monomial x_{i+1} of functions f_i cannot be moved, where "+" is addition modulo n.

Second, sources of a connection can be moved k stages left/right if the functions f_i of the k stages on the left/right of each source do shifts only (i.e. no source crosses any of the sinks of connections related to g_is during its move). This condition makes sure that time dependencies in the computation are preserved. For example, the monomial x_3 of f_3 in the mapping (3) can be moved one stage left to f_0, or two stages left to f_1, but not one stage right to f_2 since $f_2 = x_3 \oplus x_1 x_2$. Due to the circular structure of the ring, we can always reach any stage either from the left or from the right. It is sufficient that the condition is satisfied only in one of the directions. For example, although x_3 cannot be moved to f_1 from the right, it can be moved to f_1 from the left. So, we can move x_3 to f_1.

Third, the sink of a connection can be moved k stages left/right if k stages on the left/right of the sink do not serve as sources of any other connection of any g_i, $i \in \{0, 1, \ldots, n-1\}$ (i.e. the sink does not cross any of the sources of connections related to g_is during its move). This condition makes sure that values of variables participating in the computation are correct. For example, the monomial $x_1 x_2$ of f_2 in the mapping (3) cannot be moved to f_3 because x_3 is a variable of a monomial of g_3.

Suppose that, in addition to preserving the cycle structure of the STG of a mapping, we want to preserve the binary sequence generated by one of its functions, say f_i, for any $i \in \{0, 1, \ldots, n-1\}$. This might be desirable because, for example, this sequence is used as a keystream and we do not want to change its properties. Then, in addition to the three conditions above, we need to add a condition that neither the sink nor the sources of a shifted connection cross the border between ith and $i-1$st modulo n stage of the ring.

For example, if the value computed by the function f_0 of the mapping (3) is used as an output, then, in order to preserve the output sequence after the

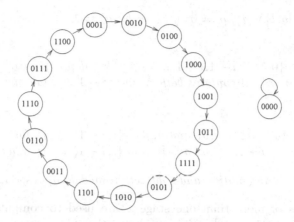

Fig. 5. The state transition graph of the mapping (5).

transformation, neither the sink nor the sources of a shifted connection should cross the border between 0th and 3rd stage. This holds for the transformation from (3) to (4). Indeed, we can see from Figs. 1 and 4 that, for the initial state $(x_0, x_1, x_2, x_3) = (0001)$, the functions f_0 of both mappings generate the periodic sequence[3] 000110101111001. However, this is not the case for the mapping (5). From its STG in Fig. 5, we can see that the sequence generated by its function f_0, namely 000111101100101 is different from the sequence above for any initial state. This is because the shifted connection crosses the border between 0th and 3rd stage.

5   Formal Description

In this section, we give a formal description of the presented transformation.

Definition 1. *The shifting, denoted by $f_i \xrightarrow{m} f_j$, $i, j \in \{0, 1, \ldots, n-1\}$, $i \neq j$, transforms an n-variate mapping of type (1) to another n-variate mapping in which the ANF monomial m of f_i is moved to f_j and each index $a \in dep(m)$ is changed to b defined by*

$$b = (a - i + j) \bmod n \qquad (6)$$

For example, by applying shifting $f_2 \xrightarrow{x_1 x_2} f_1$ to the 4-variate mapping (3), we get the mapping (4).

Given a shifting $g_i \xrightarrow{m} g_j$, we denote by g_i^* the function $g_i^* = g_i \oplus m$.

Definition 2. *Given an n-variate mapping of type (1) in which the values computed by f_z, $z \in \{0, 1, \ldots, n-1\}$ are used as an output sequence, a shifting $g_i \xrightarrow{m} g_j$, $i, j \in \{0, 1, \ldots, n-1\}$, $i \neq j$, is valid if for each $a \in dep(m)$ and for b defined by (6) the following three conditions hold:*

---

[3] Note that in this case the initial states are the same but generally they can be different [22].

1. For each $c \in [a,b] \setminus \{i\}$, $g_c = 0$; if $i \in [a,b]$, $g_i^* = 0$.
2. For all $k \in [i,j]$:
   (a) $k \notin dep(g_i^*)$;
   (b) for all $p \in \{0,1,...,i-1,i+1,...n-1\}$, $k \notin dep(g_p)$.
3. Neither $[a,b]$ nor $[i,j]$ contains both, z and $(z-1)$ modulo n.

where

$$[a,b] = \{a, a-1, ..., b\} \text{ and } [i,j] = \{i, i-1, ..., j\} \text{ for } i > j$$
$$[a,b] = \{a, a+1, ..., b\} \text{ and } [i,j] = \{i, i+1, ..., j\} \text{ otherwise}$$

and "+" and "−" are addition and subtraction modulo n, respectively.

If the values of more than one stage z are used to compute the output sequence (e.g. as in Grain [6], Trivium [7], or other filter generators), then the condition 3 should hold for each pair z and $(z-1)$ modulo n.

For example, for the mapping (3) with f_0 as an output, shifting $g_2 \xrightarrow{x_1 x_2} g_1$ is valid. However, shiftings $g_3 \xrightarrow{x_3} g_2$ and $g_2 \xrightarrow{x_1 x_2} g_3$ are not valid since the former violates the condition 1 and the latter violates the condition 2 of Definition 2.

In the theorem below, we use $f(s)$ to denote the value of the function f evaluated for the vector s. We also use $f|_j$ ($f|_{-j}$) to denote the function obtained from f by adding (subtracting) j modulo n to (from) indexes of all variables of f. For example, if $f = x_1 x_2 \oplus x_3$, then $f|_2 = x_3 x_4 \oplus x_5$ and $f|_{-1} = x_0 x_1 \oplus x_2$.

Theorem 1. *Let F be a mapping of type (1) and F' be a mapping obtained from F by applying a valid shifting $g_i \xrightarrow{m} g_j$, $i,j \in \{0,1,...,n-1\}$, $i \neq j$. If F is initialized to the state $s = (s_0, s_1, ..., s_{n-1})$ and F' is initialized to the state $r = (r_0, r_1, ..., r_{n-1})$ such that*

$$
\begin{aligned}
&\text{if } i > j, \text{ then } r_k = s_k \oplus m|_{k-i-1}(s) \text{ for } k \in \{i, i-1, ..., j+1\} \\
&\text{if } i < j, \text{ then } r_k = s_k \oplus m|_{k-j-1}(s) \text{ for } k \in \{i+1, i+2, ..., j\}
\end{aligned}
\tag{7}
$$

and $r_k = s_k$ for all remaining $k \in \{0,1,...,n-1\}$, then sequences of states generated by F and F' may differ only in bit positions $i, i-1, ..., j+1$ if $i > j$ and only in bit positions $i+1, i+2, ..., j$ if $i < j$.

Proof. First we show that Theorem 1 holds for the case of $i = j+1$. In this case, the Eq. (7) is reduced to $r_k = s_k \oplus m|_{-1}(s)$ for $k = j+1$.

Suppose that $m = x_{a_1} x_{a_2} ... x_{a_t}$, where $a_l \in \{0,1,...,n-1\}$, for all $l \in \{1,2,...,t\}$, and $a_1 > a_2 > ... > a_t$. For simplicity, let us assume that the values computed by f_0 are used as an output sequence of F. If the shifting $g_i \xrightarrow{m} g_j$ is valid, then, from the condition 3 of Definition 2, we can conclude that $a_t > 0$. Thus, after shifting, m changes to $x_{a_1-1} x_{a_2-1} ... x_{a_t-1}$. Furthermore, from the condition 2 of Definition 2 we can conclude that $\{j+1, j\} \not\subset dep(g_{j+1}^*)$ and $\{j+1, j\} \not\subset dep(g_p)$ for all $p \in \{0,1,...,j,j+2,...n-1\}$. Therefore, F is of type

$$\begin{pmatrix} x_0 \\ \cdots \\ x_j \\ x_{j+1} \\ \cdots \\ x_{n-1} \end{pmatrix} \rightarrow \begin{pmatrix} x_1 \oplus g_0(x_0, \ldots, x_{j-1}, x_{j+2}, \ldots, x_{n-1}) \\ \cdots \\ x_{j+1} \oplus g_j(x_0, \ldots, x_{j-1}, x_{j+2}, \ldots, x_{n-1}) \\ x_{j+2} \oplus g_{j+1}^*(x_0, \ldots, x_{j-1}, x_{j+2}, \ldots, x_{n-1}) \oplus x_{a_1} x_{a_2} \ldots x_{a_t} \\ \cdots \\ x_{n-1} \oplus g_{n-1}(x_0, \ldots, x_{j-1}, x_{j+2}, \ldots, x_{n-1}) \end{pmatrix}$$

and F' is of type

$$\begin{pmatrix} x_0 \\ \cdots \\ x_j \\ x_{j+1} \\ \cdots \\ x_{n-1} \end{pmatrix} \rightarrow \begin{pmatrix} x_1 \oplus g_0(x_0, \ldots, x_{j-1}, x_{j+2}, \ldots, x_{n-1}) \\ \cdots \\ x_{j+1} \oplus g_j(x_0, \ldots, x_{j-1}, x_{j+2}, \ldots, x_{n-1}) \oplus x_{a_1-1} x_{a_2-1} \ldots x_{a_t-1} \\ x_{j+2} \oplus g_{j+1}^*(x_0, \ldots, x_{j-1}, x_{j+2}, \ldots, x_{n-1}) \\ \cdots \\ x_{n-1} \oplus g_{n-1}(x_0, \ldots, x_{j-1}, x_{j+2}, \ldots, x_{n-1}) \end{pmatrix}$$

Note that, due to the restriction imposed on the function g_i in Eq. (2), $j+2 \notin \{a_1, a_2, \ldots, a_t\}$ and therefore $j+1 \notin \{a_1 - 1, a_2 - 1, \ldots, a_t - 1\}$. In addition, from the condition 1 of Definition 2 we can conclude that, for all $l \in \{1, 2, \ldots, t\}$, $g_{c_l} = 0$ for $c_l \in \{a_l, a_l - 1\}$.

Suppose that F is initialized to a state $s = (s_0, s_1, \ldots, s_{n-1})$ and F' is initialized to a state $r = (s_0, s_1, \ldots, s_j, s_{j+1} \oplus s_{a_1-1} s_{a_2-1} \ldots s_{a_t-1}, s_{j+2}, \ldots, s_{n-1})$. On one hand, for F, the next state $s^+ = (s_0^+, s_1^+, \ldots, s_{n-1}^+)$ is given by:

$$s_0^+ = s_1 \oplus g_0(s_0, \ldots, s_{j-1}, s_{j+2}, \ldots, s_{n-1})$$
$$\cdots$$
$$s_j^+ = s_{j+1} \oplus g_j(s_0, \ldots, s_{j-1}, s_{j+2}, \ldots, s_{n-1})$$
$$s_{j+1}^+ = s_{j+2} \oplus g_{j+1}^*(s_0, \ldots, s_{j-1}, s_{j+2}, \ldots, s_{n-1}) \oplus s_{a_1} s_{a_2} \ldots s_{a_t}$$
$$\cdots$$
$$s_{n-1}^+ = s_0 \oplus g_{n-1}(s_0, \ldots, s_{j-1}, s_{j+2}, \ldots, s_{n-1})$$

On the other hand, for F', the next state $r^+ = (r_0^+, r_1^+, \ldots, r_{n-1}^+)$ is given by:

$$r_0^+ = s_1 \oplus g_0(s_0, \ldots, s_{j-1}, s_{j+2}, \ldots, s_{n-1})$$
$$\cdots$$
$$r_j^+ = s_{j+1} \oplus s_{a_1-1} s_{a_2-1} \ldots s_{a_t-1} \oplus g_j(s_0, \ldots, s_{j-1}, s_{j+2}, \ldots, s_{n-1})$$
$$\qquad \oplus s_{a_1-1} s_{a_2-1} \ldots s_{a_t-1}$$
$$\quad = s_{j+1} \oplus g_j(s_0, \ldots, s_{j-1}, s_{j+2}, \ldots, s_{n-1})$$
$$r_{j+1}^+ = s_{j+2} \oplus g_{j+1}^*(s_0, \ldots, s_{j-1}, s_{j+2}, \ldots, s_{n-1})$$
$$\cdots$$
$$r_{n-1}^+ = s_0 \oplus g_{n-1}(s_0, \ldots, s_{j-1}, s_{j+2}, \ldots, s_{n-1})$$

We can see that the next states of F and F' can potentially differ in the bit position $j+1$ only. They are the same for all other bits.

In order to extend this conclusion to a sequence of states, it remains to show that r_{j+1}^+ can be expressed as $r_{j+1}^+ = s_{j+1}^+ \oplus s_{a_1-1}^+ s_{a_2-1}^+ \ldots s_{a_t-1}^+$. From

$$s_{j+1}^+ = s_{j+2} \oplus g_{j+1}^*(s_0, \ldots, s_{j-1}, s_{j+2}, \ldots, s_{n-1}) \oplus s_{a_1} s_{a_2} \ldots s_{a_t}$$

we can derive $s_{j+2} = s_{j+1}^{+} \oplus g_{j+1}^{*}(s_0, \ldots, s_{j-1}, s_{j+2}, \ldots, s_{n-1}) \oplus s_{a_1} s_{a_2} \cdots s_{a_t}$. By substituting this expression into

$$r_{j+1}^{+} = s_{j+2} \oplus g_{j+1}^{*}(s_0, \ldots, s_{j-1}, s_{j+2}, \ldots, s_{n-1})$$

and eliminating the double occurrence of $g_{j+1}^{*}(s_0, \ldots, s_{j-1}, s_{j+2}, \ldots, s_{n-1})$, we get

$$r_{j+1}^{+} = s_{j+1}^{+} \oplus s_{a_1} s_{a_2} \cdots s_{a_t}.$$

Since $s_{a_1} s_{a_2} \cdots s_{a_t} = s_{a_1-1}^{+} s_{a_2-1}^{+} \cdots s_{a_t-1}^{+}$, we obtain

$$r_{j+1}^{+} = s_{j+1}^{+} \oplus s_{a_1-1}^{+} s_{a_2-1}^{+} \cdots s_{a_t-1}^{+}.$$

By exchanging the roles of r and s and of i and j in the proof above, we can show that the result also applies for the case of $i = j - 1$. Since any shifting can be performed by repeatedly applying either $g_{j+1} \xrightarrow{m} g_j$ or $g_{j-1} \xrightarrow{m} g_j$ as many steps as required, Theorem 1 holds for the general case.

□

The following result follows directly from Theorem 1.

Lemma 1. *Let F be a mapping of type (1). Any mapping F' obtained from F by applying a sequence of valid shiftings generates a set of output sequences equivalent to the one of F.*

6   Transforming Trivium

In this section, we show how the presented transformation can be applied to Trivium stream cipher.

Trivium [7] is defined by a 287-variate mapping in which all but 3 out of 287 of functions are of type $f_i = x_{i+1}$. The remaining 3 functions are given by:

$$f_{287} = x_0 \oplus x_1 x_2 \oplus x_{45} \oplus x_{219}$$
$$f_{194} = x_{195} \oplus x_{196} x_{197} \oplus x_{117} \oplus x_{222}$$
$$f_{110} = x_{111} \oplus x_{112} x_{113} \oplus x_{24} \oplus x_{126}$$

The structure of 287-bit ring representing Trivium is shown in Fig. 6. The outputs from stages 110, 94 and 287 are added to get the keystream:

$$f_{output} = f_{287} \oplus f_{194} \oplus f_{110}.$$

There are many different possibilities for modifying Trivium. If the target is to minimize the propagation delay, then one possible solution obtained by applying the presented transformation is:

$$f_{287} = x_0 \oplus x_{219} \qquad f_{194} = x_{195} \oplus x_{222} \qquad f_{110} = x_{111} \oplus x_{24}$$
$$f_{218} = x_{219} \oplus x_{120} x_{121} \qquad f_{131} = x_{132} \oplus x_{133} x_{134} \qquad f_{21} = x_{22} \oplus x_{23} x_{24}$$
$$f_{210} = x_{211} \oplus x_{133} \qquad f_{118} = x_{119} \oplus x_{134} \qquad f_{17} = x_{18} \oplus x_{63}$$

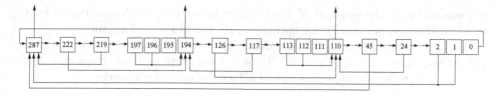

Fig. 6. The structure of Trivium.

Table 1. Propagation delays for a typical 90 nm CMOS technology.

Gate	Delay, ps
2 input AND	87
2 input XOR	115
flip-flop	221

and the remaining functions of type $f_i = x_{i+1}$. The keystream is computed as previously. By Theorem 1, it is equivalent to the keystream generated by the original Trivium. The reader can easily see that, in the original Trivium, the propagation delay is given by:

$$d_{original} = 2d_{XOR} + d_{AND} + d_{FF}$$

where d_{XOR}, d_{AND} and d_{FF} are the delays of the 2-input XOR, the 2-input AND, and the flip-flop, respectively. On the other hand, for the modified Trivium:

$$d_{modified} = d_{XOR} + d_{AND} + d_{FF}$$

By substituting d_{XOR}, d_{AND} and d_{FF} by values shown in Table 1, we get $d_{original} = 538$ ps and $d_{modified} = 423$ ps.

A shift register with the propagation delay of 538 ps can support data rates up to 1.86 Gbits/s. A shift register with the propagation delay of 423 ps can support data rates up to 2.36 Gbits/s. Note that 0.5 Gbits/s improvement (27 %) comes without any penalty in area, since the number of gates before and after the transformation remains the same.

It should be noted that the transformation reduces the maximum possible degree of parallelization of Trivium from the original 64 to 8. The modified Trivium can generate up to 8 bits per clock cycle because no variables are taken from 7 consecutive stages after each sink and after outputs 110, 94 and 287. The modified Trivium with the degree of parallelization 8 can support data rates up to 18.88 Gbits/s. The original Trivium with the degree of parallelization 8 can support data rates up to 14.88 Gbits/s.

7   Conclusion

We presented a transformation which can be applied to an n-bit shift register to construct other n-bit shift registers which generate the same set of output sequences. Using the example of Trivium stream cipher, we demonstrated that this transformation is useful for optimizing its hardware performance.

Being able to construct different shift registers generating equivalent sets of output sequences might be potentially useful for cryptanalysis. Exploring this opportunity to cryptanalyze Trivium is a focus of our future works.

Acknowledgements. This work was supported in part the research grant No 621-2010-4388 from the Swedish Research Council and in part by the research grant No SM12-0005 from the Swedish Foundation for Strategic Research.

References

1. Good, T., Benaissa, M.: ASIC hardware performance. In: Robshaw, M., Billet, O. (eds.) New Stream Cipher Designs. LNCS, vol. 4986, pp. 267–293. Springer, Heidelberg (2008)
2. Ericsson, 5G radio access - research and vision. White paper (2013). http://www.ericsson.com/news/130625-5g-radio-access-research-and-vision_244129228_c
3. Golomb, S.: Shift Register Sequences. Aegean Park Press, Laguna Hills (1982)
4. Klimov, A., Shamir, A.: A new class of invertible mappings. Revised Papers from the 4th International Workshop on Cryptographic Hardware and Embedded Systems, CHES'02, pp. 470–483. Springer, London (2002)
5. Xu, A.B., He, D.K., Wang, X.M.: An implementation of the GSM general data encryption algorithm A5. In: Proceedings of CHINACRYPT'94 (1994)
6. Hell, M., Johansson, T., Maximov, A., Meier, W.: The grain family of stream ciphers. In: Robshaw, M., Billet, O. (eds.) New Stream Cipher Designs. LNCS, vol. 4986, pp. 179–190. Springer, Heidelberg (2008)
7. De Cannière, C., Preneel, B.: TRIVIUM. In: Robshaw, M., Billet, O. (eds.) New Stream Cipher Designs. LNCS, vol. 4986, pp. 244–266. Springer, Heidelberg (2008)
8. Gammel, B., Göttfert, R., Kniffler, O.: Achterbahn-128/80: Design and analysis. In: SASC'2007: Workshop Record of the State of the Art of Stream Ciphers, pp. 152–165 (2007)
9. Gittins, B., Landman, H.A., O'Neil, S., Kelson, R.: A presentation on VEST hardware performance, chip area measurements, power consumption estimates and benchmarking in relation to the AES, SHA-256 and SHA-512. Cryptology ePrint Archive, Report 2005/415 (2005). http://eprint.iacr.org/2005/415
10. Gammel, B.M., Göttfert, R., Kniffler, O.: An NLFSR-based stream cipher. In: ISCAS (2006)
11. Aumasson, J.-P., Henzen, L., Meier, W., Naya-Plasencia, M.: Quark: A lightweight hash. J. Cryptol. **26**(2), 313–339 (2013)
12. Dubrova, E.: A transformation from the Fibonacci to the Galois NLFSRs. IEEE Trans. Inf. Theory **55**, 5263–5271 (2009)
13. Cusick, T.W., Stănică, P.: Cryptographic Boolean Functions and Applications. Academic Press, San Diego (2009)
14. Brayton, R.K., McMullen, C., Hatchel, G., Sangiovanni-Vincentelli, A.: Logic Minimization Algorithms for VLSI Synthesis. Kluwer Academic Publishers, Boston (1984)
15. Goresky, M., Klapper, A.: Fibonacci and Galois representations of feedback-with-carry shift registers. IEEE Trans. Inf. Theory **48**, 2826–2836 (2002)

16. Mrugalski, G., Rajski, J., Tyszer, J.: Ring generators - New devices for embedded test applications. Trans. Comput. Aided Des. Integr. Circuits Syst. **23**(9), 1306–1320 (2004)

17. Kagaris, D.: A similarity transform for linear finite state machines. Discrete Appl. Math. **154**, 1570–1577 (2006)

18. Arnault, F., Berger, T., Minier, M., Pousse, B.: Revisiting LFSRs for cryptographic applications. IEEE Trans. Inf. Theory **57**(12), 8095–8113 (2011)

19. Arnault, F., Berger, T.P., Pousse, B.: A matrix approach for FCSR automata. Crypt. Commun. **3**, 109–139 (2011)

20. Chabloz, J.-M., Mansouri, S.S., Dubrova, E.: An algorithm for constructing a fastest galois NLFSR generating a given sequence. In: Carlet, C., Pott, A. (eds.) SETA 2010. LNCS, vol. 6338, pp. 41–54. Springer, Heidelberg (2010)

21. Massey, J., Liu, R.-W.: Equivalence of nonlinear shift-registers. IEEE Trans. Inf. Theory **10**(4), 378–379 (1964)

22. Dubrova, E.: Finding matching initial states for equivalent NLFSRs in the Fibonacci to the Galois configurations. IEEE Trans. Inf. Theory **56**, 2961–2967 (2010)

A Lattice Rational Approximation Algorithm for AFSRs Over Quadratic Integer Rings

Weihua Liu[✉] and Andrew Klapper

University of Kentucky, Lexington, USA
weihua.liu@uky.edu

Abstract. Algebraic feedback shift registers (AFSRs) [10] are pseudo-random sequence generators that generalize linear feedback shift registers (LFSRs) and feedback with carry shift registers (FCSRs). With a general setting, AFSRs can result in sequences over an arbitrary finite field. It is well known that the sequences generated by LFSRs can be synthesized by either the Berlekamp-Massey algorithm or the extended Euclidean algorithm. There are three approaches to solving the synthesis problem for FCSRs, one based on the Euclidean algorithm [2], one based on the theory of approximation lattices [8] and Xu's algorithm which is also used for some AFSRs [11]. Xu's algorithm, an analog of the Berlekamp-Massey algorithm, was proposed by Xu and Klapper to solve the AFSR synthesis problem. In this paper we describe an approximation algorithm that solves the AFSR synthesis problem based on low-dimensional lattice basis reduction [14]. It works for AFSRs over quadratic integer rings $\mathbb{Z}[\sqrt{D}]$ with quadratic time complexity. Given the first $2\varphi_\pi(\mathbf{a}) + c$ elements of a sequence $\mathbf{a}$, it finds the smallest AFSR that generates $\mathbf{a}$, where $\varphi_\pi(\mathbf{a})$ is the π-adic complexity of $\mathbf{a}$ and c is a constant.

Keywords: AFSR synthesis · Rational approximation · Lattice basis reduction · π-adic complexity

1 Introduction

Pseudo-random sequences are ubiquitous in modern electronics and information technology. High speed generators of such sequences play essential roles in various engineering applications, such as stream ciphers, radar systems, multiple access systems, and quasi-Monte-Carlo simulation. Security has been a big concern in register design for many years. Given a short prefix of a sequence, it is undesirable to have an efficient algorithm that can synthesize a generator which can predict the whole sequence. Otherwise, a cryptanalytic attack can be launched against the system based on that given sequence. So finding such a synthesis algorithm

This material is based upon work supported by the National Science Foundation under Grant No. CNS-1420227. Any opinions, findings, and conclusions or recommendations expressed in this material are those of the author and do not necessarily reflect the views of the National Science Foundation.

© Springer International Publishing Switzerland 2014
K.-U. Schmidt and A. Winterhof (Eds.): SETA 2014, LNCS 8865, pp. 200–211, 2014.
DOI: 10.1007/978-3-319-12325-7_17

is an interesting problem in cryptanalysis. For a class of generators $\mathcal{F}$ and a sequence $\mathbf{a}$, a *register synthesis algorithm* finds the smallest generator in $\mathcal{F}$ that outputs sequence $\mathbf{a}$ given a sufficiently long prefix of $\mathbf{a}$.

LFSRs are the most widely studied pseudorandom sequence generators. The most famous LFSR synthesis algorithm is the Berlekamp-Massey algorithm. It can find the smallest LFSR that generates a given sequence $\mathbf{a}$ with only $2\lambda(\mathbf{a})$ consecutive bits of $\mathbf{a}$, where $\lambda(\mathbf{a})$ is the linear complexity of $\mathbf{a}$ [13].

FCSRs were first described by Goresky and Klapper [6,9]. They have many good algebraic properties similar to those of LFSRs. They can also be implemented efficiently, especially in hardware. FCSRs are good candidates as building blocks of stream ciphers since they seem to proffer resistance to algebraic attacks. The register synthesis problem for FCSRs was solved by Klapper and Goresky using integer approximation lattices [8,9]. These were originally proposed by Mahler [12] and de Weger [4]. In the case of binary FCSRs, the lattice approximation algorithm can construct the smallest FCSR which generates the sequence $\mathbf{a}$, and it does so using only a knowledge of the first $2\varphi_2(\mathbf{a}) + 2\log\varphi_2(\mathbf{a})$ elements of $\mathbf{a}$, where $\varphi_2(\mathbf{a})$ is the 2-adic span of $\mathbf{a}$ [7].

In later work, Klapper and Xu defined a generalization of both LFSRs and FCSRs called algebraic feedback shift registers (AFSRs) [10], described in detail in Sect. 2.1 of the current paper. Based on a choice of an integral domain R and $\pi \in R$, an AFSR can produce sequences whose elements can be thought of elements of the quotient ring $R/(\pi)$. A modification of the Berlekamp-Massey algorithm, Xu's algorithm solves the synthesis problem for AFSRs over a pair (R, π) with certain algebraic properties [11].

In this paper, we introduce a new synthesis algorithm for AFSRs. It can be seen as an extension of lattice approximation approach based on low-dimensional lattice basis reduction. For AFSRs over (R, π), where $R = \mathbb{Z}[\pi]$ with $\pi^2 = D \in \mathbb{Z}$, the algorithm can find the smallest AFSR that generates the sequence $\mathbf{a}$ given at least $2\varphi_\pi(\mathbf{a}) + 2 + \lceil \log_{|D|}(4D^2 + 2|1 + D|) \rceil$ terms of sequence $\mathbf{a}$, where $\varphi_\pi(\mathbf{a})$ is the π-adic complexity of $\mathbf{a}$. It has quadratic time complexity.

2   Preliminaries

2.1   Algebraic Feedback Shift Registers

In this section we recall the construction of AFSRs and properties of AFSR sequences. For more details on AFSRs, the reader should refer to [7, Chap. 5].

Let R be an integral domain and π be an element in R. Let S be a complete set of representatives for the quotient ring $R/(\pi)$ (i.e., the composition $S \rightarrow R \rightarrow R/(\pi)$ is a one to one correspondence). For any $u \in R$ denote its image in $R/(\pi)$ by $\tilde{u} = u \pmod{\pi}$. Given S, every element $a \in R$ has a unique expression $a = a_0 + b\pi$, where $a_0 \in S$. The element a_0 is the representative of $\tilde{a}$ in S, and $a - a_0$ is divisible by π. We write $a_0 = a \pmod{\pi}$ and $b = a(\text{div } \pi) = (a - a_0)/\pi$.

Definition 1 *[7]. Let $q_0, q_1, q_2, \cdots, q_m \in R$ and assume that q_0 is invertible* $\pmod{\pi}$. *An algebraic feedback shift register (or AFSR) over (R, π, S) of length*

m with multipliers or taps $q_1, q_2, \cdots, q_m$ is a sequence generator whose states are elements $s = (a_0, a_1, \cdots, a_{m-1}; z) \in S^m \times R$ consisting of cell contents a_i and memory z. The output is $\mathbf{out}(s) = a_0$. The state change operation is:

1. Compute
$$\sigma = \sum_{i=1}^{m} q_i a_{m-i} + z.$$

2. Find $a_m \in S$ such that $-q_0 a_m \equiv \sigma \pmod{\pi}$. That is, $\tilde{a}_m = -\tilde{q}_0^{-1}\tilde{\sigma}$.
3. Replace $(a_0, a_1, \cdots, a_{m-1})$ by $(a_1, a_2, \cdots, a_m)$ and replace z by $\sigma(\mathrm{div}\pi) = (\sigma + q_0 a_m)/\pi$.

The register outputs an infinite sequence $a_0, a_1, \ldots$ of elements in S. The state change rules may be summarized by saying that this sequence satisfies a *linear recurrence with carry*,

$$-q_0 a_n + \pi z_n = q_1 a_{n-1} + \cdots + q_m a_{n-m} + z_{n-1} \tag{1}$$

for all $n \geq m$. Here, z_i represent the sequence of memory values, with $z = z_{m-1}$ being the initial value. The procedure of an AFSR is illustrated in Fig. 1.

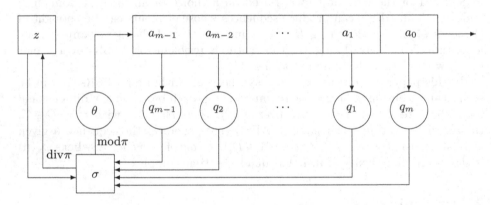

Fig. 1. An algebraic feedback shift register of length m

The element
$$q = \sum_{i=0}^{m} q_i \pi^i \in R$$
plays a central role in the analysis of AFSRs and is referred to as the *connection element*. To simplify the analysis, we suppose all the q_i's are in S.

An LFSR over a field (or even an integral domain) K, is an AFSR where $R = K[x]$ is the ring of all polynomials with coefficients in K, $\pi = x$ and $S = K$ is the set of polynomials of degree 0, which also is the quotient ring $R/(\pi) = K[x]/(x)$. An FCSR with elements in $\mathbb{Z}/(N)$ is an AFSR with $R = \mathbb{Z}$, $\pi = N$, and $S = \{0, \ldots, N-1\}$.

In this paper, we focus on AFSRs over quadratic extensions of $\mathbb{Z}$. That is, fix $\pi \in \mathbb{Z}$ such that $\pi^2 = D$, where $D \in \mathbb{Z}$ is square free. So $x^2 - D$ is irreducible over the rational numbers $\mathbb{Q}$. Let $R = \mathbb{Z}[\pi]$, a quadratic extension of $\mathbb{Z}$. It is an integral domain in which every prime ideal is maximal. It can be proved that $S = \{0, 1, \ldots, D-1\}$ is a complete set of representatives for the quotient ring $R/(\pi)$ [7, p.102].

An infinite sequence, $\mathbf{a} = a_0, a_1, \cdots$, generated by an LFSR can be identified with its generating function

$$a(x) = \sum_{i=0}^{\infty} a_i x^i,$$

an element of the ring of formal power series. An N-adic integer of the form $\sum_{i=0}^{\infty} a_i N^i$ is associated with the sequence $\mathbf{a} = a_0, a_1, \cdots$ generated by an FCSR [8,9,11]. Similarly, a sequence $\mathbf{a} = a_0, a_1, \cdots$ over R generated by an AFSR has an associated π-adic integer, defined below.

Definition 2 *[7]. Let R be an integral domain, let $\pi \in R$ and let S be a complete set of representatives for R modulo π. Then the ring of π-adic integers, R_π, is the set of expressions $\alpha = \sum_{i=0}^{\infty} a_i \pi^i$ with all $a_i \in S$.*

Take $R = \mathbb{Z}[\pi]$ with $\pi^2 = D$ as above for example. The ring $\alpha \in R_\pi$ consists of elements $\alpha = a_0 + a_1 \pi + \cdots$ with coefficients $a_i \in S = \{0, 1, \ldots, D-1\}$. In carrying out algebraic operations we must remember that $D = \pi^2$. The addition of π-adic integers may be described as term-wise addition with a "delayed carry": each carried quantity is delayed 2 steps before adding it back in. In other words, if $\beta = b_0 + b_1 \pi + \cdots \in R_\pi$ then

$$\alpha + \beta = \sum_{i=0}^{\infty} e_i \pi^i,$$

with $0 \le e_i \le D-1$, means that there exist $c_0, c_1, \ldots \in \{0, 1\}$ with $a_i + b_i + c_i = e_i + D c_{i+2}$. That is, c_i is the carry to the ith position.

Theorem 1 *(Fundamental Theorem on AFSRs [10]). Let the output sequence $\mathbf{a} = a_0, a_1, \ldots$ of an AFSR with connection element q and initial state $(a_0, a_1, \cdots, a_{m-1}; z)$ have associated π-adic integer $\alpha = \sum_{i=0}^{\infty} a_i \pi^i$. Then*

$$\alpha = \frac{\sum_{n=0}^{m-1} \sum_{i=0}^{n} q_i a_{n-i} \pi^n - z \pi^m}{q} = \frac{u}{q} \in R_\pi. \tag{2}$$

The expression u/q is called a rational expression of α.

If (u, q) is found, then the AFSR that generates sequence $\mathbf{a}$ can be constructed by Eq. (2). So our goal is to find a rational expression u/q using as few terms of sequence $\mathbf{a}$ as we can.

2.2   Size and π-Adic Complexity

From here on, we suppose $R = \mathbb{Z}[\pi] = \{a_0 + a_1\pi : a_0, a_1 \in \mathbb{Z}\}$ and $\pi^2 = D \in \mathbb{Z}$ unless otherwise mentioned. To measure the size of the elements of R, let size function $\varphi_{R,\pi} : R \to \mathbb{Z}$ be $\varphi_{R,\pi}(q) = q_0^2 + q_1^2$, where $q = q_0 + q_1\pi$ and $q_0, q_1 \in \mathbb{Z}$.

Proposition 1. *For any $u, q \in R$, we have*

1. $\varphi_{R,\pi}(u \pm q) \leq 2(\varphi_{R,\pi}(u) + \varphi_{R,\pi}(q))$ *and*
2. $\varphi_{R,\pi}(uq) \leq (D^2 + |1 + D|/2)\varphi_{R,\pi}(u)\varphi_{R,\pi}(q).$

Proof. Let $u = u_0 + u_1\pi$ and $q = q_0 + q_1\pi$ where $u_0, u_1, q_0, q_1 \in \mathbb{Z}$. We have $u \pm q = (u_0 \pm q_0) + (u_1 \pm q_1)\pi$, so

$$
\begin{aligned}
\varphi_{R,\pi}(u \pm q) &= (u_1 \pm q_1)^2 + (u_0 \pm q_0)^2 \\
&= u_1^2 + q_1^2 + u_0^2 + q_0^2 \pm 2u_1q_1 \pm 2u_0q_0 \\
&\leq 2(u_1^2 + q_1^2 + u_0^2 + q_0^2) \\
&= 2(\varphi_{R,\pi}(u) + \varphi_{R,\pi}(q)).
\end{aligned}
$$

We have $uq = (u_0 + q_0\pi)(u_1 + q_1\pi) = (u_0q_0 + Du_1q_1) + (u_0q_1 + u_1q_0)\pi$, so

$$
\begin{aligned}
\varphi_{R,\pi}(uq) &= (u_0q_0 + Du_1q_1)^2 + (u_0q_1 + u_1q_0)^2 \\
&= u_0^2q_0^2 + D^2u_1^2q_1^2 + u_0^2q_1^2 + u_1^2q_0^2 + (2 + 2D)u_0u_1q_0q_1 \\
&\leq u_0^2q_0^2 + D^2u_1^2q_1^2 + u_0^2q_1^2 + u_1^2q_0^2 + |(2 + 2D)| \cdot |u_0u_1q_0q_1|
\end{aligned}
$$

Since $\varphi_{R,\pi}(u)\varphi_{R,\pi}(q) = u_0^2q_0^2 + u_1^2q_1^2 + u_0^2q_1^2 + u_1^2q_0^2 \geq 4|u_0u_1q_0q_1|$, we have

$$
\begin{aligned}
\varphi_{R,\pi}(uq) &\leq D^2\varphi_{R,\pi}(u)\varphi_{R,\pi}(q) + \frac{|2 + 2D|}{4}\varphi_{R,\pi}(u)\varphi_{R,\pi}(q) \\
&= (D^2 + \frac{|1 + D|}{2})\varphi_{R,\pi}(u)\varphi_{R,\pi}(q)
\end{aligned}
$$

$\square$

For any $u, q \in R$, let $\Phi_{R,\pi}(u, q) = \log_{|D|}(\varphi_{R,\pi}(u) + \varphi_{R,\pi}(q))$. $\Phi_{R,\pi}(u, q)$ is defined to be the *size* of the AFSR reconstructed by Eq. (2). That is, u/q is a rational expression of α, the associated π-adic integer of sequence **a**. Then the π-*adic complexity* of **a** is $\varphi_\pi(\mathbf{a}) = \min\{\Phi_{R,\pi}(u, q) : \alpha = u/q\}$.

The AFSR synthesis problem can be rephrased as follows:

- **Given** A prefix of the eventually periodic sequence $\mathbf{a} = a_0, a_1, \cdots$ over $R/(\pi)$.
- **Find** $u, q \in R$ satisfying $\alpha = u/q$ and minimizing $\Phi_{R,\pi}(u, q)$.

This problem was studied by Xu and Klapper using Xu's rational approximation algorithm, which is a modification of the Berlekamp-Massey algorithm. They defined the π-adic complexity a little differently. For the imaginary extensions, that is $D < 0$, the size function of AFSRs was defined to be $\max(\varphi_D(N(u)), \varphi_D(N(q)))$, where $N(\cdot)$ is the rational norm function on $\mathbb{Z}[\pi]$, and $\varphi_D(\cdot)$ is the length of the D-adic representation of an integer. For real extensions, they used $\max(\varphi(u), \varphi(q))$, where $\varphi(q) = \max(2\lfloor \log_D |q_0| \rfloor, 2\lfloor \log_D |q_1| \rfloor + 1)$. Although the form of size function is different from what we use, they are related. E.g., when $D > 0$, we have

$$\begin{aligned}
\Phi_{R,\pi}(u, q) &\leq \log_D(4 \max(u_0^2, u_1^2, q_0^2, q_1^2)) \\
&\leq \log_D 4 + 2\max\left(\lfloor \log_D |u_0| \rfloor, \lfloor \log_D |u_1| \rfloor, \lfloor \log_D |q_0| \rfloor, \lfloor \log_D |q_1| \rfloor\right) + 2 \\
&\leq \max(\varphi(u), \varphi(q)) + 2 + \log_D 4.
\end{aligned}$$

2.3 Low-Dimensional Lattice Basis Reduction

A *lattice* L of rank d is a discrete additive subgroup of $\mathbb{R}^n$ of the form

$$L(\mathbf{b}_1, \mathbf{b}_2, \cdots, \mathbf{b}_d) := \sum_{i=0}^{d} \mathbf{b}_i \mathbb{Z},$$

where $\mathbf{b}_1, \mathbf{b}_2, \ldots, \mathbf{b}_d \in \mathbb{R}^n$ are linearly independent vectors over $\mathbb{R}$. We call $(\mathbf{b}_1, \mathbf{b}_2, \ldots, \mathbf{b}_d)$ a basis of lattice L. Usually, the basis of a lattice is not unique. For arbitrary vectors $\mathbf{b}_1, \mathbf{b}_2, \ldots, \mathbf{b}_d \in \mathbb{R}^n$, let

$$span(\mathbf{b}_1, \mathbf{b}_2, \cdots, \mathbf{b}_d) := \sum_{i=0}^{d} \mathbf{b}_i \mathbb{R}$$

be the space spanned by $\mathbf{b}_1, \mathbf{b}_2, \ldots, \mathbf{b}_d$.

Here is some notation we use. Let $\|\cdot\|$ and $\langle \cdot, \cdot \rangle$ be the Euclidean norm and inner product of $\mathbb{R}^n$ respectively. The notation $[\mathbf{b}_1, \mathbf{b}_2, \ldots, \mathbf{b}_d]_\leq$ means $\|\mathbf{b}_1\| \leq \|\mathbf{b}_2\| \leq \cdots \leq \|\mathbf{b}_d\|$ which is to say the $\mathbf{b}_i$s are ordered. The Gram matrix, denoted by $G(\mathbf{b}_1, \mathbf{b}_2, \ldots, \mathbf{b}_d)$, is a $d \times d$ symmetric matrix with entries given by $G_{ij} = \langle \mathbf{b}_i, \mathbf{b}_j \rangle$. The Voronoï cell is $\mathrm{Vor}(\mathbf{b}_1, \mathbf{b}_2, \ldots, \mathbf{b}_d) = \{\mathbf{x} \mid \|\mathbf{x} - \mathbf{v}\| \geq \|\mathbf{x}\|, \ \forall \ \mathbf{v} \in L\}$.

Loosely speaking, the lattice reduction problem is: given an arbitrary lattice basis, obtain a basis of shortest possible vectors which are mutually orthogonal. Finding a good reduced basis has many important applications in mathematics, computer science, and cryptography [5]. For two dimensional lattices, Gauss's basis reduction algorithm, which is a generalization of the Euclidean Algorithm, can be used. For higher dimensions, there are many different kinds of basis reduction, such as Hermite, Minkowski, Hermite-Korkine-Zolotarev(HKZ), and Lenstra-Lenstra-Lovász(LLL). The one we use here is Minkowski reduction because it can reach all the successive minima of a lattice.

Definition 3 *[3]* **(Successive Minima** $\lambda_1, \lambda_2, \ldots, \lambda_n$**).** *For every lattice $L \in \mathbb{R}^n$ of rank d the successive minima $\lambda_1, \lambda_2, \ldots, \lambda_n$ are defined as:*

$$\lambda_i = \lambda_i(L) := \min\left\{ r > 0 \,\middle|\, \begin{array}{l} \exists \ \text{linearly independent} \\ c_1, c_2, \ldots, c_i \in L \text{ with} \\ \|c_j\| \leq r \text{ for } j = 1, 2 \ldots, i \end{array} \right\}, \ \text{for } i = 1, 2, \ldots, n.$$

Definition 4 *[14]* *(*Minkowski **reduction).** *A basis* $[\mathbf{b}_1, \mathbf{b}_2, \ldots, \mathbf{b}_d]_\leq$ *of a lattice L is Minkowski-reduced if for all* $1 \leq i \leq d$, $\mathbf{b}_i$ *has minimal norm among all lattice vectors* $\mathbf{b}_i$ *such that* $[\mathbf{b}_1, \mathbf{b}_2, \ldots, \mathbf{b}_i]_\leq$ *can be extended to a basis of L.*

Notice that the first vector in a Minkowski-reduced basis is the shortest nonzero vector in lattice L. Given a basis of a lattice L, finding a lattice vector whose norm is exactly $\lambda_1(L)$ is one of the most famous lattice problems. It is called the shortest vector problem (SVP). It has been proved to be NP-hard if the dimension is unrestricted [1]. Nguyen and Stehlé [14] proposed a greedy algorithm that generalizes Lagrange's algorithm for lattice reduction to arbitrary dimension. They showed that up to dimension four, their algorithm computes a Minkowski-reduced basis in quadratic time without fast arithmetic but as the dimension increases, the analysis becomes more complex. Figure 2 is an iterative description of Nguyen and Stehl's greedy algorithm from [14].

1: **procedure** GREEDYLATTICEREDUCTION($b_1, b_2, \ldots, b_d$)
2: **Input:** A basis $[\mathbf{b}_1, \mathbf{b}_2, \ldots, \mathbf{b}_d]_\leq$ with its Gram matrix
3: **Output:** An ordered basis of $L(\mathbf{b}_1, \mathbf{b}_2, \ldots, \mathbf{b}_d)$ with its Gram matrix
4:     $m := 2$
5:     **while** $m \leq d$ **do**
6:         Compute a vector $\mathbf{c} \in L(\mathbf{b}_1, \mathbf{b}_2, \ldots, \mathbf{b}_{m-1})$ closest to $\mathbf{b}_m$
7:     **end while**
8:     $\mathbf{b}_m := \mathbf{b}_m - \mathbf{c}$ and update the Gram matrix
9:     **if** $\|\mathbf{b}_m\| \geq \|\mathbf{b}_{m-1}\|$ **then**
10:         $m := m + 1$
11:     **else**
12:         insert $\mathbf{b}_m$ between $\mathbf{b}_{m'-1}$ and $\mathbf{b}_{m'}$ such that $\|\mathbf{b}_{m'-1}\| \leq \|\mathbf{b}_m\| < \|\mathbf{b}'_m\|$.
13:         update the Gram matrix and set $m := m' + 1$.
14:     **end if**
15: **end procedure**

Fig. 2. Lattice reduction greedy algorithm

Theorem 2 *[14]. Let* $d \leq 4$. *Given as input an ordered basis* $[\mathbf{b}_1, \mathbf{b}_2, \ldots, \mathbf{b}_d]_\leq$ *and its Gram matrix, the greedy algorithm of Fig. 2 outputs a Minkowski-reduced basis of* $L(\mathbf{b}_1, \mathbf{b}_2, \ldots, \mathbf{b}_d)$, *with bit complexity in* $O(\log \|\mathbf{b}_d\| \cdot [1 + \log \|\mathbf{b}_d\| - \log \lambda_1(L)])$, *where the* $O()$ *constant is independent of the lattice. Moreover, in dimension five, the output basis may not be Minkowski-reduced.*

We use the greedy algorithm in four dimensions, i.e., $d = 4$, to find the shortest vector in L in our Rational Approximation algorithm. More exactly, the closest vector problem in step 6 of GREEDYLATTICEREDUCTION can be found as follows.

1. Let $\mathbf{h} = \sum_{i=1}^{m-1} y_i \mathbf{b}_i$ be the orthogonal projection of $\mathbf{b}_m$ on $span(\mathbf{b}_1, \mathbf{b}_2, \ldots, \mathbf{b}_{m-1})$. Then

$$G(\mathbf{b}_1, \mathbf{b}_2, \ldots, \mathbf{b}_{m-1}) \begin{pmatrix} y_1 \\ y_2 \\ \vdots \\ y_{m-1} \end{pmatrix} = \begin{pmatrix} \langle \mathbf{b}_1, \mathbf{b}_m \rangle \\ \langle \mathbf{b}_2, \mathbf{b}_m \rangle \\ \vdots \\ \langle \mathbf{b}_{k-1}, \mathbf{b}_m \rangle \end{pmatrix}.$$

2. Let $\mathbf{c}$ be the closest vector to $\mathbf{h}$ in $L(\mathbf{b}_1, \mathbf{b}_2, \ldots, \mathbf{b}_{m-1})$. Then $\mathbf{h} - \mathbf{c} \in \mathrm{Vor}(\mathbf{b}_1, \mathbf{b}_2, \ldots, \mathbf{b}_{m-1})$. With Theorem 3, $\mathbf{c}$ can be found by a suitable exhaustive search.

Theorem 3 *[14]*

1. *Let* $[\mathbf{b}_1, \mathbf{b}_2]_\leq$ *be a Minkowski-reduced basis and* $\mathbf{u} \in \mathrm{Vor}(\mathbf{b}_1, \mathbf{b}_2)$. *Write* $\mathbf{u} = x\mathbf{b}_1 + y\mathbf{b}_2$. *Then* $|x| < 3/4$ *and* $|y| \leq 2/3$.
2. *Let* $[\mathbf{b}_1, \mathbf{b}_2, \mathbf{b}_3]_\leq$ *be a Minkowski-reduced basis and* $\mathbf{u} \in \mathrm{Vor}(\mathbf{b}_1, \mathbf{b}_2, \mathbf{b}_3)$. *Write* $\mathbf{u} = x\mathbf{b}_1 + y\mathbf{b}_2 + z\mathbf{b}_3$. *Then* $|x| < 3/4$, $|y| \leq 2/3$ *and* $|z| \leq 1$.

3 k-th Approximation Lattices

Definition 5. *Let* $\pi = \sqrt{D}$, *where* $D \in \mathbb{Z}$ *is square free. Let* $R = \mathbb{Z}[\pi]$ *and let* R_π *be the ring of* π-*adic integers. Suppose* $\alpha = a_0 + a_1\pi + a_2\pi^2 + \cdots$ *is an element in* R_π. *The* kth *approximation lattice of* α *is defined as*

$$L_k = L_k(\alpha) := \{(u_1, u_2, u_3, u_4) \in \mathbb{Z}^4 : \alpha(u_3 + u_4\pi) - (u_1 + u_2\pi) \equiv 0 \ (\mathrm{mod} \ \pi^k)\}$$

Notice that for every element (u_1, u_2, u_3, u_4) in $L_k(\alpha)$, we have

$$\alpha \equiv \frac{u_1 + u_2\pi}{u_3 + u_4\pi} \ (\mathrm{mod} \ \pi^k) \quad \text{if} \quad u_3 + u_4\pi \neq 0.$$

Thus the pair (u, q) with $u = u_1 + u_2\pi$ and $q = u_3 + u_4\pi$ represents a fraction u/q whose π-adic expansion agrees with α in the first k places. We call (u, q) a rational approximation of α up to k terms. If $\alpha_k = \sum_{i=0}^{k-1} a_i\pi^i = a + b\pi$, where $a, b \in \mathbb{Z}$, then $\mathbf{u}_1 = (a, b, 1, 0) \in L_k$. Also, it can be verified that $\mathbf{u}_2 = (Db, a, 0, 1) \in L_k$. Suppose

$$\pi^k = c + d\pi = \begin{cases} D^{\frac{k-1}{2}}\pi, & \text{if k is odd;} \\ D^{\frac{k}{2}}, & \text{if k is even.} \end{cases}$$

Then $\mathbf{u}_3 = (c, d, 0, 0) \in L_k$ and $\mathbf{u}_4 = (Dd, c, 0, 0) \in L_k$.

Theorem 4. $L_k(\alpha)$ *is a four dimensional lattice and* $(\mathbf{u}_1, \mathbf{u}_2, \mathbf{u}_3, \mathbf{u}_4)$ *is a basis of* $L_k(\alpha)$. L_{i+1} *is a sublattice of* L_i *for any* $i \in \mathbb{Z}$.

Proof. If $\mathbf{u} = (u_1, u_2, u_3, u_4) \in L_k$ and $\mathbf{v} = (v_1, v_2, v_3, v_4) \in L_k$, then $\mathbf{u} + \mathbf{v} \in L$. So L_k is a lattice. The four vectors $\mathbf{u}_1, \mathbf{u}_2, \mathbf{u}_3, \mathbf{u}_4$ are linearly independent elements of L_k. Now suppose that $\mathbf{x} = (x_1, x_2, x_3, x_4)$ is an arbitrary vector in L_k. So $\alpha_k(x_3 + x_4\pi) - (x_1 + x_2\pi) = \gamma\pi^k$ for some $\gamma = r_1 + r_2\pi \in R$. Making corresponding terms equal, we have

$$\begin{cases} ax_3 + bx_4D - x_1 = r_1c + r_2dD \\ bx_3 + ax_4 - x_2 = r_2c + r_1d. \end{cases}$$

This also means that $\mathbf{x} = x_3\mathbf{u}_1 + x_4\mathbf{u}_2 - r_1\mathbf{u}_3 - r_2\mathbf{u}_4$. So $(\mathbf{u}_1, \mathbf{u}_2, \mathbf{u}_3, \mathbf{u}_4)$ is a basis of L_k.

For any $(y_1, y_2, y_3, y_4) \in L_{i+1}$ and any $i \in \mathbb{Z}$ we have $\alpha(y_3 + y_4\pi) - (y_1 + y_2\pi) \equiv 0 \pmod{\pi^{i+1}}$. So $\alpha(y_3 + y_4\pi) - (y_1 + y_2\pi) \equiv 0 \pmod{\pi^i}$. That is, $(y_1, y_2, y_3, y_4) \in L_i$. So L_{i+1} is a sublattice of L_i for any $i \in \mathbb{Z}$. □

4   The Approximation Algorithm

In this section we give an approximation algorithm based on the algorithm GREEDYLATTICEREDUCTION. Let $\mathbf{a}$ be a sequence with associated π-adic integer α. Given a sufficiently large prefix of $\mathbf{a}$, this algorithm finds the rational expression of α that realizes the π-adic complexity of $\mathbf{a}$. With the help of GREEDY-LATTICEREDUCTION, we can find the shortest vector of the kth approximation lattice which gives the best rational approximation of α up to k terms. Suppose the π-adic complexity is known. Theorem 5 shows that if k is chosen big enough, then such a rational approximation is exactly the rational expression we want. The algorithm shown in Fig. 3 is just for the case when k is even. The odd case is similar, so details are omitted here.

Theorem 5. *Let $\mathbf{a}$ be a π-adic sequence with associated π-adic integer α. Suppose the size of the AFSR that generates $\mathbf{a}$ is less than or equal to n. That is, the π-adic complexity of $\mathbf{a}$, $\varphi_\pi(\mathbf{a})$, is less than or equal to n. Let APPROXLATTICE (Fig. 3) be executed with $k \geq 2n + 2 + \lceil \log_{|D|}(4D^2 + 2|1 + D|) \rceil$. Suppose the algorithm outputs a pair (u, q) of elements of R. Then*

$$\alpha = \sum_{i=0}^{\infty} a_i\pi^i = \frac{u}{q}.$$

Proof. Let u'/q' be a rational expression of α with $\Phi_{R,\pi}(u', q') = \varphi_\pi(\mathbf{a})$. That is

$$\alpha = \sum_{i=0}^{\infty} a_i\pi^i = \frac{u'}{q'}.$$

It follows that $\Phi_{R,\pi}(u', q') \leq n$. Suppose $\mathbf{v}_1 = (v_1, v_2, v_3, v_4)$ where $u' = v_1 + v_2\pi, q' = v_3 + v_4\pi$. So $\mathbf{v}_1 \in L_k(a)$.

Let (u, q) be the output of APPROXLATTICE. Then Theorem 2 shows that $\mathbf{u}_1 = (u_1, u_2, u_3, u_4)$ in step 14 is the minimal vector in $L_k(\alpha)$.

1: **procedure** APPROXLATTICE$(a_0, a_1, \ldots, a_{k-1})$
2: **Input:** first k terms of sequence **a**
3: **Output:** $u, q \in R$ satisfying $\alpha = u/q$ and minimizing $\Phi_{R,\pi}(x, y)$
4: $a := \sum\limits_{0 \le i \le k/2} a_{2i} D^i$
5: $b := \sum\limits_{0 \le i \le (k-2)/2} a_{2i+1} D^i$
6: $c := D^{k/2}$
7: $\mathbf{u}_1 := (a, b, 1, 0)$
8: $\mathbf{u}_2 := (Db, a, 0, 1)$
9: $\mathbf{u}_3 := (c, 0, 0, 0)$
10: $\mathbf{u}_4 := (0, c, 0, 0)$
11: Sort $\mathbf{u}_1, \mathbf{u}_2, \mathbf{u}_3, \mathbf{u}_4$ by their norm $\| \cdot \|$. Let $(\mathbf{u}_1, \mathbf{u}_2, \mathbf{u}_3, \mathbf{u}_4)$ be ordered.
12: Compute the Gram matrix G so that $G_{ij} = \langle \mathbf{u}_i, \mathbf{u}_j \rangle$.
13: $(\mathbf{u}_1, \mathbf{u}_2, \mathbf{u}_3, \mathbf{u}_4) := $GREEDYLATTICEREDUCTION$(\mathbf{u}_1, \mathbf{u}_2, \mathbf{u}_3, \mathbf{u}_4)$
14: Suppose $\mathbf{u}_1 = (u_0, u_1, q_0, q_1)$
15: **return** $(u_0 + u_1\pi, q_0 + q_1\pi)$
16: **end procedure**

Fig. 3. Lattice Rational Approximation Algorithm for AFSRs over quadratic extension

We have $u = u_1 + u_2\pi$ and $q = u_3 + u_4\pi$. So

$$\|\mathbf{u}_1\| = \sqrt{u_1^2 + u_2^2 + u_3^2 + u_4^2} \le \sqrt{v_1^2 + v_2^2 + v_3^2 + v_4^2} = \|\mathbf{v}_1\|.$$

Since

$$\Phi_{R,\pi}(u, q) = \log_{|D|}(u_1^2 + u_2^2 + u_3^2 + u_4^2)$$
$$\le \log_{|D|}(v_1^2 + v_2^2 + v_3^2 + v_4^2)$$
$$= \Phi_{R,\pi}(u', q') \le n.$$

This shows that $\varphi_{R,\pi}(u'), \varphi_{R,\pi}(q'), \varphi_{R,\pi}(u), \varphi_{R,\pi}(q)$ are all less than or equal to $|D|^n$. We have

$$\frac{u}{q} \equiv \frac{u'}{q'} \pmod{\pi^k},$$

so

$$\pi^k \left| \frac{uq' - u'q}{qq'} \right..$$

Thus there exists $t \in R$ such that $tqq'\pi^k = uq' - u'q$. From Proposition 1,

$$\varphi_{R,\pi}(uq' - u'q) \le 2(\varphi_{R,\pi}(uq') + \varphi_{R,\pi}(u'q))$$
$$\le (2D^2 + |1 + D|)(\varphi_{R,\pi}(u)\varphi_{R,\pi}(q') + \varphi_{R,\pi}(u')\varphi_{R,\pi}(q'))$$
$$\le (4D^2 + 2|1 + D|)|D|^{2n}.$$

For any $e = e_1 + e_2\pi \neq 0 \in \mathbb{Z}[\pi]$, we have

$$e\pi^k = \begin{cases} e_1 D^{\frac{k}{2}} + e_2 D^{\frac{k}{2}}\pi & \text{if } k \text{ is even} \\ e_2 D^{\frac{k+1}{2}} + e_1 D^{\frac{k-1}{2}}\pi & \text{if } k \text{ is odd}. \end{cases}$$

Therefore $\varphi_{R,\pi}(e\pi^k) > |D|^{k-2}$. This is to say, $\varphi_{R,\pi}(tqq'\pi^k) > |D|^{k-2}$ if $t \neq 0$. But from $k \geq 2n+2+\lceil \log_{|D|}(4D^2+2|1+D|) \rceil$ we have $|D|^{k-2} \geq (4D^2+2|1+D|)|D|^{2n}$. So t must be 0, which also means $uq' - u'q = 0$. This proves that

$$\frac{u}{q} = \frac{u'}{q'} = \sum_{i=0}^{\infty} a_i \pi^i.$$

From the proof we also know that $\Phi_{R,\pi}(u,q)$ reaches the π-adic complexity of sequence **a** which means that we find the smallest AFSR that generates **a**.   □

Theorem 6. *The Lattice Rational Approximation Algorithm,* APPROXLATTICE, *runs in time $O(k^2)$ if k elements of **a** are used.*

Proof. The time complexity of getting $\mathbf{u}_1, \mathbf{u}_2, \mathbf{u}_3, \mathbf{u}_4$ from step 4 to step 10 in Fig. 3 is $O(k \log k)$. Since

$$|a| = \left| \sum_{0 \leq i \leq k/2} a_{2i} D^i \right| \leq |D|^{k/2+1},$$

$$|b| = \left| \sum_{0 \leq i \leq (k-2)/2} a_{2i} D^i \right| \leq |D|^{k/2}, \text{ and}$$

$$|c| \leq |D|^{k/2},$$

we have $\max(\|\mathbf{u}_1\|, \|\mathbf{u}_2\|, \|\mathbf{u}_3\|, \|\mathbf{u}_4\|) \leq \sqrt{2}|D|^{(k+3)/2}$.

In step 11, to compute and sort $\|\mathbf{u}_1\|, \|\mathbf{u}_2\|, \|\mathbf{u}_3\|, \|\mathbf{u}_4\|$ takes time $O(k^2)$ because the dimension of L_k is fixed. Also, the time complexity for computing the Gram matrix G is $O(k^2)$.

The most costly step in APPROXLATTICE is Step 13 that calls GREEDY-LATTICEREDUCTION. According to Theorem 2, the time complexity is bounded by $O\left(\log(\sqrt{2}|D|^{\frac{k+3}{2}})[1 + \log(\sqrt{2}|D|^{\frac{k+3}{2}}) - \log \lambda_1(L_k)] \right) = O(k^2)$, where $\lambda_1(L)$ is the smallest vector in L_k. To sum up, the time complexity of APPROXLATTICE is $O(k^2)$.   □

Xu's algorithm [11] has worst case time complexity $O(\sum_{k=1}^{\varphi(\mathbf{a})} \sigma(k))$, where $\sigma(k)$ is the time needed to add two elements $a, b \in \mathbb{Z}[\pi]$ with the length of π-adic expansion at most k. So it also runs in quadratic time. The number of terms needed to get the exact rational expression is $52\varphi(\mathbf{a}) + c$, with some constant c. APPROXLATTICE only needs $2\varphi_\pi(\mathbf{a}) + c'$ terms to get the exact rational expression, with some constant c'.

5   Conclusions

In this paper we proposed a synthesis algorithm for AFSRs over $\mathbb{Z}[\sqrt{D}]$ based on low-dimensional lattice reduction. It has the same time complexity as Xu's algorithm but needs fewer terms of the sequence to get the exact rational expression.

With the same idea, we may extend this approach to cubic or higher extensions of $\mathbb{Z}$. This becomes complicated because of the complexity of the lattice reduction problem. In further work we will try to use other lattice reduction algorithms, such as the LLL lattice basis reduction algorithm, to reduce the basis of kth approximation lattice so that we can find the shortest possible vectors.

References

1. Ajtai, M.: The shortest vector problem in L^2 is NP-hard for randomized reductions. In: Proceedings of the Thirtieth Annual ACM Symposium on Theory of Computing, pp. 10–19. ACM (1998)
2. Arnault, F., Berger, T.P., Necer, A.: Feedback with carry shift registers synthesis with the Euclidean algorithm. IEEE Trans. Inf. Theor. **50**(5), 910–917 (2004)
3. Dwork, C.: Lattices and their application to cryptography. Stanford University, Lecture Notes (1998)
4. de Weger, B.M.M.: Approximation lattices of p-adic numbers. J. Number Theor. **24**(1), 70–88 (1986)
5. Goldreich, O., Goldwasser, S., Halevi, S.: Public-key cryptosystems from lattice reduction problems. In: Kaliski Jr., B.S. (ed.) CRYPTO 1997. LNCS, vol. 1294, pp. 112–131. Springer, Heidelberg (1997)
6. Goresky, M., Klapper, A.: Feedback registers based on ramified extensions of the 2-adic numbers. In: De Santis, A. (ed.) EUROCRYPT 1994. LNCS, vol. 950, pp. 215–222. Springer, Heidelberg (1995)
7. Goresky, M., Klapper, A.: Algebraic Shift Register Sequences. Cambridge University Press, Cambridge (2012)
8. Klapper, A., Goresky, M.: Cryptanalysis based on 2-adic rational approximation. In: Coppersmith, D. (ed.) CRYPTO 1995. LNCS, vol. 963, pp. 262–273. Springer, Heidelberg (1995)
9. Klapper, A., Goresky, M.: Feedback shift registers, 2-adic span, and combiners with memory. J. Crypt. **10**(2), 111–147 (1997)
10. Klapper, A., Xu, J.: Algebraic feedback shift registers. Theor. Comput. Sci. **226**(1), 61–92 (1999)
11. Klapper, A., Xu, J.: Register synthesis for algebraic feedback shift registers based on non-primes. Des. Codes Crypt. **31**(3), 227–250 (2004)
12. Mahler, K.: On a geometrical representation of p-adic numbers. Ann. Math. **41**(1), 8–56 (1940)
13. Massey, J.L.: Shift register synthesis and BCH decoding. IEEE Trans. Inf. Theor. **15**(1), 122–127 (1969)
14. Nguyen, P.Q., Stehlé, D.: Low-dimensional lattice basis reduction revisited. ACM Trans. Algorithms (TALG) **5**(4), 46 (2009)

On the Lattice Structure of Inversive PRNG via the Additive Order

Domingo Gómez-Pérez[✉] and Ana Gómez

University of Cantabria, Avd. Los Castros s/n, Santander, Spain
domingo.gomez@unican.es
http://personales.unican.es/gomezd

Abstract. One of the main contributions which Harald Niederreiter made to mathematics is related to pseudorandom sequences theory. In this article, we improve on a bound on one of the pseudorandom number generators (PRNGs) proposed by Harald Niederreiter and Arne Winterhof and study its lattice structure. We obtain that this generator passes general lattice tests for arbitrary lags for high dimensions.

Keywords: Lattice tests · Inversive methods · Additive order

1 Introduction

Pseudorandom numbers are used in many fields, like cryptography, financial mathematics, simulations, etc. The diversity among methods comes from the different nature of requirements, citing a famous sentence "what is appropiate for a video game is not appropiate for a nuclear reactor". Linear methods are the most popular choice for generating pseudorandom sequences and are implemented in the API of the java language. Inversive methods are popular and competitive alternatives to the linear method for generating pseudorandom numbers, see [7] and the surveys [8,9,16,17].

In this paper we analyze the lattice structure of *digital explicit inversive pseudorandom numbers* introduced in [10] and further analyzed in [6,11,12,14]. To introduce this class of generators we need some notation.

Let $q = p^r$ be a prime power and $\mathbb{F}_q$ the finite field of order q. Let

$$\overline{\gamma} = \begin{cases} \gamma^{-1}, \text{ if } \gamma \in \mathbb{F}_q^*, \\ 0, \quad \text{ if } \gamma = 0. \end{cases}$$

We order the elements of $\mathbb{F}_q = \{\xi_0, \xi_1, \ldots, \xi_{q-1}\}$ using an ordered basis $\{\gamma_1, \ldots, \gamma_r\}$ of $\mathbb{F}_q$ over $\mathbb{F}_p$ for $0 \leq n < q$,

$$\xi_n = n_1\gamma_1 + n_2\gamma_2 + \cdots + n_r\gamma_r,$$

Dedicated to Harald Niederreiter on the occasion of his 70th birthday.

© Springer International Publishing Switzerland 2014
K.-U. Schmidt and A. Winterhof (Eds.): SETA 2014, LNCS 8865, pp. 212–219, 2014.
DOI: 10.1007/978-3-319-12325-7_18

if

$$n = n_1 + n_2 p + \cdots + n_r p^{r-1}, \quad 0 \leq n_i < p, \quad i = 1, \ldots, r.$$

For $n \geq 0$ we define $\xi_{n+q} = \xi_n$. Then the *digital explicit inversive pseudorandom number generator* of period q is defined by

$$\rho_n = \overline{\alpha \xi_n + \beta}, \quad n = 0, 1, \ldots$$

for some $\alpha, \beta \in \mathbb{F}_q$ with $\alpha \neq 0$. Digital explicit inversive pseudorandom number generators are used for generating low discrepancy sequences. If

$$\rho_n = c_{n,1} \gamma_1 + c_{n,2} \gamma_2 + \cdots + c_{n,r} \gamma_r$$

with all $c_{n,i} \in \mathbb{F}_p$, we derive *digital explicit inversive pseudorandom numbers of period q* in the interval $[0, 1)$ by defining

$$y_n = \sum_{j=1}^{r} c_{n,j} p^{-j}, \quad n = 0, 1, \ldots .$$

Bounds on the discrepancy of points generated from these sequences appear in [10] and in [1,2]. Also, inversive methods were considered by Hu and Gong in [5] where it was proven a bound on the autocorrelation of this family of sequences.

Our goal in this paper is to study the behaviour of the digital explicit inversive pseudorandom number generator under a generalized test introduced in [13]. For the convenience of the reader, we give here a brief description of this test.

For given integers $L \geq 1, 0 < d_1 < \cdots < d_{L-1} < T$ and (s_n) a sequence of elements in $\mathbb{F}_q$, (s_n) passes the L-dimensional *N-Lattice Test* with lags $d_1, \ldots, d_{L-1}$ if the vectors

$$\{ s_n - s_0 \; : \; s_n = (s_n, s_{n+d_1}, \ldots, s_{n+d_{L-1}}), \quad \text{for } 0 \leq n < N \},$$

span $\mathbb{F}_q^L$. The greatest dimension L such that (s_n) passes the L-dimensional N-lattice test for all lags $d_1, \ldots, d_{L-1}$ is denoted by $T((s_n), N)$.

The authors in [14] studied the lattice test for digital explicit inversive generators and they obtained bounds on $T((\rho_n), N)$, even in parts of the sequence. We cite here part of their main result.

Lemma 1 (Theorem 1 and 2 in [14]). *Let (ρ_n) be a sequence arising from a digital explicit inverse pseudorandom number generator defined over $\mathbb{F}_q$ with $q = p^r$, then we have that,*

$$T((\rho_n), N) \geq \frac{\log N - \log \log N - 1}{r - 1} - 1,$$

for $2 \leq N < q$ if $r > 1$. For $r = 1$ the inequality

$$T((\rho_n), N) \geq \frac{N}{2} - 1,$$

holds for $2 \leq N < q$.

We want to stress the different nature of both results. For $r = 1$, the bound is linear in N whereas only a logarithmic lower bound is given for $r > 1$. Indeed, the bound for $r > 2$ can be obtained when $N = q$ for any sequence (s_n) of period q with sufficiently high linear complexity, see [4].

Here, we show that this bound can be improved using hyperplane arrangements.

2   Hyperplane Arrangements

Hyperplane arrangements is a concept well studied in the field of combinatorial geometry, see [3]. We only introduce enough theory to understand the proof of the main result and follow the nice introduction given in [15].

Let d be a positive integer and $\mathbb{R}$ the field of real numbers. We denote by

$$\boldsymbol{a} = (a_1, \ldots, a_d), \quad a_1, \ldots, a_d \in \mathbb{R}$$

elements of $\mathbb{R}^d$, where $\mathbb{R}^d$ is a vector space of dimension d over the field $\mathbb{R}$. We also consider matrices with the usual operations involving matrices, namely multiplication, addition and transposition. Also, it is needed the topological concept of dimension of a set of points in $\mathbb{R}^d$. Vectors in $\mathbb{R}^d$ are matrices with d rows and 1 column. The notation for the transposition of a matrix $\mathbf{A}$ is $\mathbf{A}^{\mathbf{T}}$.

Definition 1. *Given $\boldsymbol{a} \in \mathbb{R} - \{0\}$ and $b \in \mathbb{R}$, the set $\{\boldsymbol{x} \in \mathbb{R}^d : \boldsymbol{a}^T \boldsymbol{x} = b\}$ is called a hyperplane.*

We also use $\boldsymbol{a} \cdot \boldsymbol{x}$ to denote $\boldsymbol{a}^T \boldsymbol{x}$, which correspond to the standard dot product, and the matrix form $\mathbf{A}\boldsymbol{x} = \boldsymbol{b}$ to encode the finite set of hyperplanes $\mathcal{H} = \{\mathcal{H}_1, \ldots, \mathcal{H}_m\}$, where

$$\mathcal{H}_i = \{\boldsymbol{x} \in \mathbb{R}^d : \sum_{j=1}^{d} a_{i,j} x_j = b_i\}. \tag{1}$$

Definition 2. *A set of hyperplanes in $\mathbb{R}^d$ partitions the space into relatively open convex polyhedral regions, called faces, of all dimensions. This partition is called a hyperplane arrangement.*

We make a distinction between the two sides of a hyperplane. A point $\boldsymbol{p} \in \mathbb{R}^d$ is on the positive side of hyperplane $\mathcal{H}_i$, denoted by $\mathcal{H}_i^+$, if

$$\sum_{j=1}^{d} a_{i,j} p_j > b_i.$$

Similarly, we define $\boldsymbol{p} \in \mathbb{R}^d$ is on the negative side of hyperplane $\mathcal{H}_i$ and we denote it by $\mathcal{H}_i^-$.

For each point $p \in \mathbb{R}^d$ we define a sign vector of length m consisting of $1, 0, -1$ signs as follows:

$$sv(p)_i = \begin{cases} 1 & \text{if } p \in \mathcal{H}_i^+, \\ -1 & \text{if } p \in \mathcal{H}_i^-, \\ 0 & \text{if } p \in \mathcal{H}_i, \end{cases}$$

where $i = 1, \ldots, m$ and m is the number of hyperplanes.

Definition 3. *A face is a set of points with the same sign vector. It is called a i-face if its dimension is $i \leq d$ and a cell if the dimension is d.*

As a small comment, the dimension of a face is at least d minus the number of ceros in the sign vector of any of the points of the face. The number of faces of given dimension in a hyperplane arrangement is given in the following result

Lemma 2 (Theorem 1.3 in [3]). *Given a set of hyperplanes $\mathcal{H} = \{H_1, \ldots, H_m\}$ in $\mathbb{R}^d$, then the number of i-faces in the correspondent hyperplane arrangement can be bounded by,*

$$\sum_{j=0}^{i} \binom{d-j}{i-j} \binom{m}{d-j}.$$

3   Main Result

Now, we have all the technical tools to prove the main result. The proof is a minor modification of the one in [14, Theorem 1] and the only difference is the estimate for the number of possible carries. Nevertheless, for the sakeness of completeness, we include it here without claiming any priority over it.

Theorem 1. *For the sequence of elements (ρ_n) defined by an inversive pseudo-random number generator of period $q = p^r$, we have*

$$6T((\rho_n), N) \geq \left(\frac{N}{(r)^{r-1}}\right)^{1/r},$$

for $2 \leq N \leq q$.

Proof. The case $r = 1$ is stated in Lemma 1 so assume that $r \geq 2$ and the sequence (ρ_n) does not pass the L-dimensional N-lattice test for some lags $0 < d_1 < d_2 < \cdots < d_{L-1} < q$. Put

$$\boldsymbol{\rho}_n = (\rho_n, \rho_{n+d_1}, \ldots, \rho_{n+d_{L-1}}), \quad n \geq 0,$$

and let V be the subspace of $\mathbb{F}_q^L$ spanned by all $\boldsymbol{\rho}_n - \boldsymbol{\rho}_0$ for $0 \leq n < N$. Consider the orthogonal space of V, i.e. $\{u : u \cdot v = 0, \forall v \in V\}$, whose dimension is different from 0. So, there exits $\alpha \neq 0$ such that,

$$\boldsymbol{\rho}_n \cdot \boldsymbol{\alpha} = \boldsymbol{\rho}_0 \cdot \boldsymbol{\alpha}, \quad \text{for } 0 \leq n < N.$$

Calling $\delta = \rho_0 \cdot \boldsymbol{\alpha}$ and j the smallest index with $\alpha_j \neq 0$ we have[1]

$$\alpha_j \rho_{n+d_j} + \alpha_{j+1} \rho_{n+d_{j+1}} + \cdots + \alpha_{L-1} \rho_{n+d_{L-1}} = \delta, \quad \text{for } 0 \leq n < N. \quad (2)$$

For all $1 \leq i < L$ and $0 \leq d_i, n < q$, let

$$d_i = \sum_{j=1}^{r} d_{i,j} p^{j-1}, \quad 0 \leq d_{i,1}, \dots, d_{i,r} < p,$$

and

$$n = \sum_{j=1}^{r} n_j p^{j-1}, \quad 0 \leq n_1, \dots, n_r < p,$$

be the p-adic expansions of d_i and n, respectively. We now define the vectors of the carries that occur in the additions of $n + d_i$. Let $w_{i,1} = 0$ and define for $1 \leq h \leq r$ recursively

$$w_{i,h+1} = \begin{cases} 1, & \text{if } d_{i,h} + n_h + w_{i,h} \geq p, \\ 0, & \text{otherwise.} \end{cases}$$

Then we have

$$n + d_i = \sum_{j=1}^{r} z_{i,j} p^{j-1}, \quad 0 \leq z_{i,1}, \dots, z_{i,r} < p,$$

with

$$z_{i,j} = d_{i,j} + n_j + w_{i,j} - w_{i,j+1} p, \quad 1 \leq j \leq r,$$

and

$$\xi_{n+d_i} = \xi_n + \xi_{d_i} + w_i, \quad \text{where } w_i = \sum_{j=1}^{r} w_{i,j} \gamma_j.$$

Previously only trivial estimates were used to count the number of possible choices for $w_j, \dots, w_{L-1}$. Now, we are going to use hyperplane arrangements to bound this number. Consider the following sets of hyperplanes in $\mathbb{R}^r$,

$$\{H_{i,j}^1 : 1 \leq i \leq L, \ 1 \leq j \leq r\} \cup \{H_{i,j}^2 : 1 \leq i \leq L, \ 1 \leq j \leq r\},$$

where

$$H_{i,j}^1 = \{\boldsymbol{x} \in \mathbb{R}^r : x_j + d_{i,j} = p - 0.1\}, \quad H_{i,j}^2 = \{\boldsymbol{x} \in \mathbb{R}^r : x_j + d_{i,j} = p - 1.1\}.$$

It is easy to encode the union of these two sets of hyperplanes by $\mathbf{A}\boldsymbol{x} = \boldsymbol{b}$ as in Eq. (1). Matrix $\mathbf{A}$ is a matrix with $2Lr$ rows and r columns that it is constructed

---

[1] If $j = 0$, we will denote $d_0 = 0$, although the lags are $d_1, \dots, d_{L-1}$.

by stacking $2L$ identity matrices of dimension r. The first L components of vector b are just joining the following L vectors,

$$(p - 0.1, \ldots, p - 0.1), (p - d_{1,1} - 0.1, \ldots, p - d_{1,r} - 0.1),$$
$$\ldots, (p - d_{L-1,1} - 0.1, \ldots, p - d_{L-1,r} - 0.1),$$

and the next L components are,

$$(p - 1.1, \ldots, p - 1.1), (p - d_{1,1} - 1.1, \ldots, p - d_{1,r} - 1.1),$$
$$\ldots, (p - d_{L-1,1} - 1.1, \ldots, p - d_{L-1,r} - 1.1).$$

Using the previous notation, it is trivial that if n, n' are two different integers satisfying

$$\xi_{n+d_i} = \xi_n + \xi_{d_i} + w_i, \quad \xi_{n'+d_i} = \xi_{n'} + \xi_{d_j} + w_i',$$

with $w_i \neq w_i'$ for some $i \in 1, \ldots, r$, then the sign vectors of the points $(n_1, \ldots, n_r)$, $(n_1', \ldots, n_r') \in \mathbb{R}^r$ are different, where

$$n = \sum_{j=1}^{r} n_j p^{j-1}, \quad n' = \sum_{j=1}^{r} n_j' p^{j-1}, \quad 0 \leq n_1, \ldots, n_k, n_1', \ldots, n_k' < p.$$

The reason is the following, if $sv((n_1, \ldots, n_r)) = sv((n_1', \ldots, n_r'))$, then both points must be in the same side of the hyperplanes $H_{i,1}^1, H_{i,1}^2$ for $i = 1, \ldots, L$, which is equivalent to,

$$d_{i,1} + n_1 > p - 0.1 \iff d_{i,1} + n_1' > p - 0.1, \implies w_{i,2} = w_{i,2}'.$$

In general, $w_{i,h} = w_{i,h}'$ because

- $w_{i,h} = w_{i,h}' = 1$ and the points lie in the same side of $H_{i+L,h+1}^2$.
- $w_{i,h} = w_{i,h}' = 0$ and the points lie in the same side of $H_{i,h+1}^1$.

We are only interested in the faces of dimension greater or equal than $r - 1$[2] Using Lemma 2, we get that the number of $(r - 1)$-faces plus the number of r-faces is less than

$$(r + 1) \sum_{j=0}^{r-1} \binom{2rL}{r - j} \leq (6rL)^{r-1}.$$

So there exists a vector $(w_j, \ldots, w_{r-1})$ such that for at least

$$\frac{N}{(6rL)^{r-1}},$$

---

[2] Because we always consider $w_{i,1} = 0$. It is also equivalent to discard x_1, i. e. working in $\mathbb{R}^{r-1}$.

different n with $0 \le n < N$ we have $\xi_{n+d_i} = \xi_n + \xi_{d_i} + w_i$, $j \le i < r$. We have $\rho_{n+d_i} = 0$ for some value $1 \le i < r$ for at most $r - j$ different n. If $\rho_{n+d} \ne 0$ then we can write $\rho_{n+d} = \overline{\alpha \xi_{n+d} + \beta}$. By Eq. (2), we have

$$\alpha_j \overline{\alpha \xi_n + \xi_{d_j} + w_j + \beta} + \ldots + \alpha_{L-1} \overline{\alpha \xi_n + \xi_{d_{L-1}} + w_{L-1} + \beta} = \delta,$$

for at least $N/(6rL)^{r-1} - L$ different elements ξ_n. Operating and using Lagrange theorem, the number of solutions of the previous equation is less than L, so $2L \ge N/(6rL)^{r-1}$ or, $6L \ge \left(\frac{N}{(r)^{r-1}} \right)^{1/r}$ and this finishes the proof.

Final Comments

No effort has been put in getting the best possible constant in the theorem. The reason is to avoid technical details as much as possible and focus on hyperplane arrangements. The new idea in this paper is using hyperplane arrangements, which seems to be new to study sequences via additive order. We think that this could lead to improvements to study distribution of sequences via additive order. However, new ideas are needed to be added. For example, hyperplane arrangements applied to the results in [2], give better constants in the results but not significant improvements. Also, the result in this paper applies only when p is sufficiently large. It would certainly be very interesting to see how to apply this technique for $p = 2$.

Acknowledgement. This work is supported in part by the Spanish Ministry of Science, project MTM2011-24678.

References

1. Chen, Z.: Finite binary sequences constructed by explicit inversive methods. Finite Fields Appl. **14**(3), 579–592 (2008)
2. Chen, Z., Gomez, D., Winterhof, A.: Distribution of digital explicit inversive pseudorandom numbers and their binary threshold sequence. In: Carlo, M., Carlo, Q.-M. (eds.) Methods 2008, pp. 249–258. Springer, Heidelberg (2009)
3. Edelsbrunner, H.: Algorithms in Combinatorial Geometry, vol. 10. Springer, New York (1987)
4. Gomez-Perez, D., Gutierrez, J.: On the linear complexity and lattice test of nonlinear pseudorandom number generators (2013)
5. Hu, H., Gong, G.: A study on the pseudorandom properties of sequences generated via the additive order. In: Golomb, S.W., Parker, M.G., Pott, A., Winterhof, A. (eds.) SETA 2008. LNCS, vol. 5203, pp. 51–59. Springer, Heidelberg (2008)
6. Meidl, W., Winterhof, A.: On the linear complexity profile of explicit nonlinear pseudorandom numbers. Inf. Process. Lett. **85**(1), 13–18 (2003)
7. Niederreiter, H.: Random Number Generation and Quasi-Monte Carlo methods, vol. 63. SIAM, Philadelphia (1992)
8. Niederreiter, H., Shparlinski, I.E.: Recent advances in the theory of nonlinear pseudorandom number generators. In: Carlo, M., Carlo, Q.-M. (eds.) Methods 2000, pp. 86–102. Springer, Heidelberg (2002)

9. Niederreiter, H., Shparlinski, I.E.: Dynamical systems generated by rational functions. In: Fossorier, M.P.C., Høholdt, T., Poli, A. (eds.) AAECC 2003. LNCS, vol. 2643, pp. 6–17. Springer, Heidelberg (2003)
10. Niederreiter, H., Winterhof, A.: Incomplete exponential sums over finite fields and their applications to new inversive pseudorandom number generators. Acta Arithmetica **93**(4), 387–399 (2001)
11. Niederreiter, H., Winterhof, A.: On a new class of inversive pseudorandom numbers for parallelized simulation methods. Periodica Mathematica Hungarica **42**(1), 77–87 (2001)
12. Niederreiter, H., Winterhof, A.: On the lattice structure of pseudorandom numbers generated over arbitrary finite fields. Appl. Algebra Eng. Commun. Comput. **12**(3), 265–272 (2001)
13. Niederreiter, H., Winterhof, A.: On the structure of inversive pseudorandom number generators. In: Boztaş, S., Lu, H.-F.F. (eds.) AAECC 2007. LNCS, vol. 4851, pp. 208–216. Springer, Heidelberg (2007)
14. Pirsic, G., Winterhof, A.: On the structure of digital explicit nonlinear and inversive pseudorandom number generators. J. Complex. **26**(1), 43–50 (2010)
15. Sleumer, N.H.: Hyperplane arrangements: Construction, visualization and applications. Master's thesis, Swiss Federal Institute of Technology (2000)
16. Topuzoğlu, A., Winterhof, A.: Pseudorandom sequences. In: Topics in Geometry. Coding Theory and Cryptography, vol. 6, pp. 135–166. Springer, Dordrecht (2007)
17. Winterhof, A.: Recent results on recursive nonlinear pseudorandom number generators. In: Carlet, C., Pott, A. (eds.) SETA 2010. LNCS, vol. 6338, pp. 113–124. Springer, Heidelberg (2010)

Weaknesses in the Initialisation Process of the Common Scrambling Algorithm Stream Cipher

Harry Bartlett[1]([⊠]), Ali Alhamdan[1,2], Leonie Simpson[1],
Ed Dawson[1], and Kenneth Koon-Ho Wong[1]

[1] Institute for Future Environments and Science and Engineering Faculty,
Queensland University of Technology, Brisbane, Australia
{h.bartlett,lr.simpson,e.dawson,kk.wong}@qut.edu.au
[2] National Information Center, Riyadh, Saudi Arabia
alhamdan@nic.gov.sa

Abstract. The Common Scrambling Algorithm Stream Cipher (CSA-SC) is a shift register based stream cipher designed to encrypt digital video broadcast. CSA-SC produces a pseudo-random binary sequence that is used to mask the contents of the transmission. In this paper, we analyse the initialisation process of the CSA-SC keystream generator and demonstrate weaknesses which lead to state convergence, slid pairs and shifted keystreams. As a result, the cipher may be vulnerable to distinguishing attacks, time-memory-data trade-off attacks or slide attacks.

Keywords: Common scrambling algorithm · Stream cipher · Initialisation · State convergence · Slid pairs · Shifted keystream

1 Introduction

European digital television signals are encrypted using the Digital Video Broadcasting Common Scrambling Algorithm (CSA) specified by the European Telecommunications Standards Institute (ETSI) [12]. CSA consists of a cascade of block and stream ciphers. To encrypt, the block cipher is applied first, followed by the stream cipher. To decrypt, the stream cipher is applied first, followed by the block cipher. Both block and stream ciphers use the same 64-bit key, although Tews et al. [17] note that many applications use keys with only 48 bits of entropy. The stream cipher component of the Common Scrambling Algorithm is referred to as CSA-SC, as in [4]. CSA-SC consists of a keystream generator which produces a pseudo-random binary sequence. This sequence is combined with the underlying digital video stream using bitwise XOR (binary addition), so the security of CSA-SC depends solely on the keystream generator.

CSA-SC has been described in several different ways, with different internal state sizes. Bewick's patents [3,4] have an internal state of 107 bits. Weinmann and Wirt [18] reduced this to 103 bits by removing a 4-bit memory. Simpson et al. [16] model CSA-SC with an 89-bit internal state, by shifting the positions

© Springer International Publishing Switzerland 2014
K.-U. Schmidt and A. Winterhof (Eds.): SETA 2014, LNCS 8865, pp. 220–233, 2014.
DOI: 10.1007/978-3-319-12325-7_19

of the inputs to the S-boxes by one stage, to remove additional memories. These models are functionally equivalent. This paper uses the 89-bit model from [16].

Keystream generators are finite state machines and require an initial state to be set before keystream can be produced. This paper analyses the initialisation process of CSA-SC; previous analyses [16,18] relate only to the keystream generation process. Two security flaws in the initialisation process of CSA-SC are investigated, which lead to state convergence, slid pairs and the production of phase-shifted keystreams. State convergence results in two generators with different inputs producing the same output sequence. Slid pairs occur when the same internal state is obtained in two different generators at slightly different times; this can lead to keystream sequences where one sequence is simply a shifted version of the other. We conclude by discussing the security implications of these weaknesses.

2   Description of CSA-SC

CSA-SC uses word based registers (with a 4-bit word size) and bit based state update functions. There are two feedback shift registers (denoted A and B), a combiner with memory (with stages denoted E, F and c) and seven 5×2 S-boxes, as shown in Fig. 1. Registers A and B each have ten stages; each stage stores a nibble (4-bit word). The combiner memories E and F each store a nibble, and c stores only 1 bit. Figure 1 shows the structure and operation of CSA-SC during both keystream generation and initialisation. (Dashed lines apply only during initialisation.) The least significant bit or stage of each nibble is indexed by 0, $\oplus$ denotes bitwise XORs, $\boxplus$ denotes addition modulo 2^4 and $\lll i$ represents a left rotation by i bits. Note also that all S-Box outputs in this model are constrained to be zero for the first iteration of initialisation [16].

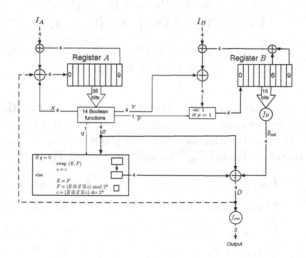

Fig. 1. Common Scrambling Algorithm Stream Cipher (CSA-SC)

Let A_i^t and B_i^t denote the contents of the ith stage of registers A and B respectively, for $i \in \{0, \ldots, 9\}$, at time t. Let $a_{i,j}^t$ and $b_{i,j}^t$ denote the content of the jth bit position of the ith stage at time t for registers A and B, respectively, for $i \in \{0, \ldots, 9\}$ and $j \in \{0, \ldots, 3\}$. Let E^t, F^t and c^t represent the contents of the memory stages E, F and c at time t, respectively. Let e_j^t and f_j^t represent the jth bit of the memories E and F at time t for $j \in \{0, \ldots, 3\}$.

CSA-SC uses a 64-bit secret key $K = k_0, \ldots, k_{63}$ and a 64-bit initial value (IV) $V = v_0, \ldots, v_{63}$, where k_i and v_i are the ith key and IV bits, respectively. During initialisation the Key and IV are spread across the entire internal state. Let I_A^t and I_B^t denote 4-bit words taken from the IV, and used at time t as input to registers A and B respectively (details are given in the following section). Let $i_{A,j}^t$ and $i_{B,j}^t$ represent the jth bits of I_A^t and I_B^t respectively for $j \in \{0, 1, 2, 3\}$.

We treat the seven 5×2 S-boxes as 14 5-input Boolean functions, each of which takes inputs from register A. Let $S_j(i_0, i_1, i_2, i_3, i_4)$ represent the jth Boolean function with respect to 5 inputs $(i_0, i_1, i_2, i_3, i_4)$ for $j \in \{1, \ldots, 14\}$. Consistent with previous references [16,18], we denote the outputs of the 14 Boolean functions as $X = (x_0^t, x_1^t, x_2^t, x_3^t)$, $Y = (y_0^t, y_1^t, y_2^t, y_3^t)$, $Z = (z_0^t, z_1^t, z_2^t, z_3^t)$, p^t and q^t. For more detail, see Tables 6 and 7 (in Appendix A).

2.1   Initialisation Process

During initialisation, the key and IV are introduced into the 89-bit internal state in two phases: *key loading* and *diffusion*. Unlike most stream ciphers, loading of the IV in CSA-SC is performed during the diffusion phase.

Key Loading Phase: All registers and memories are first set to zero and the key is transferred to specified positions in registers A and B as follows:

$$a_{i,j}^{-32} = \begin{cases} k_{4 \cdot i + j} & \text{for } i \in \{0, \ldots, 7\}, \ j \in \{0, \ldots, 3\} \\ 0 & \text{for } i \in \{8, 9\}, \ j \in \{0, \ldots, 3\} \end{cases}$$

$$b_{i,j}^{-32} = \begin{cases} k_{32 + 4 \cdot i + j} & \text{for } i \in \{0, \ldots, 7\}, \ j \in \{0, \ldots, 3\} \\ 0 & \text{for } i \in \{8, 9\}, \ j \in \{0, \ldots, 3\} \end{cases}$$

Diffusion Phase: After loading the key, 32 iterations of the initialisation state update function are performed (starting at $t = -31$) [18] while simultaneously loading the IV. At each iteration two different nibbles of the IV, denoted I_A^t and I_B^t, are input to stages A_0^t and B_0^t. The nibbles I_A^t and I_B^t are defined as:

$$I_A^t = \begin{cases} v_x, v_{x+1}, v_{x+2}, v_{x+3} & \text{for } t \in \{-31, -29, \ldots\} \\ v_{x+4}, v_{x+5}, v_{x+6}, v_{x+7} & \text{for } t \in \{-30, -28, \ldots\} \end{cases}$$

$$I_B^t = \begin{cases} v_{x+4}, v_{x+5}, v_{x+6}, v_{x+7} & \text{for } t \in \{-31, -29, \ldots\} \\ v_x, v_{x+1}, v_{x+2}, v_{x+3} & \text{for } t \in \{-30, -28, \ldots\} \end{cases}$$

$$\text{for } x = 8 \times [7 + \lceil t/4 \rceil]$$

Thus, every byte of the IV is used in four consecutive iterations of the state update function, with the high and low nibbles of the byte being input alternately to each of A_0 and B_0.

The state update functions for shift registers A and B during the initialisation process (shown in Fig. 1, including the dashed lines) are as follows:

$$A_i^t = A_{i-1}^{t-1} \qquad \text{for } i \in \{1, \dots, 9\} \tag{1a}$$

$$A_0^t = A_9^{t-1} \oplus X^{t-1} \oplus D^{t-1} \oplus I_A^t \tag{1b}$$

$$B_i^t = B_{i-1}^{t-1} \qquad \text{for } i \in \{1, \dots, 9\} \tag{2a}$$

$$B_0^t = (B_6^{t-1} \oplus B_9^{t-1} \oplus Y^{t-1} \oplus I_B^t) \lll p^{t-1} \tag{2b}$$

The state update functions for memories E, F and c are:

$$E^t = F^{t-1} \tag{3a}$$

$$F^t = \begin{cases} E^{t-1} & \text{if } q^{t-1} = 0 \\ E^{t-1} \boxplus Z^{t-1} \boxplus c^{t-1} \bmod 2^4 & \text{if } q^{t-1} = 1 \end{cases} \tag{3b}$$

$$c^t = \begin{cases} c^{t-1} & \text{if } q^{t-1} = 0 \\ (E^{t-1} \boxplus Z^{t-1} \boxplus c^{t-1}) \text{ div } 2^4 & \text{if } q^{t-1} = 1 \end{cases} \tag{3c}$$

Selected bits from register B are XORed with E^t and Z^t to give a 4-bit word denoted D^t:

$$D^t = E^t \oplus Z^t \oplus B_{out}^t \tag{4}$$

where
$$\begin{aligned} B_{out}^t &= (b_{0,out}^t \parallel b_{1,out}^t \parallel b_{2,out}^t \parallel b_{3,out}^t) \\ &= (b_{8,2}^t \oplus b_{5,3}^t \oplus b_{2,1}^t \oplus b_{7,0}^t \parallel b_{4,3}^t \oplus b_{7,2}^t \oplus b_{3,0}^t \oplus b_{4,1}^t \parallel \\ &\quad b_{5,0}^t \oplus b_{7,1}^t \oplus b_{2,3}^t \oplus b_{3,2}^t \parallel b_{2,0}^t \oplus b_{5,1}^t \oplus b_{6,2}^t \oplus b_{8,3}^t) \end{aligned}$$

and $\parallel$ denotes concatenation. During initialisation, D is used in updating A, as per Eq. 1b.

At the end of the initialisation process ($t = 0$), the cipher is in its *initial state* and keystream generation can begin.

2.2 Keystream Generation

During this phase, there is no feedback from pre-output word D to register A and no IV input. The state update functions for A and B are simply:

$$A_0^t = A_9^{t-1} \oplus X^{t-1}$$

$$B_0^t = (B_6^{t-1} \oplus B_9^{t-1} \oplus Y^{t-1}) \lll p^{t-1}$$

At time t, the keystream generator generates a 2-bit output word z^t from D^t by combining bits as follows: $z^t = ((d_2^t \oplus d_3^t)||(d_0^t \oplus d_1^t))$.

3   State Convergence in CSA-SC

Before considering state convergence in CSA-SC, we note a related issue. CSA-SC uses a 64-bit secret key and a 64-bit IV to form an 89-bit internal state. Thus, the 2^{128} possible Key-IV combinations are mapped to at most 2^{89} internal states, so clearly multiple Key-IV pairs map to the same internal state. On average, each internal state corresponds to 2^{39} Key-IV pairs.

State convergence occurs when two or more distinct states at time t are mapped to the same state at time $t + 1$; that is, when the state update function is not one-to-one. As the state size is fixed, this implies that some states at time $t + 1$ must have no pre-image, that is, these states cannot occur at time $t + 1$. Thus state convergence reduces the effective state size, which may leave the stream cipher vulnerable to attacks such as distinguishing attacks [15] or time-memory-data trade-off attacks [5].

State convergence does not occur in CSA-SC during keystream generation, as the state update function is one-to-one during this process; however, it does occur during initialisation (specifically, in the diffusion phase). To show this, we assume that the state contents are known at time t and consider possible pre-image states at time $t - 1$. Most of the state contents at time $t - 1$ are directly transferred to known locations at time t; these cannot take multiple values at time $t - 1$. The exceptions are A_9^{t-1}, B_9^{t-1} and, when $q^{t-1} = 1$, E^{t-1} and c^{t-1}; we consider possible contents for these locations.

We present below the analysis for the case $q^{t-1} = 0$. (As the Boolean function S_{14} is balanced, this case applies to exactly half of the possible internal states.) State convergence can also occur in CSA-SC when $q^{t-1} = 1$, as demonstrated by the example in Table 1, but we omit details of that analysis.

Table 1. Two pre-images for a given state when $q^{t-1} = 1$

	A	B	E	F	c
State	0010010101010111000100001101110101010001	1001111001011110010101101110110101110011	0010	0011	1
1st pre-image	0101010101110001000011011101010100010111	1110010111100101011011101101011100111000	0101	0010	0
2nd pre-image	0101010101110001000011011101010100010100	1110010111100101011011101101011100111100	0101	0010	1

3.1   Bit-Level Analysis ($q^{t-1} = 0$)

Assume the state contents at time t are known and consider the possible contents of A_9^{t-1}. The contents of A_9^{t-1} affect the values of A_0^t and B_0^t, but are not in the state at time t. We show below that multiple distinct values for A_9^{t-1} may result in the same value for each of A_0^t and B_0^t.

After substituting from Eq. 4, Eqs. 1b and 2b can be re-written to give the following bit-level equations for A_9^{t-1} and B_9^{t-1}:

$$a_{9,0}^{t-1} = C_0 \oplus a_{0,0}^t \oplus S_9(a_{2,1}^{t-1} \ldots a_{9,1}^{t-1}) \tag{6a}$$

$$a_{9,1}^{t-1} = C_1 \oplus a_{0,1}^t \oplus S_{10}(a_{4,0}^{t-1} \ldots a_{9,0}^{t-1}) \tag{6b}$$

$$a_{9,2}^{t-1} = C_2 \oplus a_{0,2}^t \oplus S_3(a_{2,1}^{t-1} \ldots a_{9,1}^{t-1}) \oplus S_{11}(a_{3,1}^{t-1} \ldots a_{9,3}^{t-1}) \tag{6c}$$

$$a_{9,3}^{t-1} = C_3 \oplus a_{0,3}^t \oplus S_4(a_{4,0}^{t-1} \ldots a_{9,0}^{t-1}) \oplus S_{12}(a_{5,2}^{t-1} \ldots a_{9,2}^{t-1}) \tag{6d}$$

$$b_{9,0}^{t-1} = C_4 \oplus b_0'^t \oplus S_5(a_{3,1}^{t-1} \ldots a_{9,3}^{t-1}) \tag{7a}$$

$$b_{9,1}^{t-1} = C_5 \oplus b_1'^t \oplus S_6(a_{5,2}^{t-1} \ldots a_{9,2}^{t-1}) \tag{7b}$$

$$b_{9,2}^{t-1} = C_6 \oplus b_2'^t \tag{7c}$$

$$b_{9,3}^{t-1} = C_7 \oplus b_3'^t \tag{7d}$$

In these equations, C_0 to C_7 represent terms that are unaffected by the values of $a_{9,0}^t$ to $a_{9,3}^t$ and $b_j'^t = b_{0,j}^t(1 \oplus p^{t-1}) \oplus b_{0,(j-1)\bmod 4}^t(p^{t-1})$ for $j \in \{0,1,2,3\}$. More specifically,

$$C_0 = S_1(a_{3,3}^{t-1} \ldots a_{8,0}^{t-1}) \oplus e_0^{t-1} \oplus b_{0,out}^{t-1} \oplus i_{A,0}^t$$

$$C_1 = S_2(a_{1,3}^{t-1} \ldots a_{6,2}^{t-1}) \oplus e_1^{t-1} \oplus b_{1,out}^{t-1} \oplus i_{A,1}^t$$

$$C_2 = e_2^{t-1} \oplus b_{2,out}^{t-1} \oplus i_{A,2}^t$$

$$C_3 = e_3^{t-1} \oplus b_{3,out}^{t-1} \oplus i_{A,3}^t$$

$$C_4 = b_{6,0}^{t-1} \oplus i_{B,0}^t$$

$$C_5 = b_{6,1}^{t-1} \oplus i_{B,1}^t$$

$$C_6 = b_{6,2}^{t-1} \oplus S_7(a_{3,3}^{t-1} \ldots a_{8,0}^{t-1}) \oplus i_{B,2}^t$$

$$C_7 = b_{6,3}^{t-1} \oplus S_8(a_{1,3}^{t-1} \ldots a_{6,2}^{t-1}) \oplus i_{B,3}^t$$

Note that $a_{9,0}^{t-1}$ to $a_{9,3}^{t-1}$ appear implicitly on the right hand sides of Eqs. 6 and 7 and that $b_{9,0}^{t-1}$ to $b_{9,3}^{t-1}$ are determined uniquely once $a_{9,2}^{t-1}$ and $a_{9,3}^{t-1}$ are known. That is, every distinct solution for A_9^{t-1} from Eqs. 6a to 6d yields a unique corresponding solution for B_9^{t-1}.

We now examine Eqs. 6a–6d to identify conditions for state convergence in CSA-SC. Let $\bar{x}$ denote the complement of x. For any of the Boolean functions $S_3, S_4, S_9, S_{10}, S_{11}$ and S_{12}, the function is said to have *even parity* with respect to i_4 if $S(i_0, i_1, i_2, i_3, \bar{i_4}) = S(i_0, i_1, i_2, i_3, i_4)$, and to have *odd parity* with respect to i_4 if $S(i_0, i_1, i_2, i_3, \bar{i_4}) = \overline{S(i_0, i_1, i_2, i_3, i_4)}$. For example, $S_9(0,1,1,0,0) = S_9(0,1,1,0,1) = 0$, so S_9 has even parity with respect to i_4 when $(i_0, i_1, i_2, i_3) = (0,1,1,0)$. For brevity, we refer to these parity conditions by saying that S is even or odd. We fix the values of $a_{0,0}^t$ to $a_{0,3}^t$ in Eqs. 6a–6d and investigate possible solutions $a_{9,0}^{t-1}$ to $a_{9,3}^{t-1}$.

If S_9 is even, $a_{9,0}^{t-1}$ is determined uniquely by Eq. 6a; Eq. 6b then determines a unique value for $a_{9,1}^{t-1}$. Similarly, if S_{10} is even, $a_{9,1}^{t-1}$ is determined uniquely and leads to a unique solution for $a_{9,0}^{t-1}$. Thus $a_{9,0}^{t-1}$ and $a_{9,1}^{t-1}$ are both uniquely determined if either S_9 or S_{10} is even. For example, if $a_{0,0}^t = a_{0,1}^t = C_0 = C_1 = 0$ and the inputs $(i_0, i_1, i_2, i_3) = (0, 1, 1, 0)$ for both S_9 and S_{10}, then S_9 is even, $a_{9,0}^{t-1} = 0$ from Eq. 6a and $a_{9,1}^{t-1} = 1$ from Eq. 6b.

If S_9 and S_{10} are both odd, the solutions of Eqs. 6a and 6b depend on each other and the equations may be either inconsistent (no valid solution) or consistent (two valid solutions for $a_{9,0}^{t-1}$ and $a_{9,1}^{t-1}$). For example, if $a_{0,0}^t = a_{0,1}^t = C_0 = C_1 = 0$ but $(i_0, i_1, i_2, i_3) = (0, 1, 1, 1)$ for both S_9 and S_{10}, then Eq. 6a gives $a_{9,0}^{t-1} = S_9(0, 1, 1, 1, a_{9,1}^{t-1}) = \overline{a_{9,1}^{t-1}}$ but Eq. 6b gives $a_{9,1}^{t-1} = S_{10}(0, 1, 1, 1, a_{9,0}^{t-1}) = a_{9,0}^{t-1}$ and these equations have no consistent solution for $a_{9,0}^{t-1}$ and $a_{9,1}^{t-1}$. Conversely, if $a_{0,0}^t = a_{0,1}^t = C_0 = C_1 = 0$ and $(i_0, i_1, i_2, i_3) = (1, 0, 1, 0)$ for both S_9 and S_{10}, then Eq. 6a gives $a_{9,0}^{t-1} = S_9(1, 0, 1, 0, a_{9,1}^{t-1}) = a_{9,1}^{t-1}$, while Eq. 6b gives $a_{9,1}^{t-1} = S_{10}(1, 0, 1, 0, a_{9,0}^{t-1}) = a_{9,0}^{t-1}$, with the two solutions $a_{9,0}^{t-1} = a_{9,1}^{t-1} = 0$ and $a_{9,0}^{t-1} = a_{9,1}^{t-1} = 1$.

Once $a_{9,0}^{t-1}$ and $a_{9,1}^{t-1}$ have been determined, we consider Eqs. 6c and 6d for $a_{9,2}^{t-1}$ and $a_{9,3}^{t-1}$. For a particular solution to Eqs. 6a and 6b, the values of $a_{9,0}^{t-1}$ and $a_{9,1}^{t-1}$ determine the values of $S_3(\ldots, a_{9,1}^{t-1})$ and $S_4(\ldots, a_{9,0}^{t-1})$. These values can be included in the constants C_2 and C_3; a similar argument and analysis then shows that:

- If either S_{11} or S_{12} is even, the values of $a_{9,2}^{t-1}$ and $a_{9,3}^{t-1}$ are determined uniquely from the related equations.
- If S_{11} and S_{12} are both odd, there is either no solution for $a_{9,2}^{t-1}$ and $a_{9,3}^{t-1}$ or there are two complementary solution pairs for these variables.

Number of Solutions: Combining the above arguments, there are zero to four solutions to Eqs. 6a–6d. In particular,

- If there is no solution for $a_{9,0}^{t-1}$ and $a_{9,1}^{t-1}$, it is not possible to obtain a valid solution set for $a_{9,0}^{t-1}$ to $a_{9,3}^{t-1}$; that is, no solution exists for A_9^{t-1}.
- If there is a unique solution for $a_{9,0}^{t-1}$ and $a_{9,1}^{t-1}$, there are either no, one or two solutions for A_9^{t-1} exactly when $a_{9,2}^{t-1}$ and $a_{9,3}^{t-1}$ have no, one or two solutions.
- When there are two valid solution pairs for $a_{9,0}^{t-1}$ and $a_{9,1}^{t-1}$, then
  (a) if either S_{11} or S_{12} is even, each of these pairs has a unique solution for $a_{9,2}^{t-1}$ and $a_{9,3}^{t-1}$, so there are exactly two solutions for A_9^{t-1};
  (b) if S_{11} and S_{12} are both odd, the equations for $a_{9,2}^{t-1}$ and $a_{9,3}^{t-1}$ have either zero or two solutions for each solution pair $(a_{9,0}^{t-1}, a_{9,1}^{t-1})$. In fact,
    • If S_3 and S_4 are both even or both odd, there are either zero or four solution sets for A_9^{t-1}.
    • If one of S_3 and S_4 is even and the other is odd, there are exactly two solution sets for A_9^{t-1}.

Based on this analysis, it is possible to find states with multiple pre-images. Tables 2 and 3 show examples of states with two and four pre-images respectively. Note that the pre-image states differ in bits $a_{9,0}$ to $a_{9,3}$ and $b_{9,0}$ to $b_{9,3}$ (shown in bold). The probability that a randomly chosen state has a given number of pre-images is discussed in Sect. 3.2 and summarised in Table 4.

Table 2. Two pre-images for a given state when $q^{t-1} = 0$

	A	B	E	F	c
State	1110010101010101000100001101010101010001	00101110010111100101011011101101010101011	0010	1100	1
1st pre-image	0101010101010001000011010101010100010100	11100101111001010110111011010101010110111	1100	0010	1
2nd pre-image	0101010101010001000011010101010100010111	11100101111001010110111011010101010110111	1100	0010	1

Table 3. Four pre-images for a given state when $q^{t-1} = 0$

	A	B	E	F	c
State	0100010101010001000101010101010001110101	10011110010111100101011011101101010101011	0010	0001	1
1st pre-image	0101010100010001010101010100011101010000	11100101111001010110111011010101010110100	0001	0010	1
2nd pre-image	0101010100010001010101010100011101010011	11100101111001010110111011010101010111000	0001	0010	1
3rd pre-image	0101010100010001010101010100011101011110	11100101111001010110111011010101010110000	0001	0010	1
4th pre-image	0101010100010001010101010100011101011101	11100101111001010110111011010101010111100	0001	0010	1

3.2   Probabilities for Numbers of Pre-image States

The options discussed in the analysis above can be represented as a tree diagram, as in Fig. 2 ("Cons." and "Incons." indicate consistent and inconsistent equation pairs, respectively). Based on this, we determine the distribution of the number of pre-image states at time $t - 1$ for a randomly chosen state at time t.

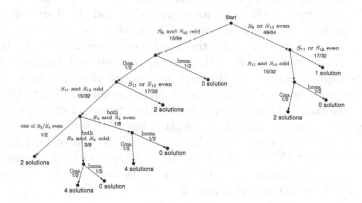

Fig. 2. Tree diagram for solution alternatives

The loading format (Sect. 2.1) dictates that A_9 contains zeros during the first two clocks of initialisation ($t = -31, -30$). Therefore, multiple valid pre-images cannot occur at these clock steps. From $t = -30$ onwards, assume a uniform distribution for the contents of register A; the probabilities of each branch in Fig. 2 can be found from the truth tables for $S_3, S_4, S_9, S_{10}, S_{11}$ and S_{12} (Table 7 in Appendix A) and by noting that the probability of consistent equations (where relevant) is $\frac{1}{2}$ (as the contents of A_9 are uniformly distributed). From the branch probabilities, we calculate the total probability of each leaf node and combine these according to the number of solutions for each node, giving the results in Table 4. Note that the number of solutions is the number of pre-images for the given state.

Table 4. Probabilities of various numbers of pre-images when $q^{t-1} = 0$

Number of pre-images	0	1	2	4
Probability	$\frac{5085}{16384}$	$\frac{833}{2048}$	$\frac{2205}{8192}$	$\frac{225}{16384}$
Probability (decimal)	0.31	0.407	0.269	0.014

3.3 Extent of Convergence

At $t = -30$, there are 2^{72} possible states, as 64 key bits and 8 IV bits have been loaded. Of these, 2^{71} states have $q^{-30} = 0$. At the next clock step, these 2^{71} states will clock to only $(1 - \frac{5085}{16384}) \cdot 2^{71} = 2^{70.46}$ distinct states, a proportional loss of $\frac{5085}{16384} \approx 0.31$. The 2^{71} states with $q^{-30} = 1$ will also clock to a smaller number of distinct states, but we have not yet determined the size of this effect. Ignoring the latter effect for the moment, the proportion of states lost overall is at least $\frac{5085}{32768} \approx 0.155$.

At $t = -30$, stages A_2 to A_9 of register A contain key bits and the first byte of the IV has determined the contents of stages A_0 and A_1. Thus, all stages of register A can take any possible value, and the above analysis of state space reduction is exact. After $t = -30$, the results of subsequent clock steps may depend on those of earlier steps, as the register contents and feedback bits clock through the input locations for the various S-boxes. However, none of the S-boxes has an input set which clocks directly onto the inputs of any other S-box, so it seems reasonable to assume that the results at later clock steps are nearly independent of those at earlier steps. On this basis, we obtain an approximate upper bound on the effective state size at $t = 0$ (the end of initialisation) of $(1 - 0.155)^{30} \approx (2^{-0.243})^{30} \approx 2^{-7.30}$ of the state size without convergence; in other words, the amount of accessible state space decreases by a minimum of approximately 7 bits during initialisation.

4   Analysis of Slid Pairs

Slide attacks were first introduced against block ciphers [7,8] and later adapted to stream ciphers [7,19]. These attacks depend on the existence of slid pairs (defined below) and on whether the slid pairs lead to phase-shifted keystreams. When a Key-IV pair (K', V') produces a loaded state that can also be obtained from another Key-IV pair (K, V) after a number of iterations α of initialisation, we refer to these two states as a slid pair. Subject to certain conditions, the keystream generated by (K', V') may then mimic the keystream generated by (K, V), but phase-shifted by a fixed multiple of α bits.

In this section we look for pairs (K, V) and (K', V') which produce phase-shifted versions of the same keystream. For CSA-SC, loaded states correspond to loaded keys only since the IV is loaded during the diffusion phase. Thus, the second state of a slid pair depends only on K' and we only need to consider V' when determining whether the slid pair gives shifted keystream.

Firstly, suppose that a pair (K, V) yields a loaded state after α iterations of diffusion. This requires that $A_8^{\alpha-31}$, $A_9^{\alpha-31}$, $B_8^{\alpha-31}$, $B_9^{\alpha-31}$, $E^{\alpha-31}$, $F^{\alpha-31}$ and $c^{\alpha-31}$ are all zeros. Equations 3a–3c show that the last three conditions require E^t, F^t and c^t to be zero for all $t < \alpha - 31$; to ensure this, the outputs of either S_9 to S_{12} (Z) or S_{14} (q) must be zeros for $-31 \leq t < \alpha - 31$. Further, Stages A_8^{α}, A_9^{α}, B_8^{α} and B_9^{α} can be zeros under the following conditions:

(1) If $\alpha = 1$: the 8 key bits $k_{28} \ldots k_{31}$ and $k_{60} \ldots k_{63}$ must be zeros.
(2) For $2 \leq \alpha \leq 8$: key bytes $k_{32-4\alpha} \ldots k_{39-4\alpha}$ and $k_{64-4\alpha} \ldots k_{71-4\alpha}$ are zeros.
(3) For $\alpha = 9$: key bits $k_0 \ldots k_3$ and $k_{32} \ldots k_{35}$ must be zeros, and Eqs. 8a and 8b (below) must be satisfied at $t = -31$.
(4) For $\alpha \geq 10$: Eqs. 8a and 8b must be satisfied for two successive iterations to generate two consecutive stages of zeros in A and B.

$$A_9^{t-1} \oplus X^{t-1} \oplus D^{t-1} \oplus I_A^t = 0 \tag{8a}$$
$$(B_6^{t-1} \oplus B_9^{t-1} \oplus Y^{t-1} \oplus I_B^t) \lll p^{t-1} = 0 \tag{8b}$$

Now, if the second loaded state corresponds to (K', V'), then K' is uniquely determined by K and the first $\lceil \frac{\alpha}{4} \rceil$ bytes of V, via Eqs. 1–4. As the registers are not autonomous during diffusion, we must also consider the conditions on V and V' in order for the states resulting from (K, V) and (K', V') to remain in step (clock identically) for the rest of the diffusion phase. These conditions depend on the value of α.

Finally, to obtain shifted keystream from a slid pair that has remained in step, we also require that the last α iterations with (K', V') satisfy

$$D^{t-1} \oplus I_A^t = 0 \tag{9a}$$
$$I_B^t = 0 \tag{9b}$$

Recall that the IV in CSA-SC is loaded during the diffusion phase and that each IV byte is loaded into registers A and B for four successive iterations. If α is not a multiple of 4, this operation imposes significant restrictions on the form of V and V' if the states of a slid pair are to remain in step. For this reason, we describe our analysis for $\alpha = 4$ but report only the results for other values of α.

For $\alpha = 4$, the three steps required for slid pairs and shifted keystream are:

(i) To obtain a slid pair, k_{16} to k_{23} and k_{48} to k_{55} must be zeros (probability 2^{-16}). Also, either q or z_0 to z_3 must be zeros for the 2nd, 3rd and 4th clocks so that E, F and c are all zeros (probability $(\frac{17}{32})^3 = 2^{-2.738}$).
(ii) Keeping the slid pair in step imposes no restrictions on V and requires only that V' is a 1-byte shifted version of V.

(iii) If the slid pairs remain in step, shifted keystream occurs if the last byte of V' is zero and also $D^{t-1} = 0$ for the last four iterations of diffusion (probability 2^{-16}).

Thus, the combined probability that a randomly chosen pair (K, V) leads (via a slid pair) to shifted keystream is $2^{-16} \times 2^{-2.738} \times 2^{-16} = 2^{-34.738}$. K' is then determined by K and the first byte of V, V' is determined by the rest of V (plus a final byte of zeros) and the keystreams are out of phase by 8 bits.

The analysis for $\alpha = 4$ can be extended to $\alpha = 8, 12, \ldots, 28$. The probability of a key-IV pair leading to slid pairs and shifted keystreams in these cases is $2^{-16} \times \left(\frac{17}{32}\right)^{(\alpha-1)} \times 2^{-4\alpha} = 2^{-15.087-4.913\alpha}$. When α is not a multiple of 4, the number of IVs, V, which allow a slid pair to generate shifted keystream are extremely limited, so the overall probability of obtaining a slid pair and shifted keystream is much smaller. The results for $\alpha = 1, 2, 3, 4$ and 8 are shown in Table 5, from which it is clear that the most favourable result occurs for $\alpha = 4$.

Table 5. Probabilities of obtaining slid pairs and shifted keystreams for α clocks

α	1	2	3	4	8
(i) To obtain slid pairs	2^{-8}	$2^{-16.913}$	$2^{-17.825}$	$\mathbf{2^{-18.738}}$	$2^{-22.388}$
(ii) Proportion of valid IVs	2^{-60}	2^{-64}	2^{-64}	2^0	2^0
(iii) To satisfy Eq. 9a	2^{-4}	2^{-8}	2^{-12}	$\mathbf{2^{-16}}$	2^{-32}
Total probabilities	2^{-72}	$2^{-88.913}$	$2^{-93.825}$	$\mathbf{2^{-34.738}}$	$2^{-54.388}$

5   Conclusion

State convergence clearly occurs during the initialisation of CSA-SC, with distinct key-IV inputs producing the same output keystream sequence. This is clearly not desirable. Slid pairs leading to shifted key stream also occur; that is, distinct key-IV inputs lead to output sequences which are phase-shifted versions of one another. Both flaws may leave the cipher vulnerable to generic attacks.

In Sect. 3, state convergence was shown to occur during initialisation, beginning at the third iteration of the diffusion phase. This convergence can occur both when $q = 0$ and when $q = 1$; we investigated the case $q = 0$ in detail. At the third iteration, states with $q = 0$ clock to only 0.69 as many states. The convergence for states with $q = 1$ is undetermined but expected to be similar. In any case, the effective state space is reduced by a proportion of at least 0.155 at this iteration. We argue that the size of this effect remains approximately constant throughout diffusion, and hence that the effective state size is reduced by at least 7 bits during initialisation (more if there is significant convergence for $q = 1$).

This reduction in state space may leave CSA-SC vulnerable to time-memory-data trade-off attacks [5]. Previous attacks on the A5/1 stream cipher [10], e.g. Biryukov et al. [6] and Golić [13], are based on the reduction of state space in this cipher due to state convergence. For the case of 48-bit keys, CSA is already vulnerable to specific attacks [17]; state convergence on the scale indicated by our analysis would further compound this vulnerability.

In Sect. 4, we showed that slid pairs occur during the initialisation of CSA-SC and may lead to a phase-shifted version of the same keystream. The best probability of finding slid pairs leading to shifted keystream is for $\alpha = 4$, where a Key-IV pair generates a key loaded state after four iterations. This case has a probability of $2^{-18.74}$ of obtaining slid pairs, with a further probability of 2^{-16} that a slid pair leads to shifted key stream.

The existence of slid pairs and shifted keystream may leave this cipher vulnerable to slide attacks. A number of stream ciphers have previously been attacked in this way, including LEX [19], WAKE-ROFB [7], Grain [11,20] and Trivium [14]. More recently, Alhamdan et al. [1,2] showed that the existence of slid pairs can be used to recover the secret key of A5/1 [10] and Sfinks [9] stream ciphers using ciphertext-only attacks.

A   CSA-SC S-Boxes

Table 6. Input and output bits of the 14 Boolean functions

Functions	X				Y				Z					
	S_1	S_2	S_3	S_4	S_5	S_6	S_7	S_8	S_9	S_{10}	S_{11}	S_{12}	S_{13}	S_{14}
Output bits	x_0^t	x_1^t	x_2^t	x_3^t	y_0^t	y_1^t	y_2^t	y_3^t	z_0^t	z_1^t	z_2^t	z_3^t	p^t	q^t
Input $\;i_0^t$	$a_{3,3}$	$a_{1,3}$	$a_{2,1}$	$a_{4,0}$	$a_{3,1}$	$a_{5,2}$	$a_{3,3}$	$a_{1,3}$	$a_{2,1}$	$a_{4,0}$	$a_{3,1}$	$a_{5,2}$	$a_{2,2}$	$a_{2,2}$
i_1^t	$a_{1,1}$	$a_{2,0}$	$a_{3,2}$	$a_{1,2}$	$a_{4,1}$	$a_{4,3}$	$a_{1,1}$	$a_{2,0}$	$a_{3,2}$	$a_{1,2}$	$a_{4,1}$	$a_{4,3}$	$a_{3,0}$	$a_{3,0}$
bits $\;i_2^t$	$a_{2,3}$	$a_{5,1}$	$a_{6,3}$	$a_{6,1}$	$a_{5,0}$	$a_{6,0}$	$a_{2,3}$	$a_{5,1}$	$a_{6,3}$	$a_{6,1}$	$a_{5,0}$	$a_{6,0}$	$a_{7,1}$	$a_{7,1}$
i_3^t	$a_{4,2}$	$a_{5,3}$	$a_{7,0}$	$a_{7,3}$	$a_{7,2}$	$a_{8,1}$	$a_{4,2}$	$a_{5,3}$	$a_{7,0}$	$a_{7,3}$	$a_{7,2}$	$a_{8,1}$	$a_{8,2}$	$a_{8,2}$
i_4^t	$a_{8,0}$	$a_{6,2}$	$a_{9,1}$	$a_{9,0}$	$a_{9,3}$	$a_{9,2}$	$a_{8,0}$	$a_{6,2}$	$a_{9,1}$	$a_{9,0}$	$a_{9,3}$	$a_{9,2}$	$a_{8,3}$	$a_{8,3}$

Table 7. The truth table of the 14 Boolean functions

i_0^t	0000000000000000	1111111111111111
i_1^t	0000000011111111	0000000011111111
i_2^t	0000111100001111	0000111100001111
i_3^t	0011001100110011	0011001100110011
i_4^t	0101010101010101	0101010101010101
S_1	1000111011100001	0110110011010010
S_2	0011011010001101	0110001100011110
S_3	1001111001100001	1001110100011010
S_4	1100011011010010	1101100000100111
S_5	1001111000011010	0100101110010110
S_6	0010011111011000	1101100000100111
S_7	1011010101001001	0010110101100110
S_8	1101001010011001	1011010101001001
S_9	1000111100110010	1110010000111001
S_{10}	0001101001111001	0100011110101100
S_{11}	0011011000101101	1101100010010110
S_{12}	0101100001100111	0100101101100110
S_{13}	0111100010010110	0011000111001101
S_{14}	0100100110111001	1011011001100100

References

1. Alhamdan, A., Bartlett, H., Dawson, E., Simpson, L., Wong, K.K.: Slide attacks on the sfinks stream cipher. In: 2012 6th International Conference on Signal Processing and Communication Systems (ICSPCS), December 2012, pp. 1–10 (2012)
2. Alhamdan, A., Bartlett, H., Dawson, E., Simpson, L., Wong, K.K.: Slid pairs in the initialisation of the A5/1 stream cipher. In: Thomborson, C., Parampalli, U. (eds.) Information Security 2013 (AISC 2013). CRPIT, vol. 138, pp. 3–12. ACS, Adelaide (2013)
3. Bewick, S.: Descrambling DVB data according to ETSI common scrambling specification, UK Patent GB2322994A, September 1998
4. Bewick, S.: Descrambling DVB data according to ETSI common scrambling standard, UK Patent GB2322995A, September 1998
5. Biryukov, A., Shamir, A.: Cryptanalytic time/memory/data tradeoffs for stream ciphers. In: Okamoto, T. (ed.) ASIACRYPT 2000. LNCS, vol. 1976, pp. 1–13. Springer, Heidelberg (2000)
6. Biryukov, A., Shamir, A., Wagner, D.: Real time cryptanalysis of A5/1 on a PC. In: Goos, G., Hartmanis, J., van Leeuwin, J., Schneier, B. (eds.) FSE 2000. LNCS, vol. 1978, pp. 1–18. Springer, Heidelberg (2001)
7. Biryukov, A., Wagner, D.: Slide attacks. In: Knudsen, L.R. (ed.) FSE 1999. LNCS, vol. 1636, pp. 245–259. Springer, Heidelberg (1999)
8. Biryukov, A., Wagner, D.: Advanced slide attacks. In: Preneel, B. (ed.) EURO-CRYPT 2000. LNCS, vol. 1807, pp. 589–606. Springer, Heidelberg (2000)
9. Braeken, A., Lano, J., Mentens, N., Preneel, B., Verbauwhede, I.: SFINKS: a synchronous stream cipher for restricted hardware environments. eSTREAM, ECRYPT Stream Cipher Project, Report 2005/026 (2005). www.ecrypt.eu.org/stream/ciphers/sfinks/sfinks.ps

10. Briceno, M., Goldberg, I., Wagner, D.: A pedagogical implementation of A5/1 (1999). http://cryptome.org/jya/a51-pi.htm
11. De Cannière, C., Küçük, Ö., Preneel, B.: Analysis of Grain's initialization algorithm. In: Vaudenay, S. (ed.) AFRICACRYPT 2008. LNCS, vol. 5023, pp. 276–289. Springer, Heidelberg (2008)
12. European Standards Organization - European Union. European Telecommunications Standards Institute. http://www.etsi.org/
13. Golić, J.: Cryptanalysis of three mutually clock-controlled stop/go shift registers. IEEE Trans. Inf. Theor. **46**(3), 1081–1090 (2000)
14. Priemuth-Schmid, D., Biryukov, A.: Slid pairs in Salsa20 and Trivium. In: Chowdhury, D.R., Rijmen, V., Das, A. (eds.) INDOCRYPT 2008. LNCS, vol. 5365, pp. 1–14. Springer, Heidelberg (2008)
15. Rose, G., Hawkes, P.: On the applicability of distinguishing attacks against stream ciphers. In: Proceedings of the 3rd NESSIE Workshop, p. 6. Citeseer (2002)
16. Simpson, L., Henricksen, M., Yap, W.-S.: Improved cryptanalysis of the common scrambling algorithm stream cipher. In: Boyd, C., Gonzálcz Nicto, J. (eds.) ACISP 2009. LNCS, vol. 5594, pp. 108–121. Springer, Heidelberg (2009)
17. Tews, E., Wälde, J., Weiner, M.: Breaking DVB-CSA. In: Armknecht, F., Lucks, S. (eds.) WEWoRC 2011. LNCS, vol. 7242, pp. 45–61. Springer, Heidelberg (2012)
18. Weinmann, R., Wirt, K.: Analysis of the DVB common scrambling algorithm. In: Chadwick, D., Preneel, B. (eds.) Communications and Multimedia Security. IFIP - The International Federation for Information Processing, vol. 175, pp. 195–207. Springer, US (2005)
19. Wu, H., Preneel, B.: Resynchronization attacks on WG and LEX. In: Robshaw, M. (ed.) FSE 2006. LNCS, vol. 4047, pp. 422–432. Springer, Heidelberg (2006)
20. Zhang, H., Wang, X.: Cryptanalysis of stream ciphcr grain family. Cryptology ePrint Archive, Report 2009/109 (2009). http://cprint.iacr.org/

Distribution Properties of Half-ℓ-Sequence

Ting Gu$^{(\boxtimes)}$ and Andrew Klapper

Department of Computer Science, University of Kentucky,
Lexington 40506-0046, USA
ting.gu@uky.edu
http://www.cs.uky.edu/~klapper/

Abstract. Speed is essential in stream cipher design. In 2011, Lee and Park [5] proposed a software implementation for word-based FCSRs. The sequences generated by the FCSRs using Lee and Park's implementation methods are half-ℓ-sequences. In this paper, we investigate the imbalance properties of half-ℓ-sequences. Bounds on the numbers of occurrences of one and two consecutive symbols are given. The experimental results show how tight our bound is.

Keywords: FCSRs · Half-ℓ-sequence · Stream cipher · Pseudo-random sequence

1 Introduction

Classical stream ciphers are often based on linear feedback shift registers (LFSRs), filtered or combined by non-linear functions. However, this type of stream cipher is vulnerable to algebraic attacks [1–3]. In 1994, Klapper and Goresky proposed a new type of feedback shift register called a feedback with carry shift register (FCSR) [4, p. 69]. An FCSR is a feedback shift register with a small amount of auxiliary memory. The critical reason that LFSR based stream ciphers are vulnerable to algebraic attacks is that a degree d multinomial composed with a linear state change gives a degree d multinomial. However, this fails for FCSRs due to the nonlinearity. An FCSR is described by an associated integer q, the connection integer. The analysis of FCSRs is based largely on the algebra of N-adic numbers. If the output is the N-ary sequence $\mathbf{a} = a_0, a_1, ...$, then the associated N-adic number is $\alpha = a_0 + a_1 N + a_2 N^2 + ...$ It can be seen that $\alpha = -p/q$ where $p \in Z$. The period of $\mathbf{a}$ is the multiplicative order of N modulo q. Also, $a_i = q^{-1}(pN^{-i} \mod q) \mod N$. An important characteristic of an FCSR is that the elementary additions are not additions modulo 2, but additions with propagation of carries. This leads to proved results on period and

This material is based upon work supported by the National Science Foundation under Grant No. CNS-1420227. Any opinions, findings, and conclusions or recommendations expressed in this material are those of the author and do not necessarily reflect the views of the National Science Foundation.

© Springer International Publishing Switzerland 2014
K.-U. Schmidt and A. Winterhof (Eds.): SETA 2014, LNCS 8865, pp. 234–245, 2014.
DOI: 10.1007/978-3-319-12325-7_20

non-degeneration of internal states. Also, the sequence that achieves this maximum period is called an ℓ-sequence. The period of an ℓ-sequence with connection integer q is $q - 1$. An ℓ-sequence has many good statistical properties, just as an m-sequence does.

Speed is essential in stream cipher design. In 2011, Lee and Park [5] proposed a software implementation for word-based FCSRs. They extended the size of register cells from 1 bit to k bits where k is the size of words in the given CPU (typically $k = 32$ or 64) to produce k bits at every clocking. Lee and Park proposed two types of implementations. One uses full-size words (in this case $N = 2^{32}$). The other one uses half-size words (in this case $N = 2^{16}$). The two implementations were claimed to have better efficiency than methods using conditional operators. However, there is an intrinsic problem with the choices $N = 2^{32}$ and 2^{16}. We show that if $N = 2^k$ and connection integer $q \equiv -1$ mod N (as Lee and Park assumed), then the multiplicative order of N modulo q is at most $(q-1)/2$. Thus there are no ℓ-sequences with such connection integers. The largest possible period is $(q-1)/2$, in which case we call the output sequence a half-ℓ-sequence. In this paper, we investigate the imbalance properties of half-ℓ-sequences. Bounds on the numbers of occurrences of one and two consecutive symbols are given.

The paper is organized as follows. In Sect. 2, we give upper bounds on the imbalance of half-ℓ-sequences in the one symbol and two consecutive symbol cases. In Sect. 3, we discuss an exceptional case for a binary half-ℓ-sequence. In Sect. 4, we show experimental results that give some sense of how tight our bounds are.

2   Bounds on the Imbalance of a Half-ℓ-Sequence

Lee and Park proposed software implementations for FCSRs over the N-adic number where $N = 2^k$ with $k = 32$ or 16. The corresponding connection integer q is expressed as $q = -1 + Nr \equiv -1$ mod 8 for some r. According to quadratic reciprocity, if $q \equiv -1$ mod 8, then 2 is a quadratic residue (QR) modulo q. Hence also $N = 2^k$ is a QR modulo q. So every power of N is a QR modulo q. It follows from the exponential representation described in the introduction that the period is at most $(q - 1)/2$. In this section we discuss the imbalance properties of sequences associated with such qs. Note that we do not restrict our discussion to $N = 2^{16}$ and 2^{32} but reason in a general manner.

Let $N = 2^t$ for some t and $q = q_0 + mN$ for $q_0, m \in \mathbb{Z}$ with $0 \le q_0 < N$ and $\gcd(q_0, N) = 1$. We consider an N-adic sequence $\mathbf{a}$. I.e., $\mathbf{a}$ is generated by an FCSR Γ whose connection integer is q.

Definition 1. *A sequence $\mathbf{a}$ with prime connection integer q is called a half-ℓ-sequence if the period of $\mathbf{a}$ is $(q-1)/2$.*

Actually, if $\mathbf{a}$ has period $(q - 1)/2$, then it can be shown that q is prime. Assume that the associated N-adic number for sequence $\mathbf{a}$ is the rational number $-h/q$. Then $a_0 \equiv q^{-1}h$ mod N. We want to obtain bounds on the imbalance of

a. Let Q be the set of quadratic residues modulo q and Q' be the set of quadratic non-residues modulo q. Under those conditions the set of integers of the form N^i mod q is precisely Q if $N \in Q$. For any nonzero z, we have $zQ = Q$ if $z \in Q$, and $zQ = Q'$ if $z \in Q'$. Q corresponds to the states in one cycle of the state space. Q' corresponds to the states in a second cycle.

Let ξ be a complex primitive qth root of unity. The Fourier transform of a complex valued function $f : \mathbb{Z}_q \to \mathbb{C}$ is given by

$$\hat{f}(b) = \frac{1}{q} \sum_{c=0}^{q-1} f(c) \xi^{-bc}.$$

By the Fourier inversion formula we have

$$f(c) = \sum_{b=0}^{q-1} \hat{f}(b) \xi^{bc}.$$

2.1 The One Symbol Case

Let $F_v = \{x | 0 \le x < q, x = v \mod N\}$ and $0 \le v < N$. Each consecutive pair of elements of F_v differs by N. As a result, we have

$$F_v = \{v, v + N, v + 2N, \cdots, v + t_v N\} \subset \{0, 1, \cdots, q-1\}.$$

Then we have $q - N \le v + t_v N \le q - 1$, so

$$\frac{1+v}{N} \le \frac{q}{N} - t_v \le 1 + \frac{v}{N}.$$

Then $t_v \approx q/N$ and $|\{v + eN : 0 \le v + eN < q\}| \approx q/N$.

Let $\mathbf{a} = a_0, a_1, \cdots$ be a half-ℓ-sequence with connection integer q. For $j = 0, 1, \cdots$, let $u_j/q = \sum_{i=0}^{\infty} a_{i+j} N^i$. Then u_j and u_{j+1} are related by the equation $u_j = qa_j + N u_{j+1}$. Thus $u_{j+1} \equiv N^{-1} u_j \mod q$ and $a_j \equiv q^{-1} u_j \mod N$. From the first congruence it follows that either all u_j are quadratic residues, or all are non-quadratic residues. Suppose they are quadratic residues. From the second congruence it follows that for any $a \in \{0, 1, \cdots, N-1\}$, the number of occurrences of a in one period of the sequence equals the number of quadratic residues u with $u \equiv qa \mod N$. Let $\mu(v)$ be the number of occurrences of $q^{-1}v$ mod N in one period of $\mathbf{a}$. We have $\mu(v) = |Q \cap F_v|$.

Define

$$f_v(x) = \begin{cases} 1 & \text{if } x \in F_v \\ 0 & \text{otherwise.} \end{cases}$$

Thus

$$\mu(v) = \sum_{c \in Q} f_v(c) = \sum_{c \in Q} \sum_{b=0}^{q-1} \hat{f}_v(b) \xi^{bc} = \sum_{b=0}^{q-1} \hat{f}_v(b) \sum_{c \in Q} \xi^{bc}$$

$$= \sum_{b=0}^{q-1} \hat{f}_v(b) \sum_{d=1}^{(q-1)/2} \xi^{bd^2} = \sum_{b=0}^{q-1} \hat{f}_v(b) \sigma(b).$$

First consider the term $b = 0$. Then $\sigma(0) = (q-1)/2$ and

$$\hat{f}_v(0) = \frac{1}{q}\sum_{c=0}^{q-1} f_v(c) = \frac{1}{q}|\{v + eN : 0 \le v + eN < q\}|$$

$$\approx \frac{1}{q}\frac{q}{N} = \frac{1}{N}.$$

Thus we want to bound

$$B = \sum_{b=1}^{q-1} \hat{f}_v(b)\sigma(b) \le \left(\sum_{b=1}^{q-1} |\hat{f}_v(b)|\right) \max_{b \ne 0} |\sigma(b)|.$$

By the preceding discussion, $\sigma(b)$ depends only on whether $b \in Q$.

Theorem 1 *(Weil's Theorem [6, p. 223]). Let r be a power of a prime. Let $g \in \mathbb{F}_r[x]$ be of degree $1 \le n < r$ with $\gcd(n, r) = 1$ and let χ be a nontrivial additive character of $\mathbb{F}_r$. Then*

$$\left|\sum_{c \in \mathbb{F}_r} \chi(g(c))\right| \le (n-1)r^{1/2}.$$

In our case $r = q \in \mathbb{Z}$ is prime, $\chi(x) = \xi^x$, $n = 2$, and $g(x) = bx^2$. Thus by Weil's theorem we have

$$\left|\sigma(b)\right| - \frac{1}{2}\left|\sum_{c=1}^{q-1} \xi^{bc^2}\right| = \frac{1}{2}\left|\sum_{c=0}^{q-1}\xi^{bc^2} - 1\right| \le \frac{1}{2}\left(q^{1/2} + 1\right).$$

Lemma 1. *If $0 \le v \le N-1$, then the following inequality holds:*

$$\sum_{b=1}^{q-1} |\hat{f}_v(b)| \le 1 + \ln\left(\frac{q-1}{2}\right).$$

Proof

$$|\hat{f}_v(b)| = \frac{1}{q}\left|\sum_{c=0}^{q-1} f_v(c)\xi^{-bc}\right| = \frac{1}{q}\left|\sum_{d=0}^{t_v} \xi^{b(v+Nd)}\right| = \frac{1}{q}\left|\sum_{d=0}^{t_v} \xi^{bNd}\right|$$

$$= \frac{1}{q}\left|\frac{\xi^{bN(t_v+1)} - 1}{\xi^{bN} - 1}\right| = \frac{1}{q}\left|\frac{\xi^{bN(t_v+1)/2} - \xi^{-bN(t_v+1)/2}}{\xi^{bN/2} - \xi^{-bN/2}}\right|$$

$$= \frac{1}{q}\left|\frac{\sin(\pi bN(t_v+1)/q)}{\sin(\pi bN/q)}\right|.$$

Thus

$$\sum_{b=1}^{q-1} |\hat{f}_v(b)| \le \frac{1}{q}\sum_{b=1}^{q-1}\left|\frac{1}{\sin(\pi bN/q)}\right| = \frac{1}{q}\sum_{c=1}^{q-1}\left|\frac{1}{\sin(\pi c/q)}\right| = \frac{2}{q}\sum_{c=1}^{(q-1)/2}\left|\frac{1}{\sin(\pi c/q)}\right|.$$

For $0 \le x \le \pi/2$ we have $\sin(x) \ge 2x/\pi$, so

$$\sum_{b=1}^{q-1} |\hat{f}_v(b)| \le \frac{2}{q} \sum_{c=1}^{(q-1)/2} \frac{q}{2c} = \sum_{c=1}^{(q-1)/2} \frac{1}{c} \le 1 + \ln\left(\frac{q-1}{2}\right).$$

$\square$

Theorem 2. *If $0 \le v \le N-1$, then*

$$\left| \mu(v) - \frac{q-1}{2N} \right| \le \frac{1}{2}\left(1 + \ln\left(\frac{q-1}{2}\right)\right)\left(q^{1/2} + 1\right).$$

2.2   The Two Symbol Case

Let $G_v = \{x | 0 \le x < q, (Nx \mod q) \equiv v \mod N\}$ and $0 \le v < N$. We want to investigate the number of occurrences of two consecutive symbols. The number of occurrences of $(q^{-1}v_1 \mod N), (q^{-1}v_2 \mod N)$ in one period is $\mu(v_1, v_2) = |Q \cap F_{v_2} \cap G_{v_1}|$.

For $0 \le x < q/N$, we have $0 \le Nx < q$, so $Nx \mod q = Nx \equiv 0 \mod N$. These xs are in G_0. For $q/N \le x < 2q/N$, we have $q \le Nx < 2q$, so $Nx \mod q = Nx - q \equiv -q \mod N$. These xs are in $G_{-q \mod N}$. For $2q/N \le x < 3q/N$, we have $2q \le Nx < 3q$, so $Nx \mod q = Nx - 2q \equiv -2q \mod N$. These xs are in $G_{-2q \mod N}$. We continue in this way. In general $G_{-iq \mod N} = \{x : iq/N \le x < (i+1)q/N\}$. Thus $G_{v_1} = \{x : iq/N \le x < (i+1)q/N\}$ where $i = (-v_1 q^{-1} \mod N)$. Each G_v is an interval of length about q/N.

Assume that $-iq \mod N = v_1$ and $q = q_0 + mN$ for some m. We have $-iq \mod N = v_1$, so $iq = N - v_1 + k_1 N$ for some k_1. We also have $(i+1)q = iq + q = N - v_1 + k_1 N + q_0 + mN = (k_1 + m)N + N - v_1 + q_0$.

Then we have

$$\left\lceil \frac{iq}{N} \right\rceil = \frac{iq - N + v_1}{N} + 1 = \frac{iq + v_1}{N}$$

and

$$\left\lfloor \frac{(i+1)q}{N} \right\rfloor = \begin{cases} \frac{(i+1)q - N + v_1 - q_0}{N} & \text{if } q_0 < v_1 \\ \frac{(i+1)q + v_1 - q_0}{N} & \text{otherwise.} \end{cases}$$

Thus

$$\left\lfloor \frac{(i+1)q}{N} \right\rfloor - \left\lceil \frac{iq}{N} \right\rceil = \begin{cases} \frac{q - N - q_0}{N} & \text{if } q_0 < v_1 \\ \frac{q - q_0}{N} & \text{otherwise.} \end{cases}$$

We have $\lfloor (i+1)q/N \rfloor - \lceil iq/N \rceil \approx q/N$ since $|\lfloor (i+1)q/N \rfloor - \lceil iq/N \rceil - q/N| \le 1 + q_0/N < 2$.

Define

$$g_v(x) = \begin{cases} 1 & \text{if } x \in G_v \\ 0 & \text{otherwise.} \end{cases}$$

Then $g_{v_1}(x) = 1$ when $\lceil iq/N \rceil \leq x \leq \lfloor (i+1)q/N \rfloor$ and

$$\hat{g}_{v_1}(d) = \frac{1}{q} \sum_{c=0}^{q-1} g_{v_1}(c)\xi^{-dc}.$$

We have

$$\mu(v_1, v_2) = \sum_{x \in Q} f_{v_2}(x)g_{v_1}(x) = \sum_{x \in Q} \sum_{b=0}^{q-1} \hat{f}_{v_2}(b)\xi^{bx} \sum_{d=0}^{q-1} \hat{g}_{v_1}(d)\xi^{dx}$$

$$= \sum_{b=0}^{q-1} \hat{f}_{v_2}(b) \sum_{d=0}^{q-1} \hat{g}_{v_1}(d) \sum_{x \in Q} \xi^{(b+d)x}$$

$$= \sum_{b=0}^{q-1} \hat{f}_{v_2}(b) \sum_{d=0}^{q-1} \hat{g}_{v_1}(d) \sum_{k=1}^{(q-1)/2} \xi^{(b+d)k^2}$$

$$= \sum_{b=0}^{q-1} \hat{f}_{v_2}(b) \sum_{d=0}^{q-1} \hat{g}_{v_1}(d)\sigma(b+d).$$

First consider the term $b + d \equiv 0 \mod q$. We have $\sigma(b+d) = (q-1)/2$.

Case 1: If $b = 0$ (equivalently $d = 0$), let

$$B_0 = \hat{f}_{v_2}(0)\hat{g}_{v_1}(0)\frac{q-1}{2} = \frac{1}{q}\sum_{c=0}^{q-1} f_{v_2}(c)\frac{1}{q}\sum_{h=0}^{q-1} g_{v_1}(h)\frac{q-1}{2}$$

$$= \frac{q-1}{2q^2}\left(\sum_{c=0}^{q-1} f_{v_2}(c)\right)\left(\sum_{h=0}^{q-1} g_{v_1}(h)\right)$$

$$= \frac{q-1}{2q^2} \mid \{v_2 + eN : 0 \leq v_2 + eN < q\} \mid \left(\left\lfloor \frac{(i+1)q}{N} \right\rfloor - \left\lceil \frac{iq}{N} \right\rceil\right)$$

$$\approx \frac{q-1}{2q^2}\frac{q}{N}\frac{q}{N} = \frac{q-1}{2N^2}.$$

Note that this is the average number of occurrences of a pair of symbols.
Case 2: If $b \neq 0$ (equivalently, $d = q - b \neq 0$), let

$$B_1 = \frac{q-1}{2}\sum_{b=1}^{q-1} \hat{f}_{v_2}(b)\hat{g}_{v_1}(q-b)$$

$$= \frac{q-1}{2}\sum_{b=1}^{q-1}\frac{1}{q}\sum_{d=0}^{t_{v_2}} \xi^{b(v_2+Nd)}\frac{1}{q}\sum_{c=\lceil iq/N \rceil}^{\lfloor (i+1)q/N \rfloor} \xi^{-(q-b)c}$$

$$= \frac{q-1}{2q^2}\sum_{d=0}^{t_{v_2}}\sum_{c=\lceil iq/N \rceil}^{\lfloor (i+1)q/N \rfloor}\sum_{b=1}^{q-1} \xi^{b(c+v_2+Nd)}.$$

When $c + v_2 + Nd \equiv 0 \mod q$,

$$\sum_{b=1}^{q-1} \xi^{b(v_2+Nd+c)} = \sum_{b=1}^{q-1} 1 = q-1.$$

When $c + v_2 + Nd \neq 0 \mod q$,

$$\sum_{b=1}^{q-1} \xi^{b(v_2+Nd+c)} = \frac{\xi^{v_2+Nd+c}(1-\xi^{(q-1)(v_2+Nd+c)})}{1-\xi^{v_2+Nd+c}} = -1.$$

Define a set

$$D = \{d : 0 \leq d \leq t_{v_2} \wedge \lceil iq/N \rceil \leq -(v_2 + Nd) \mod q \leq \lfloor (i+1)q/N \rfloor\}.$$

Then, in the case when $c + v_2 + Nd \equiv 0$, we can sum over $d \in D$ and the number of pairs of c, d in the given ranges which make $c + v_2 + Nd \equiv 0 \mod q$ is $|D|$. Let

$$W = \sum_{d=0}^{t_{v_2}} \sum_{c=\lceil iq/N \rceil}^{\lfloor (i+1)q/N \rfloor} \sum_{b=1}^{q-1} \xi^{b(c+v_2+Nd)}.$$

Then we have

$$W = \left((q-1)|D| - (t_{v_2}+1)\left(\left\lfloor \frac{(i+1)q}{N} \right\rfloor - \left\lceil \frac{iq}{N} \right\rceil + 1\right) + x\right)$$

$$= \left(q|D| - (t_{v_2}+1)\left(\left\lfloor \frac{(i+1)q}{N} \right\rfloor - \left\lceil \frac{iq}{N} \right\rceil + 1\right)\right),$$

and

$$B_1 = \frac{q-1}{2q^2}W = \frac{q-1}{2q^2}\left(q|D| - (t_{v_2}+1)\left(\left\lfloor \frac{(i+1)q}{N} \right\rfloor - \left\lceil \frac{iq}{N} \right\rceil + 1\right)\right).$$

where $q/N - 1 - v_2/N \leq t_{v_2} \leq q/N - (1+v_2)/N$ and $(q-N-q_0)/N \leq \lfloor (i+1)q/N \rfloor - \lceil iq/N \rceil \leq (q-q_0)/N$.

Except in the case when $c = 0$ and $v_2 + Nd = 0$, we have $-(v_2 + Nd) \mod q = q - v_2 - Nd$. So we have

$$\left\lfloor \frac{\lfloor (i+1)q/N \rfloor - \lceil iq/N \rceil + 1}{N} \right\rfloor \leq |D| \leq \left\lceil \frac{\lfloor (i+1)q/N \rfloor - \lceil iq/N \rceil + 1}{N} \right\rceil.$$

Since

$$|D| \leq \left\lceil \frac{\lfloor (i+1)q/N \rfloor - \lceil iq/N \rceil + 1}{N} \right\rceil$$

$$t_{v_2} \geq \frac{q}{N} - 1 - \frac{v_2}{N}$$

$$\left\lfloor \frac{(i+1)q}{N} \right\rfloor - \left\lceil \frac{iq}{N} \right\rceil \geq \frac{q-N-q_0}{N},$$

B_1 is upper bounded by

$$(B_1)_{max} = \frac{q-1}{2q^2}\left(q\frac{q-q_0+N}{N^2} - \left(\frac{q}{N}-1-\frac{v_2}{N}+1\right)\left(\frac{q-q_0-N}{N}+1\right)\right)$$

$$= \frac{(q-1)(qN+v_2q-q_0v_2)}{2q^2N^2}$$

$$< \frac{1}{N}.$$

Since

$$|D| \geq \left\lfloor \frac{\lfloor(i+1)q/N\rfloor - \lceil iq/N\rceil + 1}{N} \right\rfloor$$

$$t_{v_2} \leq \frac{q}{N} - \frac{1+v_2}{N}$$

$$\left\lfloor \frac{(i+1)q}{N} \right\rfloor - \left\lceil \frac{iq}{N} \right\rceil \leq \frac{q-q_0}{N},$$

B_1 is lower bounded by

$$(B_1)_{min} = \frac{q-1}{2q^2}\left(q\frac{q-q_0}{N^2} - \left(\frac{q}{N}-\frac{v_2+1}{N}+1\right)\left(\frac{q-q_0}{N}+1\right)\right)$$

$$= \frac{(q-1)(q(v_2+1-2N)+(v_2+1-N)(N-q_0))}{2q^2N^2}$$

$$> -\left(\frac{1}{N}+\frac{1}{2q}\right).$$

So $(B_1)_{min} \leq B_1 \leq (B_1)_{max}$ and $|B_1| \leq \max\{|(B_1)_{min}|,|(B_1)_{max}|\} = 1/N + 1/2q$.

Next consider the term $b + d \neq 0 \mod q$. For $0 \leq z < q$ let $\mu_z(v_1, v_2) = \sum_{x \in Q} \xi^{zx} \sum_{b=0}^{q-1} \hat{f}_{v_2}(b)\hat{g}_{v_1}(z-b)$. Then we have

$$\mu(v_1, v_2) = B_0 + B_1 + \sum_{z=1}^{q-1}\mu_z(v_1, v_2).$$

We can bound $\mu_z(v_1, v_2)$ by

$$\mu_z(v_1, v_2) = \sum_{x \in Q} \xi^{zx} \sum_{b=0}^{q-1} \hat{f}_{v_2}(b)\hat{g}_{v_1}(z-b)$$

$$= \sum_{x \in Q} \xi^{zx} \sum_{b=0}^{q-1} \frac{1}{q}\sum_{d=0}^{t_{v_2}}\xi^{b(v_2+Nd)}\frac{1}{q}\sum_{c=\lceil iq/N\rceil}^{\lfloor(i+1)q/N\rfloor}\xi^{-c(z-b)}$$

$$= \frac{\sigma(z)}{q^2}\xi^{-cz}\sum_{d=0}^{t_{v_2}}\sum_{c=\lceil iq/N\rceil}^{\lfloor(i+1)q/N\rfloor}\sum_{b=0}^{q-1}\xi^{b(v_2+Nd+c)}.$$

First,

$$\sum_{b=0}^{q-1} \xi^{b(v_2+Nd+c)} = \begin{cases} q & \text{if } d \in D \\ 0 & \text{otherwise.} \end{cases}$$

Second, by applying Weil's theorem, we can get $|\sigma(z)| \leq (q^{1/2}+1)/2$. Notice here if $d \in D$, then we can take $c = q - v_2 - Nd$. Then

$$|\mu_z(v_1, v_2)| = \left| \frac{\sigma(z)}{q^2} \xi^{-cz} \sum_{d=0}^{t_{v_2}} \sum_{c=\lceil iq/N \rceil}^{\lfloor (i+1)q/N \rfloor} \sum_{b=0}^{q-1} \xi^{b(v_2+Nd+c)} \right|$$

$$\leq \frac{q^{1/2}+1}{2q} \left| \sum_{d \in D} \xi^{z(-q+v_2+Nd)} \right|$$

$$= \frac{q^{1/2}+1}{2q} \left| \sum_{d \in D} \xi^{zNd} \right|$$

$$\leq \frac{q^{1/2}+1}{2q} \left| \sum_{d=0}^{t_{v_2}} \xi^{zNd} \right|$$

$$\leq \frac{q^{1/2}+1}{2q} \left| \frac{\sin(\pi z N(t_{v_2}+1)/q)}{\sin(\pi z N/q)} \right|.$$

Using similar techniques to those in the proof of Lemma 1, we can get

$$|\mu(v_1, v_2)| = \left| B_0 + B_1 + \sum_{z=1}^{q-1} \mu_z(v_1, v_2) \right|$$

$$< \frac{q-1}{2N^2} + \frac{1}{N} + \frac{1}{2q} + \frac{q^{1/2}+1}{2q} \sum_{z=1}^{q-1} \left| \frac{\sin(\pi z N(t_{v_2}+1)/q)}{\sin(\pi z N/q)} \right|$$

$$= \frac{q-1}{2N^2} + \frac{1}{N} + \frac{1}{2q} + \frac{q^{1/2}+1}{2} \left(1 + \ln\left(\frac{q-1}{2} \right) \right).$$

Theorem 3. *If* $0 \leq v_1, v_2 \leq N - 1$, *then*

$$\left| \mu(v_1, v_2) - \frac{q-1}{2N^2} \right| < \frac{1}{N} + \frac{1}{2q} + \frac{q^{1/2}+1}{2} \left(1 + \ln\left(\frac{q-1}{2} \right) \right).$$

3   A Sharper Bound When $N = 2$

Theorem 4. *Let* $\mathbf{a} = a_0, a_1, a_2, \ldots$ *be a binary half-ℓ-sequence with* $q \equiv 1 \mod 8$ *and* q *an odd prime. Then* $\mathbf{a}$ *is balanced.*

Proof. Since $q \equiv 1 \mod 8$ and q is an odd prime, the order of 2 is $(q-1)/2$. Then we have $2^{(q-1)/2} \equiv 1 \mod q$. As a result, $2^{(q-1)/4} \equiv \pm 1 \mod q$. Because the order of 2 is $(q-1)/2$, $2^{(q-1)/4} \not\equiv 1 \mod q$, we have $2^{(q-1)/4} \equiv -1 \mod q$.

There is an integer h so that for all $i \geq 0$ we have

$$a_i \equiv 2^{-i}h \mod q \mod 2.$$

Consider a_j where $j \in [0, (q-1)/2)$. Then we have

$$a_j \equiv 2^{-j}h \mod q \mod 2.$$

and

$$a_{\frac{q-1}{4}+j} \equiv 2^{-(\frac{q-1}{4}+j)}h \mod q \mod 2$$
$$\equiv \left(2^{-\frac{q-1}{4}}2^{-j}\right)h \mod q \mod 2$$
$$\equiv -2^{-j}h \mod q \mod 2$$
$$\equiv \left(q - 2^{-j}h\right) \mod q \mod 2,$$

which is the complementary bit to a_j. So the first half of half-ℓ-sequence **a** is the bit-wise complement of its second half. Then the numbers of 1's and 0's in **a** are equal. So **a** is balanced. □

4   Experimental Results for the One Symbol Case

In this section we analyze the imbalance properties of half-ℓ-sequences by experiments. In other words, we investigate how tight the bound is in Theorem 2. Let

$$\mathbf{var}_v = |\mu(v) - (q-1)/2N|, \; v \in \{0, 1, \cdots, N-1\}$$

$$\mathbf{bound} = (1 + \ln((q-1)/2)(q^{1/2} + 1)/2.$$

The quantity **var**$_v$ is the difference between the number of occurrence of v in one period of sequence **a** and the average occurrence. Let **max**$_{var}$ = max$\{$**var**$_v$: $0 \leq v < N\}$. We define **max_ratio** = **max**$_{var}$/**bound**.

The smaller **max_ratio** is, the more balanced the sequence is. Ideally, for a pseudo-random sequence, we would like the **max_ratio** to be close to zero. We generated the sequences for corresponding q and calculated the **max_ratio** for these qs. It is impractical to calculate **var**$_v$ for half-ℓ-sequences with big qs and $N = 2^{32}$ or 2^{16} as discussed in Lee and Park's paper. As a result, we choose half-ℓ-sequences with smaller Ns and qs for investigation.

In the experiment, we generated some connection integers q of the form $q = 2p + 1$ with p and q prime and $q \equiv -1 \mod 8$. The sequences generated by an FCSR with those connection integers are half-ℓ-sequences or ℓ-sequences. Note that when $N \geq 2^3$ the sequences generated are all half-ℓ-sequences and these may not be the only half-ℓ-sequences, they are just the easiest to find. We would like to see how **max_ratio** changes as the connection integers increase for a particular N. We have done experiments for $N = 2^3, 2^4$ and 2^5. For each value of N, we generate the **max_ratio** with FCSR sizes 2, 3 and 4. Note that if the size of an FCSR is m, then the corresponding connection integer $q \in (N^m, N^{m+1})$. Figure 1 shows that the **max_ratio** for $N = 8$ with FCSR size 2, 3 and 4 is greater than 0.02. Figure 2 shows that the **max_ratio** for $N = 16$ with FCSR size 2, 3 and 4 is greater or equal to 0.02. Figure 3 shows that the **max_ratio** for $N = 32$ with FCSR size 2, 3 and 4 is greater than 0.01. As we can observe from the three figures, there is no increase or decrease pattern as q increases. It also shows that there are many qs with **max_ratio** much higher. There is no known way to find the best qs if the period is large enough to be useful. **max**$_{var}$ is the product of **max_ratio** and **bound**. As the connection integer q increases, the **bound** will increase accordingly. For a specific **max_ratio**, the **max**$_{var}$ will increase accordingly, and the more imbalanced the sequence will be.

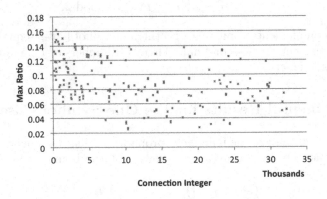

Fig. 1. Max ratio for N = 8 with FCSR size 2, 3 and 4

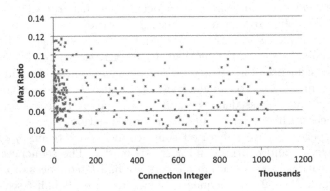

Fig. 2. Max ratio for N = 16 with FCSR size 2, 3 and 4

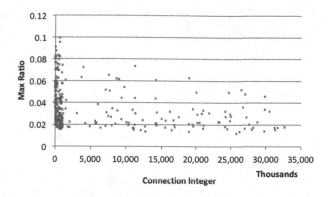

Fig. 3. Max ratio for N = 32 with FCSR size 2, 3 and 4

5 Conclusion and Open Questions

In this paper, we discuss the imbalance properties of half-ℓ-sequences. We see that this type of sequences is not uniformly distributed and their period is only half of the connection integer. It is preferred to have uniformly distributed and long period sequences such as ℓ-sequences. However, there is a tradeoff between speed and statistical properties, since the implementation of FCSRs with N primitive modulo q gives rise to new challenges. We will investigate this area in the future.

References

1. Canteaut, A.: Open problems related to algebraic attacks on stream ciphers. In: Ytrehus, Ø. (ed.) WCC 2005. LNCS, vol. 3969, pp. 120–134. Springer, Heidelberg (2006)
2. Courtois, N., Meier, W.: Algebraic attacks on stream ciphers with linear feedback. In: Biham, E. (ed.) EUROCRYPT 2003. LNCS, vol. 2656, pp. 345–359. Springer, Heidelberg (2003)
3. Courtois, N.T.: General principles of algebraic attacks and new design criteria for cipher components. In: Dobbertin, H., Rijmen, V., Sowa, A. (eds.) AES 2005. LNCS, vol. 3373, pp. 67–83. Springer, Heidelberg (2005)
4. Goresky, M., Klapper, A.: Algebraic Shift Register Sequences. Cambridge University Press, Cambridge (2012)
5. Lee, D., Park, S.: Word-based FCSRs with fast software implementations. J. Commun. Netw. **13**(1), 1–5 (2011)
6. Lidl, R., Niederreiter, H.: Finite Fields in Encyclopedia of Mathematics. Cambridge University Press, Cambridge (1983)

Crosscorrelation of Sequences

A Note on Cross-Correlation Distribution Between a Ternary m-Sequence and Its Decimated Sequence

Yongbo Xia[1][✉], Tor Helleseth[2], and Gaofei Wu[3]

[1] Department of Mathematics and Statistics,
South-Central University for Nationalities, Wuhan 430074, China
xia@mail.scuec.edu.cn
[2] Department of Informatics, University of Bergen, 5020 Bergen, Norway
Tor.Helleseth@ii.uib.no
[3] State Key Laboratory of Integrated Service Networks,
Xidian University, Xi'an 710071, China
gaofei.wu@student.uib.no

Abstract. Let r be an arbitrary positive integer greater than 1 and $n = 3r$. For the decimation $d = 3^r + 2$ or $3^{2r} + 2$, the cross-correlation distribution between a ternary m-sequence and its d-decimated sequence is completely determined. The result presented in this paper generalizes the recent work of Zhang, Li, Feng and Ge, and settles a conjecture proposed by them.

Keywords: m-Sequence · Cross-correlation distribution · Decimated sequence · Exponential sum

1 Introduction

Throughout this paper, we always assume that p is an odd prime unless otherwise stated. Let $\{s(t)\}_{t=0}^{p^n-2}$ be a p-ary m-sequence of period $p^n - 1$, where n is a positive integer. For a positive integer d satisfying $\gcd(d, p^n - 1) = 1$, the d-decimated sequence of $\{s(t)\}$, denoted by $\{s(dt)\}$, is also a p-ary m-sequence. Here d is said to be a decimation. The cross-correlation function between $\{s(t)\}$ and its d-decimated sequence $\{s(dt)\}$ is defined by

$$C_d(\tau) = \sum_{t=0}^{p^n-2} \zeta_p^{s(t+\tau)-s(dt)}$$

where $\zeta_p = e^{\frac{2\pi\sqrt{-1}}{p}}$ is a primitive complex p-th root of unity and $0 \leq \tau < p^n - 1$.

In the theory of sequences, one is interested in finding the cross-correlation distribution between a p-ary m-sequence $\{s(t)\}$ and its d-decimated sequence $\{s(dt)\}$, i.e., determining the multiset

$$\{C_d(\tau) \,|\, 0 \leq \tau < p^n - 1\}.$$

© Springer International Publishing Switzerland 2014
K.-U. Schmidt and A. Winterhof (Eds.): SETA 2014, LNCS 8865, pp. 249–259, 2014.
DOI: 10.1007/978-3-319-12325-7_21

This problem has been extensively studied during the past decades because of its wide applications in sequence design [1,4,6,7,9–12]. For an overview on this topic, the reader is referred to [1] and references therein.

In a recent paper [12], for an integer $r \geq 2$, $n = 3r$ and $d = 3^r + 2$ or $3^{2r} + 2$, under the condition that $\gcd(r, 3) = 1$, Zhang et. al. derived the cross-correlation distribution between a ternary m-sequence of period $3^n - 1$ and its d-decimated sequence. Based on numerical experiments, they conjectured that their result also holds for $\gcd(r, 3) = 3$.

Motivated by their conjecture, in this paper we further investigate the cross-correlation distribution between a ternary m-sequence of period $3^{3r} - 1$ and its d-decimated sequence with $d = 3^r + 2$ or $3^{2r} + 2$. For arbitrary positive integer $r \geq 2$, the corresponding cross-correlation distribution is determined by a unified method. Hence, the result in [12] is generalized to any positive integer $r \geq 2$ and the conjecture proposed there is confirmed. Our proof uses the same basic idea as that in [12] but some modifications are made. The key point of our method is that we find a suitable irreducible polynomial over $\mathbb{F}_{3^r}$ of degree 3 to yield the finite field $\mathbb{F}_{3^n}$, and then we compute a direct representation of $\mathrm{Tr}_r^n(\gamma x + x^d)$ as a function of three variables over $\mathbb{F}_{3^r}$, where $x \in \mathbb{F}_{3^n}$ is a variable, γ is a given element of $\mathbb{F}_{3^n}$ and $\mathrm{Tr}_r^n(\cdot)$ is the trace function from $\mathbb{F}_{3^n}$ to $\mathbb{F}_{3^r}$ [8]. This representation makes the evaluation of the cross-correlation function possible.

The remainder of this paper is organized as follows. In Sect. 2, we introduce some preliminaries. In Sect. 3, for $r \geq 2$, $n = 3r$ and $d = 3^r + 2$ or $3^{2r} + 2$, we derive the cross-correlation distribution between a ternary m-sequence and its d-decimated sequence. The concluding remarks are given in Sect. 4.

2   Preliminaries

For a positive integer n, let $\mathbb{F}_{p^n}$ denote the finite field with p^n elements and $\mathbb{F}_{p^n}^* = \mathbb{F}_{p^n} \setminus \{0\}$. The trace function from $\mathbb{F}_{p^n}$ to its subfield $\mathbb{F}_{p^e}$ is defined by [8]

$$\mathrm{Tr}_e^n(x) = \sum_{i=0}^{\frac{n}{e}-1} x^{p^{ei}},$$

where $x \in \mathbb{F}_{p^n}$ and e is a divisor of n. For $e > 1$, it is well known that $\mathrm{Tr}_1^e(\mathrm{Tr}_e^n(x)) = \mathrm{Tr}_1^n(x)$ for all $x \in \mathbb{F}_{p^n}$. Let α be a primitive element of $\mathbb{F}_{p^n}$. After a suitable cyclic shift, a p-ary m-sequence $\{s(t)\}$ can be written in terms of the trace function as

$$s(t) = \mathrm{Tr}_1^n(\alpha^t),$$

and its d-decimated sequence $\{s(dt)\}$ is given by

$$s(dt) = \mathrm{Tr}_1^n(\alpha^{dt}).$$

Then, the cross-correlation function between $\{s(t)\}$ and $\{s(dt)\}$ can be expressed by

$$C_d(\tau) = \sum_{t=0}^{p^n-2} \zeta_p^{\operatorname{Tr}_1^n(\alpha^{t+\tau}) - \operatorname{Tr}_1^n(\alpha^{dt})}$$

$$= \sum_{x \in \mathbb{F}_{p^n}} \zeta_p^{\operatorname{Tr}_1^n(\gamma x + x^d)} - 1,$$

where $\zeta_p = e^{\frac{2\pi\sqrt{-1}}{p}}$ and $\gamma = -\alpha^\tau$. When τ runs through $\{0, 1, \cdots, p^n - 2\}$, $\gamma = -\alpha^\tau$ runs through $\mathbb{F}_{p^n}^*$. Thus, we usually investigate

$$C_d(\gamma) = \sum_{x \in \mathbb{F}_{p^n}} \zeta_p^{\operatorname{Tr}_1^n(\gamma x + x^d)} - 1, \ \gamma \in \mathbb{F}_{p^n}^* \tag{1}$$

instead of $C_d(\tau)$.

The following three lemmas are basic results about irreducible polynomials over finite fields and will be employed in the sequel.

Lemma 1. *([8, Corollary 3.79]) Let p be a prime and n be a positive integer. Let $a \in \mathbb{F}_{p^n}$. Then, the trinomial $x^p - x - a$ is irreducible in $\mathbb{F}_{p^n}[x]$ if and only if $\operatorname{Tr}_1^n(a) \neq 0$.*

The following result comes from Exercise 3.85 in [8]. We give the proof below for the sake of completeness.

Lemma 2. *Assume p is a prime and q is a power of p. If $x^p - x - a$ is irreducible over $\mathbb{F}_q$ and β is a root of this trinomial in an extension field of $\mathbb{F}_q$, then $x^p - x - a\beta^{p-1}$ is irreducible over $\mathbb{F}_q(\beta)$, where $\mathbb{F}_q(\beta)$ denotes the extension field of $\mathbb{F}_q$ obtained by adjoining the element β to $\mathbb{F}_q$, and is exactly $\mathbb{F}_{q^p}$.*

Proof. Assume $q = p^n$. Since β is a root of $x^p - x - a$, then all the roots of $x^p - x - a$ are given by the p distinct elements $\beta, \beta^q, \cdots, \beta^{q^{p-1}}$. For convenience, let $x_i = \beta^{q^i}$, $i = 0, 1, \cdots, p-1$. By $\beta^p - \beta - a = 0$, one has $\beta^{p-1} = 1 + \frac{a}{\beta}$. Thus,

$$\operatorname{Tr}_1^{np}(a\beta^{p-1}) = \operatorname{Tr}_1^n\left(a\operatorname{Tr}_n^{np}\left(1 + \frac{a}{\beta}\right)\right)$$

$$= \operatorname{Tr}_1^n\left(a^2\operatorname{Tr}_n^{np}\left(\frac{1}{\beta}\right)\right). \tag{2}$$

Note that

$$\operatorname{Tr}_n^{np}\left(\frac{1}{\beta}\right) = \frac{1}{\beta} + \frac{1}{\beta^q} + \cdots + \frac{1}{\beta^{q^{p-1}}}$$

$$= \frac{1}{x_0} + \frac{1}{x_1} + \cdots + \frac{1}{x_{p-1}}$$

$$= \frac{\sum\limits_{i=0}^{p-1} \prod\limits_{j \neq i} x_j}{\prod\limits_{i=0}^{p-1} x_i} \tag{3}$$

$$= \frac{-1}{a},$$

where the last equality holds due to the relation between roots and coefficients of the polynomial equation $x^p - x - a = 0$. By (2) and (3), we have

$$\operatorname{Tr}_1^{np}(a\beta^{p-1}) = -\operatorname{Tr}_1^n(a),$$

which is not equal to zero since $x^p - x - a$ is irreducible over $\mathbb{F}_q$. Then, from Lemma 1, the desired result follows. □

Lemma 3. *([8, Corollary 3.47]) Let p be a prime and q be a power of p. An irreducible polynomial over $\mathbb{F}_q$ of degree s remains irreducible over $\mathbb{F}_{q^k}$ if and only if $\gcd(s, k) = 1$.*

Let ψ be a multiplicative character and χ be an additive character of $\mathbb{F}_{p^n}$. Then, the Gaussian sum $G(\psi, \chi)$ is defined by [8]

$$G(\psi, \chi) = \sum_{x \in \mathbb{F}_{p^n}^*} \psi(x)\chi(x).$$

Gaussian sums are important types of exponential sums for finite fields, and only for certain special characters, the associated Gaussian sums can be evaluated explicitly. Let

$$\chi^{(n)}(x) = e^{\frac{2\pi\sqrt{-1}}{p} \operatorname{Tr}_1^n(x)}, \quad x \in \mathbb{F}_{p^n}$$

be the canonical additive character of $\mathbb{F}_{p^n}$, and $\eta^{(n)}$ be the quadratic character of $\mathbb{F}_{p^n}$ [8]. The following two results related to Gaussian sums will turn out to be useful in the sequel.

Lemma 4. *([8, Theorem 5.15]) Let p be an odd prime and n be a positive integer. Let $\chi^{(n)}$ be the canonical additive character of $\mathbb{F}_{p^n}$ and $\eta^{(n)}$ be the quadratic character of $\mathbb{F}_{p^n}$. Then, the associated Gaussian sum*

$$G(\eta^{(n)}, \chi^{(n)}) = \begin{cases} (-1)^{n-1}p^{\frac{n}{2}}, & \text{if } p \equiv 1 \ (\mathrm{mod}\ 4), \\ (-1)^{n-1}(\sqrt{-1})^n p^{\frac{n}{2}}, & \text{if } p \equiv 3 \ (\mathrm{mod}\ 4). \end{cases}$$

Lemma 5. *([8, Theorem 5.33]) Let χ be a nontrivial additive character of $\mathbb{F}_{p^n}$ with p odd, and let $f(x) = a_2 x^2 + a_1 x + a_0 \in \mathbb{F}_{p^n}[x]$ with $a_2 \neq 0$. Then,*

$$\sum_{x \in \mathbb{F}_{p^n}} \chi(f(x)) = \chi(a_0 - a_1^2(4a_2)^{-1})\eta^{(n)}(a_2)G(\eta^{(n)}, \chi),$$

where $\eta^{(n)}$ is the quadratic character of $\mathbb{F}_{p^n}$.

From the properties of the trace function, the following lemma can be obtained, and it is useful in finding the cross-correlation distribution.

Lemma 6. *([9, Theorem 2.4] and [6, Theorem 3.4]) Let $C_d(\gamma)$ be defined in (1) with $\gcd(d, p^n - 1) = 1$. Then,*

(i) $\sum_{\gamma \in \mathbb{F}_{p^n}^*} (C_d(\gamma) + 1) = p^n$;

(ii) $\sum_{\gamma \in \mathbb{F}_{p^n}^*} (C_d(\gamma) + 1)^2 = p^{2n}$;

(iii) $\sum_{\gamma \in \mathbb{F}_{p^n}^*} (C_d(\gamma) + 1)^3 = p^{2n}N$, *where N is the number of $x \in \mathbb{F}_{p^n}$ such that*

$$(x + 1)^d = x^d + 1.$$

For convenience, let

$$S_d(\gamma) = C_d(\gamma) + 1, \tag{4}$$

where $C_d(\gamma)$ is given by (1). In the sequel, we will mainly deal with $S_d(\gamma)$.

3   Main Result and Its Proof

From now on, we will focus on $p = 3$ and adopt the following notations:

- r is an arbitrary positive integer greater than 1;
- t is the maximal power of 3 that divides r. Then, r can be written as $r = 3^t k$ with $\gcd(k, 3) = 1$;
- $n = 3r$ and $d = 3^r + 2$ or $3^{2r} + 2$. Then, $\gcd(d, 3^n - 1) = 1$;
- $\{s(t)\}$ denotes a ternary m-sequence of period $3^n - 1$.

Our main theorem is stated as follows.

Theorem 1. *Let $r \geq 2$, $n = 3r$ and $d = 3^r + 2$ or $3^{2r} + 2$. Then, the cross-correlation distribution between a ternary m-sequence $\{s(t)\}$ and its d-decimated sequence $\{s(dt)\}$ is given in Table 1 if r is even and in Table 2 if r is odd.*

Compared with Theorem 2.5 in [12], our result presented in Theorem 1 generalizes their work and settles the conjecture proposed there. Note that when $r = 1$, $d = 3^r + 2$ or $3^{2r} + 2$ gives a three-valued cross-correlation function. Thus, we need $r \geq 2$ here. In order to prove Theorem 1, we need to make some preparations.

The following lemma is a consequence of Lemmas 1–3, and plays an important role in this paper.

Lemma 7. *For any nonnegative integer t, there exists $a \in \mathbb{F}_{3^{3t}}$ such that $x^3 - x - a$ is irreducible over $\mathbb{F}_{3^{3t}}$ and also irreducible over $\mathbb{F}_{3^{3tk}}$, where $\gcd(k, 3) = 1$.*

Proof. By Lemma 1, we know that $x^3 - x - 2$ is irreducible over $\mathbb{F}_3$. Applying Lemma 2 to $x^3 - x - 2$, we can obtain an element $a_1 \in \mathbb{F}_{3^3}$ such that $x^3 - x - a_1$ is irreducible over $\mathbb{F}_{3^3}$. Repeating this process $t - 1$ times, we can find an element

Table 1. Cross-correlation distribution for d if r is even

Value	Frequency
-1	$\frac{3^{3r} + 3^{2r}}{2} - 3^r - 1$
$3^{2r} - 1$	3^r
$3^{\frac{3r}{2}} - 1$	$\frac{3^{3r-1} - 3^{2r-1}}{2}$
$-3^{\frac{3r}{2}} - 1$	$\frac{3^{3r-1} - 3^{2r-1}}{2}$
$2 \cdot 3^{\frac{3r}{2}} - 1$	$\frac{3^{3r-1} - 3^{2r-1}}{4}$
$-2 \cdot 3^{\frac{3r}{2}} - 1$	$\frac{3^{3r-1} - 3^{2r-1}}{4}$

Table 2. Cross-correlation distribution for d if r is odd

Value	Frequency
-1	$2 \cdot 3^{3r-1} + 3^{2r-1} - 3^r - 1$
$3^{2r} - 1$	3^r
$3^{\frac{3r+1}{2}} - 1$	$\frac{3^{3r-1} - 3^{2r-1}}{2}$
$-3^{\frac{3r+1}{2}} - 1$	$\frac{3^{3r-1} - 3^{2r-1}}{2}$

$a \in \mathbb{F}_{3^{3t}}$ such that $x^3 - x - a$ is irreducible over $\mathbb{F}_{3^{3t}}$. Furthermore, note that $\gcd(3, k) = 1$. Then, by Lemma 3, $x^3 - x - a$ is also irreducible over $\mathbb{F}_{3^{3tk}}$.   □

For any given $r \geq 2$, recall that r can be written as $r = 3^t k$, where $\gcd(k, 3) = 1$. By Lemma 7, we can choose an element $a \in \mathbb{F}_{3^{3t}}$ such that $g(x) = x^3 - x - a$ is irreducible over $\mathbb{F}_{3^{3t}}$. Then, $g(x)$ is also irreducible over $\mathbb{F}_{3^r}$. In the sequel, once r is given, we will fix such an irreducible polynomial $g(x)$. Let ω be a root of $g(x)$. Then, adjoining ω to $\mathbb{F}_{3^r}$ yields the finite field $\mathbb{F}_{3^n}$, i.e., $\mathbb{F}_{3^n} = \mathbb{F}_{3^r}(\omega)$, and thus $\{1, \omega, \omega^2\}$ is a basis of $\mathbb{F}_{3^n}$ over $\mathbb{F}_{3^r}$. The following properties about ω are very useful in the sequel.

Lemma 8. *With the notation above, let $g(x) = x^3 - x - a$ be an irreducible polynomial in $\mathbb{F}_{3^r}[x]$ and ω be a root of $g(x)$. Then, ω has the following properties:*

(i) $\mathrm{Tr}_r^n(1) = 0$, $\mathrm{Tr}_r^n(\omega) = 0$ *and* $\mathrm{Tr}_r^n(\omega^2) \neq 0$ *(the trace function of the basis $\{1, \omega, \omega^2\}$);*
(ii) Denote $\theta = \mathrm{Tr}_r^n(\omega^2)$. *Then,* $\mathrm{Tr}_r^n(\omega^3) = 0$, $\mathrm{Tr}_r^n(\omega^4) = \theta$, $\mathrm{Tr}_r^n(\omega^5) = a\theta$ *and* $\mathrm{Tr}_r^n(\omega^6) = \theta$;
(iii) $\omega^{3^r} = \omega + \mathrm{Tr}_1^r(a)$ *and* $\omega^{3^{2r}} = \omega + \mathrm{Tr}_1^{2r}(a)$. *Moreover,* $\mathrm{Tr}_1^r(a) \neq 0$ *and* $\mathrm{Tr}_1^{2r}(a) \neq 0$.

Proof. (i) The trace function $\mathrm{Tr}_r^n(x)$ from $\mathbb{F}_{3^n}$ to $\mathbb{F}_{3^r}$ is a linear transformation and this mapping is onto. Note that $\{1, \omega, \omega^2\}$ is a basis of $\mathbb{F}_{3^n}$ over $\mathbb{F}_{3^r}$. Thus, at least one of $\mathrm{Tr}_r^n(1)$, $\mathrm{Tr}_r^n(\omega)$ and $\mathrm{Tr}_r^n(\omega^2)$ is not equal to zero. It is easily seen that $\mathrm{Tr}_r^n(1) = 0$. If we can prove $\mathrm{Tr}_r^n(w) = 0$, then the desired conclusion follows. Below we show $\mathrm{Tr}_r^n(\omega) = 0$. Since $g(x) = x^3 - x - a$ is irreducible over $\mathbb{F}_{3^r}$ and ω is a root of it, then all roots of $g(x)$ are given by the three distinct elements ω, ω^{3^r} and $\omega^{3^{2r}}$. By the relation between roots and coefficients of the equation $g(x) = 0$, $-\left(\omega + \omega^{3^r} + \omega^{3^{2r}}\right)$ is equal to the coefficient of x^2 in $g(x)$. The latter is zero. Thus, $\omega + \omega^{3^r} + \omega^{3^{2r}} = \mathrm{Tr}_r^n(\omega) = 0$.

(ii) By $\omega^3 = \omega + a$ and (i), the values of $\mathrm{Tr}_r^n(\omega^i)$, $i = 3, 4, 5, 6$, can be computed.

(iii) For any positive integer i, by $\omega^3 = \omega + a$, we have

$$\omega^{3^i} = \omega^{3^{i-1}} + a^{3^{i-1}} = \omega + a + a^3 + \cdots + a^{3^{i-1}}. \tag{5}$$

Note that $a \in \mathbb{F}_{3^r} \subseteq \mathbb{F}_{3^{2r}}$. Thus,

$$a + a^3 + \cdots + a^{3^{r-1}} = \mathrm{Tr}_1^r(a) \text{ and } a + a^3 + \cdots + a^{3^{2r-1}} = \mathrm{Tr}_1^{2r}(a). \qquad (6)$$

Combining (5) and (6), we have

$$\omega^{3^r} = \omega + \mathrm{Tr}_1^r(a) \text{ and } \omega^{3^{2r}} = \omega + \mathrm{Tr}_1^{2r}(a).$$

Furthermore, note that $x^3 - x - a$ is irreducible over $\mathbb{F}_{3^r}$. By Lemma 1, $\mathrm{Tr}_1^r(a) \neq 0$. Since $\mathrm{Tr}_1^{2r}(a) = -\mathrm{Tr}_1^r(a)$, then $\mathrm{Tr}_1^{2r}(a)$ is not equal to zero either. $\qquad \square$

Corresponding to $p = 3$, recall from (1) and (4) that

$$S_d(\gamma) = C_d(\gamma) + 1 = \sum_{x \in \mathbb{F}_{3^n}} \zeta_3^{\mathrm{Tr}_1^n(\gamma x + x^d)} = \sum_{x \in \mathbb{F}_{3^n}} \zeta_3^{\mathrm{Tr}_1^r\left(\mathrm{Tr}_r^n(\gamma x + x^d)\right)}. \qquad (7)$$

The basic technique for calculating $S_d(\gamma)$ is to transform $S_d(\gamma)$ into an exponential sum over the subfield $\mathbb{F}_{3^r}$. This technique originated from [2] and was employed to compute the Walsh spectrum of some power functions over finite fields of even characteristic [2,3,5]. Here, following the proof of Theorem 2.5 in [12], we also calculate $S_d(\gamma)$ by this technique.

Note that since $\mathbb{F}_{3^n} = \mathbb{F}_{3^r}(\omega)$. Then, each $x \in \mathbb{F}_{3^n}$ can be uniquely represented in the form

$$x = x_0 + x_1\omega + x_2\omega^2, \ x_i \in \mathbb{F}_{3^r}, \ i = 0, 1, 2.$$

For a given $\gamma = \gamma_0 + \gamma_1\omega + \gamma_2\omega^2 \in \mathbb{F}_{3^n}$ with $\gamma_i \in \mathbb{F}_{3^r}$, our first step is to compute an expression of $\mathrm{Tr}_r^n\left(\gamma x + x^d\right)$ as a function of x_i, $i = 0, 1, 2$.

Lemma 9. *With the notation above, let $\beta = \mathrm{Tr}_1^r(a)$ if $d = 3^r + 2$, and $\beta = \mathrm{Tr}_1^{2r}(a)$ if $d = 3^{2r} + 2$. Let $\theta = \mathrm{Tr}_r^n\left(\omega^2\right)$, and $x = x_0 + x_1\omega + x_2\omega^2$ with $x_i \in \mathbb{F}_{3^r}$. For a given $\gamma = \gamma_0 + \gamma_1\omega + \gamma_2\omega^2 \in \mathbb{F}_{3^n}$ with $\gamma_i \in \mathbb{F}_{3^r}$, we have*

(i) $\mathrm{Tr}_r^n\left(x^d\right) = \left(-x_0 x_2^2 + 2\beta x_1 x_2^2 + x_1^2 x_2 + \beta x_1^3 - (1 + \beta a) x_2^3\right)\theta$;
(ii) $\mathrm{Tr}_r^n\left(\gamma x\right) = \left(\gamma_0 x_2 + \gamma_1 x_1 + \gamma_2 x_0 + \gamma_2 x_2\right)\theta$.

Proof. (i) We only give the proof for $d = 3^r + 2$ and the case $d = 3^{2r} + 2$ can be proved similarly. By Lemma 8 (iii), we have

$$\begin{aligned}
x^d &= \left(x_0 + x_1\omega + x_2\omega^2\right)^{3^r+2} \\
&= \left(x_0 + x_1\omega^{3^r} + x_2\omega^{2\cdot3^r}\right)\left(x_0 + x_1\omega + x_2\omega^2\right)^2 \\
&= \left(x_0 + x_1(\omega + \beta) + x_2(\omega + \beta)^2\right)\left(x_0 + x_1\omega + x_2\omega^2\right)^2.
\end{aligned}$$

Note that $\beta \in \mathbb{F}_3^*$ and thus $\beta^2 = 1$. Expanding the right hand side of the above equation, we will get a linear combination of $1, \omega, \omega^2, \cdots, \omega^6$ with coefficients lying in $\mathbb{F}_{3^r}$. Then, with the help of Lemma 8 (i)–(ii), the desired conclusion follows.

(ii) Similarly, by expanding $\left(x_0 + x_1\omega + x_2\omega^2\right)\left(\gamma_0 + \gamma_1\omega + \gamma_2\omega^2\right)$ as a linear combination of ω^i, $i = 0, 1, \cdots, 4$, the expression for $\mathrm{Tr}_r^n(\gamma x)$ can also be calculated. $\qquad \square$

Remark 1. For any given $r \geq 2$, Lemma 9 gives a unified expression for $\mathrm{Tr}_r^n(\gamma x + x^d)$, which depends on $\mathrm{Tr}_r^n(\omega^2)$, a and β (actually depends only on a). This expression is crucial to the proof of Theorem 1, and it enables us to transform $S_d(\gamma)$ in (7) into an exponential sum over $\mathbb{F}_{3^r}$.

The proof of the following lemma is the same as that of Lemma 2.4 in [12], where the condition $\gcd(r, 3) = 1$ on r turns out to be unnecessary.

Lemma 10. *Let $r \geq 2$, $n = 3r$ and $d = 3^r + 2$ or $d = 3^{2r} + 2$. Then, $(x+1)^d = x^d + 1$ has 3^r solutions in $\mathbb{F}_{3^n}$.*

With the above preparations, now we can give the proof of Theorem 1.

The proof of Theorem 1: Due to (7), determining the cross-correlation distribution is equivalent to determining the value distribution of $S_d(\gamma)$ as γ runs through $\mathbb{F}_{3^n}^*$. By (7) and Lemma 9, we have

$$
\begin{aligned}
&S_d(\gamma) \\
&= \sum_{x_0, x_1, x_2 \in \mathbb{F}_{3^r}} \zeta_3^{\mathrm{Tr}_1^r\left(\theta\left(-x_0 x_2^2 + 2\beta x_1 x_2^2 + x_1^2 x_2 + \beta x_1^3 - (1+\beta a)x_2^3 + \gamma_0 x_2 + \gamma_1 x_1 + \gamma_2 x_0 + \gamma_2 x_2\right)\right)}.
\end{aligned}
$$
(8)

Further, using the properties of the trace function, we have

$$
\mathrm{Tr}_1^r(\theta \beta x_1^3) = \mathrm{Tr}_1^r\left(\theta^{3^{r-1}} \beta x_1\right),
$$
(9)

and

$$
\mathrm{Tr}_1^r\left(-\theta(1 + \beta a)x_2^3\right) = \mathrm{Tr}_1^r\left(-\theta^{3^{r-1}}\left(1 + \beta a^{3^{r-1}}\right)x_2\right).
$$
(10)

Substituting (9) and (10) into (8), we have

$$
\begin{aligned}
&S_d(\gamma) \\
&= \sum_{x_1, x_2 \in \mathbb{F}_{3^r}} \zeta_3^{\mathrm{Tr}_1^r\left(\theta x_2 x_1^2 + \left(2\beta\theta x_2^2 + \theta^{3^{r-1}}\beta + \theta\gamma_1\right)x_1 + \left(\theta\gamma_0 + \theta\gamma_2 - \theta^{3^{r-1}}\left(1+\beta a^{3^{r-1}}\right)\right)x_2\right)} \\
&\quad \times \sum_{x_0 \in \mathbb{F}_{3^r}} \zeta_3^{\mathrm{Tr}_1^r\left(x_0 \theta\left(\gamma_2 - x_2^2\right)\right)}.
\end{aligned}
$$
(11)

For a given $\gamma = \gamma_0 + \gamma_1 \omega + \gamma_2 \omega^2 \in \mathbb{F}_{3^n}^*$ with $\gamma_i \in \mathbb{F}_{3^r}$, $i = 0, 1, 2$, define

$$
M_\gamma = \left\{x_2 \in \mathbb{F}_{3^r} \mid x_2^2 = \gamma_2\right\}.
$$

We consider the following three cases.

Case 1: $\gamma_2 = 0$. Then, $M_\gamma = \{0\}$. By (11), we have

$$
S_d(\gamma) = 3^r \sum_{x_1 \in \mathbb{F}_{3^r}} \zeta_3^{\mathrm{Tr}_1^r\left(\left(\theta^{3^{r-1}}\beta + \theta\gamma_1\right)x_1\right)},
$$

which implies that $S_d(\gamma) = 3^{2r}$ if $\gamma_1 = -\theta^{3^{r-1}-1}\beta$ and otherwise, $S_d(\gamma) = 0$. Once r, $g(x)$ and d are given, θ and β are fixed and nonzero. Thus, when γ runs through $\mathbb{F}_{3^n}^*$, there are 3^r distinct γ such that $\gamma_2 = 0$ and $\gamma_1 = -\theta^{3^{r-1}-1}\beta$. This means that in this case there are 3^r distinct $\gamma \in \mathbb{F}_{3^n}^*$ such that $S_d(\gamma) = 3^{2r}$. Furthermore, one can conclude that in this case there are $3^{2r} - 1 - 3^r$ distinct $\gamma \in \mathbb{F}_{3^n}^*$ such that $S_d(\gamma) = 0$.

Case 2: γ_2 is a nonsquare in $\mathbb{F}_{3^r}^*$. Then, M_γ is empty. Thus, by (11), we have $S_d(\gamma) = 0$ in this case. Corresponding to this case, there are $\frac{3^r-1}{2} \cdot 3^{2r}$ distinct $\gamma \in \mathbb{F}_{3^n}^*$.

Case 3: γ_2 is a square in $\mathbb{F}_{3^r}^*$. Assume $\gamma_2 = \mu^2$ with $\mu \in \mathbb{F}_{3^r}^*$. Then, $M_\gamma = \{\mu, -\mu\}$. By (11), we have

$$S_d(\gamma) = 3^r \left(\sum_{x_1 \in \mathbb{F}_{3^r}} \zeta_3^{\mathrm{Tr}_1^r\left(\theta\mu x_1^2 + bx_1 + c\right)} + \sum_{x_1 \in \mathbb{F}_{3^r}} \zeta_3^{\mathrm{Tr}_1^r\left(-\theta\mu x_1^2 + bx_1 - c\right)} \right), \tag{12}$$

where

$$b = \left(2\beta\theta\mu^2 + \theta^{3^{r-1}}\beta + \theta\gamma_1\right) \text{ and } c = \left(\theta\gamma_0 + \theta\gamma_2 - \theta^{3^{r-1}}\left(1 + \beta a^{3^{r-1}}\right)\right)\mu.$$

Then, by Lemmas 4 and 5, (12) can be rewritten as

$$S_d(\gamma) = (-1)^{r-1}(\sqrt{-1})^r 3^{\frac{3r}{2}} \left(\chi^{(r)}\left(c - \tfrac{b^2}{\theta\mu}\right)\eta^{(r)}(\theta\mu) + \chi^{(r)}\left(-c + \tfrac{b^2}{\theta\mu}\right)\eta^{(r)}(-\theta\mu) \right),$$

where $\chi^{(r)}$ is the canonical additive character of $\mathbb{F}_{3^r}$ and $\eta^{(r)}$ is the quadratic character of $\mathbb{F}_{3^r}$. Let

$$A = \chi^{(r)}\left(c - \tfrac{b^2}{\theta\mu}\right)\eta^{(r)}(\theta\mu) + \chi^{(r)}\left(-c + \tfrac{b^2}{\theta\mu}\right)\eta^{(r)}(-\theta\mu).$$

Note that $\chi^{(r)}\left(-c + \tfrac{b^2}{\theta\mu}\right)$ is the complex conjugate of $\chi^{(r)}\left(c - \tfrac{b^2}{\theta\mu}\right)$. If r is even, $\eta^{(r)}(-1) = 1$. Then,

$$A = \eta^{(r)}(\theta\mu)\left(\chi^{(r)}\left(c - \tfrac{b^2}{\theta\mu}\right) + \overline{\chi^{(r)}\left(c - \tfrac{b^2}{\theta\mu}\right)}\right)$$

$$= \begin{cases} \pm 2, & \text{if } c = \tfrac{b^2}{\theta\mu}, \\ \pm 1, & \text{if } c \neq \tfrac{b^2}{\theta\mu}, \end{cases}$$

and thus $S_d(\gamma) \in \left\{\pm 3^{\frac{3r}{2}}, \pm 2 \cdot 3^{\frac{3r}{2}}\right\}$. If r is odd, $\eta^{(r)}(-1) = -1$. Then,

$$A = \eta^{(r)}(\theta\mu)\left(\chi^{(r)}\left(c - \tfrac{b^2}{\theta\mu}\right) - \overline{\chi^{(r)}\left(c - \tfrac{b^2}{\theta\mu}\right)}\right)$$

$$= \begin{cases} 0, & \text{if } c = \tfrac{b^2}{\theta\mu}, \\ \pm\sqrt{-3}, & \text{if } c \neq \tfrac{b^2}{\theta\mu}, \end{cases}$$

and $S_d(\gamma)$ belongs to $\left\{0, \pm 3^{\frac{3r+1}{2}}\right\}$.

Combining *Cases 1–3*, we consider the following:

- If r is even, the possible values of $S_d(\gamma)$ are 0, $3^{2r}, 3^{\frac{3r}{2}}, -3^{\frac{3r}{2}}, 2\cdot 3^{\frac{3r}{2}}, -2\cdot 3^{\frac{3r}{2}}$. When γ runs through $\mathbb{F}_{3^n}^*$, assume the numbers of occurrences of the above values are $N_1, N_2, \cdots, N_6$, respectively. First, note that the values 0 and 3^{2r} only appear in *Case 1* and *Case 2*. By the analysis there, we have

$$\begin{cases} N_1 = \left(3^{2r} - 1 - 3^r\right) + \frac{3^r-1}{2}\cdot 3^{2r} = \frac{3^{3r}+3^{2r}}{2} - 3^r - 1, \\ N_2 = 3^r. \end{cases} \tag{13}$$

Then, by Lemma 6, we have

$$\begin{cases} 3^{2r}N_2 + 3^{\frac{3r}{2}}\left(N_3 - N_4\right) + 2\cdot 3^{\frac{3r}{2}}\left(N_5 - N_6\right) = 3^{3r}, \\ 3^{4r}N_2 + 3^{\frac{6r}{2}}\left(N_3 - N_4\right) + 4\cdot 3^{\frac{6r}{2}}\left(N_5 - N_6\right) = 3^{6r}, \\ 3^{6r}N_2 + 3^{\frac{9r}{2}}\left(N_3 - N_4\right) + 8\cdot 3^{\frac{9r}{2}}\left(N_5 - N_6\right) = 3^{7r}. \end{cases} \tag{14}$$

Moreover, we also have

$$N_1 + N_2 + N_3 + N_4 + N_5 + N_6 = 3^{3r} - 1. \tag{15}$$

From (13)–(15), the value distribution of $C_d(\gamma) = S_d(\gamma) - 1$ in Table 1 is obtained.
- If r is odd, the possible values of $S_d(\gamma)$ are 0, $3^{2r}, 3^{\frac{3r+1}{2}}, -3^{\frac{3r+1}{2}}$. By similar analysis as the case where r is even, Table 2 can be obtained.   □

4   Conclusion

For any given $r \geq 2$, $n = 3r$ and the decimation $d = 3^r + 2$ or $3^{2r} + 2$, the cross-correlation distribution between a ternary m-sequence of period $3^n - 1$ and its d-decimated sequence is completely determined. The result generalizes previous work in [12], and settles a conjecture proposed there. The proof in this paper uses the same basic idea as that of Theorem 2.5 in [12], but some modifications are made. The key point of our proof is that for any given $r \geq 2$, we can find an irreducible polynomial in $\mathbb{F}_{3^r}[x]$ of the form $x^3 - x - a$ to yield the finite field $\mathbb{F}_{3^n}$. This allows us to express $\mathrm{Tr}_r^n\left(\gamma x + x^d\right)$ as a function of three variables over $\mathbb{F}_{3^r}$ with a simple form, and makes the calculation of $C_d(\gamma)$ possible.

Acknowledgment. Y. Xia was supported by the National Natural Science Foundation of China (NSFC) under Grant 11301552, and the Natural Science Foundation of Hubei Province under Grant 2012FFB07403. T. Helleseth and G. Wu were supported by the Norwegian Research Council. This work was done while the first author was visiting the Department of Informatics, University of Bergen, Norway, during Sept. 2013 to Sept. 2014. He is grateful for the hospitality and support.

References

1. Choi, S.T., No, J.S.: On the cross-correlation distributions of p-ary m-sequences and their decimated sequences. IEICE Trans. Fundam. **E95–A**(11), 1808–1818 (2012)

2. Cusick, T.W., Dobbertin, H.: Some new three-valued crosscorrelation functions for binary m-sequences. IEEE Trans. Inf. Theory **42**(4), 1238–1240 (1996)
3. Dobbertin, H.: One-to-one highly nonlinear power functions on $GF(2n)$. Appl. Algebra Engrg. Comm. Comput. **9**(2), 139–152 (1998)
4. Dobbertin, H., Helleseth, T., Kumar, P.V., Martinsen, H.: Ternary m-sequences with three-valued cross-correlation function: new decimations of Welch and Niho type. IEEE Trans. Inf. Theory **47**(4), 1473–1481 (2001)
5. Feng, T., Leung, K., Xiang, Q.: Binary cyclic codes with two primitive nonzeros. Sci. China Math. **56**(7), 1403–1412 (2013)
6. Helleseth, T.: Some results about the cross-correlation function between two maximal linear sequences. Discr. Math. **16**, 209–232 (1976)
7. Helleseth, T., Kumar, P.V.: Sequences with low correlation. In: Pless, V., Huffman, C. (eds.) Handbook of Coding Theory, pp. 1767–1853. Elsevier Science, Amsterdam (1998)
8. Lidl, R., Niederreiter, H.: Finite Fields. Encyclopedia of Mathematics, vol. 20. Cambridge University Press, Cambridge (1983)
9. Rosendahl, P.: Niho type cross-correlation functions and related equations. Ph.D. dissertation, Univ. Turku, Turku, Finland (2004)
10. Trachtenberg, H.M.: On the cross-correlation functions of maximal recurring sequences. Ph.D. dissertation, Univ. of Southern California, Log Angeles, CA (1970)
11. Xia, Y., Chen, S.: A new family of p-ary sequences with low correlation constructed from decimated sequences. IEEE Trans. Inf. Theory **58**(9), 6037–6046 (2012)
12. Zhang, T., Li, S., Feng, T., Ge, G.: Some new results on the cross correlation of m-sequences. IEEE Trans. Inf. Theory **60**(5), 3062–3068 (2014)

Prime Numbers in Sequences

Conjectures Involving Sequences and Prime Numbers

Solomon Golomb[✉]

University of Southern California, Electrical Engineering Building 504A,
3740 McClintock Avenue, Los Angeles, CA 90089-2565, USA
sgolomb@usc.edu

Abstract. A 1992 conjecture of Golomb asserts the existence of an infinite increasing sequence $A = \{a_n\}$ of positive integers for which each translate $A_k = \{a_n + k\}$ of $A = A_o$ contains no more than B primes, for some finite bound B. This conjecture is inconsistent with the "Prime k-tuples Conjecture", which asserts that for infinitely many positive integers n, all k numbers $\{n, n + a_1, n + a_2, ..., n + a_{k-1}\}$ are prime, provided that there is no prime number q for which the k integers $\{0, a_1, a_2, ..., a_{k-1}\}$ occupy all q residue classes modulo q. This paper discusses reasons for believing or disbelieving each of these conjectures.

1 Introduction

In 1992, I asked (in [1]) the following two-part question. a. Is there an infinite sequence $A = \{a_n\}$ of positive integers such that $A_k = \{a_n + k\}$ contain only a finite number of prime numbers, for every $k \epsilon Z$, that is, for every integer k, positive, negative, and zero? b. Is there such a sequence $A = \{a_n\}$ for which the number of primes in A_k is less than some finite bound B, for every $k \epsilon Z$?

The answer to question(a) is "yes". It is easy to exhibit such a sequence $A = \{a_n\}$. Specifically let $a_n = ((2n)!)^3$. Then every term in $A = A_o$ is composite. At $k = 1$, $A_1 = \{a_n + 1\} = \{(2n!)^3 + 1\}$ is composite for all n because $x^3 + 1 = (x + 1)(x^2 - x + 1)$, and both factors exceed 1 for all $n \geq 1$. Similarly, $A_{-1} = \{a_n - 1\} = \{(2n!)^3 - 1\}$ is composite for all $n > 1$, since $x^3 - 1 = (x - 1)(x^2 + x + 1)$. Finally, for all k with $|k| > 1$, $A_k = \{a_n + k\} = ((2n!)^3 + k)$ is composite for all $n \geq |k|$, since $|k| > 1$ will be a common factor of $(2n)!$ and k.

The answer to question (b) remains unknown. It is not hard to show [5] that if the answer to question (b) is "yes", then the famous "Prime k-tuples Conjecture" is false.

The "Prime k-tuples Conjecture" asserts that for infinitely many values of n, all k of the numbers $\{n, n + a_1, n + a_2, ..., n + a_{k-1}\}$ are prime, provided only that there is no prime numbers q such that the k numbers $\{0, a_1, a_2, ..., a_{k-1}\}$ occupy all q residue classes modulo q.

In [5], the existence of a sequence A satisfying (b) is called "Golomb's Conjecture." This paper explores the plausibility of that conjecture.

K.-U. Schmidt and A. Winterhof (Eds.): SETA 2014, LNCS 8865, pp. 263–266, 2014.
DOI: 10.1007/978-3-319-12325-7_22

2   The Number of Candidate Sequences

Let s_n be any infinite increasing sequence of positive integers. The number of such sequence is well-known to be uncountably infinite. Then, with $a_n = ((2s(n))!)^3$, the sequence $A = \{a_n\}$ satisfies the condition that each of its translates, $A_k = \{a_n + k\}$, for all $k \epsilon Z$, contains only finitely many primes. If, for any one of the uncountably many such sequence $A = \{a_n\}$ the number of primes in A_k is less than some finite bound B (where B can be arbitrarily large), then Golomb's Conjecture is true.

The famous "Prime Number Theorem" (PNT) states that, with $\pi(x)$ =number of primes $\leq x$, $\pi(x) \sim x/\ln x$ as $x \to \infty$, which is equivalent to the statement $\lim_{x\to\infty} \pi(x)/(x/\ln x) = 1$. (Here $\ln x$ is the logarithm of x to the base e.) The **Prime Number Theorem** justifies the statement that "the probability that an integer in the vicinity of x is prime is $1/\ln x$." From this, the expected number of primes in an infinite increasing sequence $g(n)$ of positive integers is given by $\sum_{n=1}^{\infty} 1/\ln g(n)$. If $g(n) \geq e^{2^n}$ for all $n \geq 1$, then this expected number of primes is $< \sum_{n=1}^{\infty} 1/2^n = 1$.

Among the uncountably infinite number of sequence $A = \{((2s(n))!)^3\}$, an uncountably infinite subset of them have terms growing much faster than e^{2^n}, and this will be true of the terms in the translates A_k of $A = A_o$. Hence, it should be extremely likely that there is at least one such sequence A for which each of its translates A_k contains no more than B primes, for some finite bound B, where B is allowed to be extremely large.

3   The k-tuples of Prime Numbers

The simplest instance of the Prime k-tuples Conjecture is for $k = 2$, and asserts that infinitely often both n and $n + 2$ are prime. While no one seriously doubts that this "twin prime conjecture" is true, it remains unproved.

In 2012, Jiteng Zhang proved the remarkable theorem that there is a finite bound J such that, infinitely often, there are two prime numbers p and q such that $|p - q| < J$. While Zhang's original proof had $J = 70$ Million, a world-wide collaboration for the past year has improved J to below 200, though still far above the conjectured value of "2". Zhang's result is the first to show that the Prime k-tuples Conjecture is *sometimes* true, in some very simple special cases. This is a far cry from proving, or even suggesting, that it is *always* true.

From the fact, which follows from the PNT, that the sequence of primes "thins out", and from looking at tables of primes, Hardy and Littlewood conjectured [4] that $\pi(x + y) \leq \pi(x) + \pi(y)$ for all $x > 2, y > 2$. However, if this conjecture is true, then the Prime k-tuples Conjecture is false [5].

Thus, if either the Hardy-Littlewood Conjecture or Golomb's Conjecture is *true*, then the Prime k-Tuples Conjecture is *false*. It is even logically possible that all three of these conjectures are false.

4   Crosscorrelation of Sequences

Let $S = \{s_n\}$ and $T = \{t_n\}$ be any two sequences of increasing non-negative integers, defined for each $n \epsilon Z^+$. With S we associate the sequence $\bar{S} = \{\sigma_n\}$, and with T the sequence $\bar{T} = \{\tau_n\}$, where $\sigma_n = 1$ whenever n is a value of s_n, but $\sigma_n = 0$ otherwise; and $\tau_n = 1$ whenever n is a value of t_n, but $t_n = 0$ otherwise. We define the (infinite) crosscorrelation (function) $C_{ST}(\theta)$ between the two sequences S and T to be $C_{ST}(\theta) = \sum_{n=-\infty}^{\infty} \sigma_n \tau_{n+\theta}$, where $\sigma_n = 0$ for $n \leq 0$ and $\tau_n = 0$ for $n \leq 0$.

Our interest is in pairs of sequences S and T for which $C_{ST}(\theta)$ is finite for all $\theta \epsilon Z$. Among such pairs of sequences, we may also consider stronger restrictions on $C_{ST}(\theta)$. A very strong restriction would be $0 \leq C_{ST}(\theta) \leq 1$ for all $\theta \epsilon Z$. This condition is satisfied if and only if S and T form a "Sidon Sequences Pair;" that is, all differences $|s_i = s_j|$ of two elements in S are distinct from all differences $|t_i = t_j|$ of two elements in T. In my two previous SETA papers [2,3], I considered questions about and constructions for such sequence pairs. However, it appears that the sequence of the prime numbers, $P = \{2, 3, 5, 7, 11 \ldots\}$, is too dense to be a member of a Sidon Sequences Pair. Fortunately, that is far more than is required of the sequence P for Golomb's Conjecture to be true.

Specifically, Golomb's Conjecture is true if and only if there is an infinite increasing sequence of positive integers $S = \{S_n\}$ such that $C_{SP}(\theta)$ is **bounded**, $0 \leq C_{SP}(\theta) < B$, for a specific bound B, and all $\theta \epsilon Z$, where B is allowed to be huge but finite.

One may visualize $C_{SP}(\theta)$ as the process of sliding the characteristic function $\bar{S} = \{\sigma_i\}$ past the characteristic function $\bar{P} = \{\pi_i\}$ of the prime numbers, and observing the number of "hits" for each shift amount θ. We have already seen that sequences S exist for which this number of hits is *finite* for every shift θ. The challenge is to construct, or merely to prove the existence of, such a sequence S where the number of hits will be **bounded** for every shift θ.

5   Conclusions

A plausible conjecture of Golomb is inconsistent with the "Prime k-tuples Conjecture." It may be possible to prove Golomb's Conjecture by exhibiting a specific infinite increasing sequence of positve integers, $A = \{a_n\}$, such that none of the "translates" A_k of A, where $A_k = \{a_n + k\}$, for each $k \epsilon Z$, contains no more than B primes, where B is allowed to be huge, but finite. There is a (non-rigorous) probability argument that suggests, if the terms of $A = \{a_n\}$ grow fast enough, then A is increasingly likely to have this property. The challenge is to find such a sequence A for which the fact that each of its translates contains no more than B primes can actually be proved.

Acknowledgments. 1. I am grateful to Dr. Kai-Uwe Schmidt, Faculty of Mathematics, Otto-von-Guericke University, for alerting me to the references to Sidon Sequences, including his own work on this subject. 2. Discussions with Professor Barry Masur of Harvard University helped me to focus my own presentation of this material.

References

1. Golomb, S.W.: Problem 10208. Amer. Math. Mon. **99**, 266 (1992)
2. Golomb, S.W.: Infinite sequences with finite cross-correlation. In: Carlet, C., Pott, A. (eds.) SETA 2010. LNCS, vol. 6338, pp. 430–441. Springer, Heidelberg (2010)
3. Golomb, S.W.: Infinite sequences with finite cross-correlation-II. In: Helleseth, T., Jedwab, J. (eds.) SETA 2012. LNCS, vol. 7280, pp. 110–116. Springer, Heidelberg (2012)
4. Hardy, G.H., Littlewood, J.E.: Some problems of 'Partitio numerorum'; III: on the expression of a number as a sum of primes. Acta Math. **44**(1), 1–70 (1923)
5. Ribenboim, P.: The Little Book of Bigger Primes, 2nd edn. Springer-Verlag, New York (2004)

OFDM and CDMA

Optimal Sign Patterns for a Generalized Schmidl-Cox Method

Yutaka Jitsumatsu$^{(\boxtimes)}$, Masahiro Hashiguchi, and Tatsuro Higuchi

Department of Informatics, Kyushu University, 744 Motooka,
Nishi-ku, Fukuoka 819-0395, Japan
{jitumatu,hashiguchi,higuchi}@me.inf.kyushu-u.ac.jp

Abstract. The timing synchronization method proposed by Schmidl and Cox for orthogonal frequency division multiplexing (OFDM) systems uses a reference block consisting of two identical parts, while the one proposed by Shi and Serpedin uses a reference block consisting of four parts with a sign pattern $(+1, +1, -1, +1)$. The accuracy of estimated delays of the latter method is higher than the former. In this paper, the number of partitions is generalized as an integer number M. Two criteria for optimization are proposed. Optimal codes with code length $5 \leq M \leq 30$ are investigated.

1 Introduction

Time and frequency synchronization is an important issue in Orthogonal Frequency Division Multiple Access (OFDMA) systems with a large Doppler shift [1]. A transmitted signal in an OFDMA system consists of a reference block and a data block; the former is a control signal and can be used for synchronization, while the latter is a payload.

The Schmidl-Cox (S&C) method [2] in which a reference block consists of two identical parts is used as a coarse synchronization method in OFDMA systems and in ultrawide band (UWB) communications [5]. A receiver establishes its synchronization by finding the peak value of the auto-correlation between the received signal and the half-block delayed signal. Minn et al. [3] proposed an improved version of S&C method, where a reference block consists of multiple repetitive parts with a specific sign pattern. Then, Shi and Serpedin (S&S) [4] proposed training symbols consisting of four identical parts, with a sign pattern $(+1, +1, -1, +1)$. At the receiver, the sum of auto-correlation values for every pair of four parts is calculated. Other synchronization methods for OFDM using similar training sequences have been proposed [6–8]. Most of them are designed to have low computational complexity, hence the number of partitions, M, is not so large, because it should be implemented in a low-power receiver. On the other hand, we consider a fundamental question about which sign pattern is the best choice for a given M.

This research is supported by the Aihara Project, the FIRST program from JSPS, initiated by CSTP and JSPS KAKENHI Grant Number 25820162.

© Springer International Publishing Switzerland 2014
K.-U. Schmidt and A. Winterhof (Eds.): SETA 2014, LNCS 8865, pp. 269–279, 2014.
DOI: 10.1007/978-3-319-12325-7_23

We review the definition of a timing metric $\Gamma(\theta)$ for a reference block consisting of M parts [9]. We introduce a slightly modified timing metric, denoted by $\widetilde{\Gamma}(\theta)$ to simplify the analysis. Barker sequences would seem to be good candidates for such sign patterns. We will show that the shape of the timing metric of a Barker sequence of $M = 4$ is very sharp and that the peak is steep. However, those of $M = 5$ and $M = 7$ are not sharp and have sidelobes, and those of $M = 11$ and $M = 13$ have very large sidelobes. In this paper, two criteria for determining the optimal sign pattern are defined. Their associated optimal patterns are shown for $5 \leq M \leq 30$.

2   Generalized Schmidl-Cox method

2.1   Channel model

A multi-path fading channel with a time delay and a Doppler shift can be modeled as follows: Consider a discrete-time and time-invariant system with impulse response $(h_0, h_1, \ldots h_{L_T-1})$, where L_T denotes the maximum delay. Let $\varepsilon_0 = N f_D T_s \in \{0, 1, \ldots, N-1\}$ be a normalized frequency offset, where f_D is a Doppler frequency, T_s is a sampling interval, and N is the size of a Discrete Fourier Transform (DFT) for an OFDMA that is equal to the length of a reference block. Then, the discrete-time received signal of the m-th time instance is expressed by

$$r_m = e^{j2\pi\varepsilon_0 m/N} \sum_{\ell=0}^{L_T-1} h_\ell s_{m-\theta_0-\ell} + w_m, \tag{1}$$

where s_m is a transmitted signal, θ_0 is a timing offset, and w_m is a proper[1] complex Gaussian noise with zero mean and variance σ^2.

In a Spread Spectrum (SS) system, multiple paths are resolved by a rake receiver. The channel coefficients $\{h_\ell\}$ are estimated and the received signals through each path are combined. Such a rake receiver consists of several fingers that are correlators or matched filters, where the *cross-correlation* value between the received signal and the locally generated spreading signal is calculated. On the other hand, OFDM systems are sensitive to carrier frequency offset, which causes inter-channel interference. Hence, carrier frequency offset should be compensated for. The S&C approach is a non-coherent method for OFDM systems that uses the *autocorrelation* value of the received signal and that can detect the timing offset as well as the frequency offset.

For simplicity, the multi-path effect is ignored, i.e., the received signal (1) is replaced by

$$r_m = e^{j2\pi\varepsilon_0 m/N} s_{m-\theta_0} + w_m. \tag{2}$$

---

[1] If Z is a proper complex Gaussian random variable, its real and imaginary parts are independent.

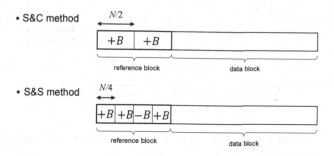

Fig. 1. Transmitted signals for the Schmidl-Cox (upper) and Shi-Serpedin (lower) methods

2.2 Timing Metric and Synchronization Method

A transmitted signal consists of a reference block and data block in an OFDMA system (See Fig. 1). A reference block, also known as a preamble, is utilized to establish time and frequency synchronization. In the S&C method [2], a reference block consists of two identical parts, denoted by (B, B). The auto-correlation value between the received signal and its half-block delayed version is calculated. If the receiver finds a peak of auto-correlations, synchronization is declared. Otherwise, the auto-correlation value of the next timing is calculated. The S&C method has a drawback in that the peak value exhibits a large plateau that greatly reduces the accuracy of the estimated delay [1]. In order to overcome this drawback, Shi-Serpedin used a reference block composed of four identical parts, with the third part being multiplied by -1. Then, the S&S method exhibits a smaller plateau than does the S&C method. This result shows that the synchronization performance of $M = 4$ is better than that of $M = 2$. Then, it is natural to ask: Can we get better performance by increasing the number of repetitive parts to more than four? We will show in a separate paper [16] that an optimal number of partitions for a multi-path fading channel with a Doppler environment is approximately given by $M = \sqrt{N/2}$ for $60 \leq N \leq 240$, where N is the length of the reference block.

In this paper, we consider a general case where the number of repetitive parts is M. Each part is called a sub-block and is multiplied by $+1$ or -1. We denote the sequence of these ± 1 by $\boldsymbol{d} = (d_0, d_1, \ldots, d_{M-1})$, while the repetitive part B consists of $X_0, X_1, \ldots, X_{L-1}$, where $L = N/M$. Then, the transmitted signal of a reference block is expressed by

$$s_{n+iL} = d_i \cdot X_n \tag{3}$$

for $0 \leq i \leq M - 1, 0 \leq n \leq L - 1$.

A timing metric is introduced by Shi and Serpedin for the case $M = 4$, which can be generalized as follows [9] (See also Fig. 2): Define an autocorrelation of the received signal between the i-th and j-th sub-block with timing offset θ as

$$Z_{i,j}(\theta) = \sum_{n=0}^{L-1} \bar{r}_{n+\theta+iL} r_{n+\theta+jL}, \tag{4}$$

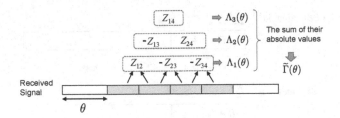

Fig. 2. How to calculate the timing metric for the S&S method.

where $\bar{r}_m$ denotes the complex conjugate of r_m. Note that if $\theta = \theta_0$ the desired signal component of $Z_{i,j}(\theta_0)$ has a sign $d_i d_j$ with a phase shift of $2\pi\epsilon_0(j-i)L/N$. Define

$$\Lambda_p(\theta) = \sum_{i=0}^{M-p-1} d_i d_{i+p} Z_{i,i+p}(\theta). \qquad (5)$$

This quantity is the sum of the autocorrelations and its desired signal component has a phase shift $2\pi\epsilon_0 pL/N$. The effect of a phase shift with a signal component of $\Lambda_p(\theta)$ is eliminated by taking the absolute value. Then, a timing metric is defined by

$$\Gamma(\theta) = \frac{\binom{M}{2}^{-1} \sum_{p=1}^{M-1} |\Lambda_p(\theta)|}{\frac{1}{M}|\Lambda_0(\theta)|}, \qquad (6)$$

where $\binom{M}{2} = \frac{M(M-1)}{2}$ is a binomial coefficient, θ is a controlled parameter for estimating θ_0, and $|z|$ is the absolute value of a complex number z. It is expected that $\Gamma(\theta)$ will have its peak value when $\theta = \theta_0$ and will have a small value for any $\theta \neq \theta_0$. Hence, we would like to find the sign pattern $\boldsymbol{d}$ to minimize $\Gamma(\theta)$ for all $\theta \neq \theta_0$. Note that $\Lambda_0(\theta)$ is equal to the energy of the received signal, i.e., $\sum_{n=0}^{N-1} |r_{n+\theta}|^2$.

In order to simplify the analysis, we introduce a modified timing metric as

$$\tilde{\Gamma}(\theta) = \sum_{p=1}^{M-1} |\Lambda_p(\theta)|. \qquad (7)$$

The parameter θ that attains the maximum value of $\Gamma(\theta)$ is selected as an estimate of θ_0, i.e.,

$$\hat{\theta}_0 = \arg\max_{\theta} \tilde{\Gamma}(\theta). \qquad (8)$$

Note that using a threshold is more practical than finding the θ that takes the maximum value. A normalization process is needed if we compare a timing metric with a fixed threshold (e.g. [10]).

We refer to the part of the timing metric for $\theta < L$ as a *mainlobe* and that for $\theta \geq L$ as a *sidelobe*. Fig. 3 shows four examples of the modified timing metric

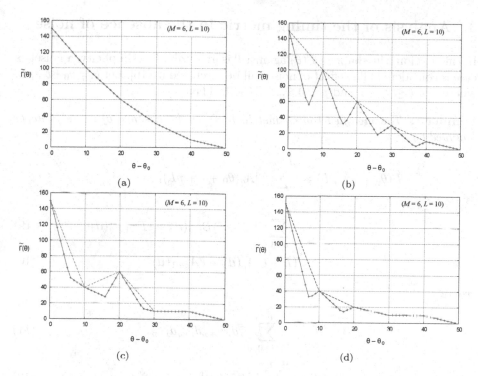

Fig. 3. An example of timing metric $\widetilde{\Gamma}(\theta)$ for $M = 6$ and $L = 10$, where (a) $\boldsymbol{d} = (1, -1, 1, -1, 1, -1)$, (b) $\boldsymbol{d} = (1, -1, -1, 1, 1, -1)$, (c) $\boldsymbol{d} = (1, -1, 1, 1, 1, -1)$, (d) $\boldsymbol{d} = (1, 1, 1, -1, -1, 1)$ are used.

$\widetilde{\Gamma}(\theta)$ for $M = 6$. Solid lines show the modified timing metric and dashed lines are upper bounds of them, which will be defined later. These examples show that the shape of $\widetilde{\Gamma}(\theta)$ largely depends on $\boldsymbol{d}$. Case (d) is the most desirable among the four cases, since the modified timing metric does not have large sidelobes and its peak is sharp. We will define two criteria for determining the optimal timing metric.

Remark: *The timing metric Eq. (6) is based on an auto-correlation of the received signal. On the other hand, Rick and Milstein [11] derived their optimal decision statistics based on a it cross-correlation function between the received signal and a replica of the transmitted signal for spread-spectrum (SS) communications in Rayleigh and Rician fading channels. Doppler shift was not taken into account in [11]. Timing estimation methods combining auto-correlation and cross-correlation functions for OFDM systems with Doppler shift have been proposed in [12, 13]. A comparison of several preamble designs were given in [14]*

It may be worth noting that this estimate has bias for a multi-path channel, that is, the estimate is likely to be larger than the true timing offset because of the channel distortion. This topic will be discussed in [16].

3   Analysis of the timing metric in the absence of noise

In this section, the shape of $\widetilde{\Gamma}(\theta)$ against θ is analyzed. For simplicity, a noiseless case is considered. The effect of noise will be discussed in [16]. Let $\theta = \theta_0 + \ell + pL$, where $0 \leq \ell \leq L - 1$ and $p = 0, 1, \ldots M - 1$. Then,

Lemma 1: *The modified timing metric $\widetilde{\Gamma}(\theta)$ in the absence of noise is upper bounded by*

$$\widetilde{\Gamma}(\theta_0 + \ell + pL) = \sum_{q=1}^{M-p-1} |A_q(\theta_0 + \ell + pL)|$$

$$= \sum_{q=1}^{M-p-1} |(L - \ell)B_{p,q}(\boldsymbol{d}) + \ell B_{p+1,q}(\boldsymbol{d})| \tag{9}$$

$$\leq (L - \ell)A_p(\boldsymbol{d}) + \ell A_{p+1}(\boldsymbol{d}) \tag{10}$$

where

$$B_{p,q}(\boldsymbol{d}) = \sum_{i=0}^{M-p-q-1} d_i d_{i+p} d_{i+q} d_{i+p+q}, \tag{11}$$

$$A_p(\boldsymbol{d}) = \sum_{q=1}^{M-p-1} |B_{p,q}(\boldsymbol{d})|. \tag{12}$$

Define $B_{p,q}(\boldsymbol{d}) = 0$ if $p + q \geq M$. Note that $A_p(\boldsymbol{d}) = 0$ for $p \geq M - 1$, $B_{0,q}(\boldsymbol{d}) = M - q$, and $A_0(\boldsymbol{d}) = \binom{M}{2}$.

Proof. Substituting Eqs. (2) and (3) into (4) under the noiseless assumption, i.e., $w_m = 0$ for every m, gives

$$Z_{ij}(\theta_0 + \ell + pL)$$

$$= \sum_{n=0}^{L-1} \bar{r}_{n+\theta_0+\ell+(p+i)L} r_{n+\theta_0+\ell+(p+j)L}$$

$$= e^{j2\pi\varepsilon_0(j-i)L} \sum_{n=0}^{L-1} s_{n+\ell+(p+i)L} s_{n+\ell+(p+j)L}$$

$$= e^{j2\pi\varepsilon_0(j-i)L} \left\{ \sum_{n=0}^{L-\ell-1} d_{i+p} d_{j+p} X_{n+\ell}^2 + \sum_{n=L-\ell}^{L-1} d_{i+p+1} d_{j+p+1} X_{n+\ell-N}^2 \right\}$$

$$= e^{j2\pi\varepsilon_0(j-i)L} \{ (L - \ell) d_{i+p} d_{j+p} + \ell d_{i+p+1} d_{j+p+1} \}.$$

Then,

$$\sum_{i=0}^{M-q-1} d_i d_{i+q} Z_{i,i+q}(\theta_0 + \ell + pL)$$

$$= e^{j2\pi\varepsilon_0(j-i)L} \left\{ (L-\ell) \sum_{i=0}^{M-q-1} d_{i+p} d_{j+p} + \sum_{i=0}^{M-q-1} \ell d_{i+p+1} d_{j+p+1} \right\}$$

$$= e^{j2\pi\varepsilon_0(j-i)L} \{ (L-\ell) B_{p,q}(\boldsymbol{d}) + B_{p+1,q}(\boldsymbol{d}) \}$$

Using (7) and (5) together with the triangular inequality $|X + Y| \le |X| + |Y|$ gives

$$\widetilde{\Gamma}(\theta_0 + \ell + pL) = \sum_{q=1}^{M-1} |(L-\ell) B_{p,q}(\boldsymbol{d}) + \ell B_{p+1,q}(\boldsymbol{d})|$$

$$\le (L-\ell) \sum_{q=1}^{M-1} |B_{p,q}(\boldsymbol{d})| + \ell \sum_{q=1}^{M-1} |B_{p+1,q}(\boldsymbol{d})|,$$

which completes the proof.

The inequality (10) shows that, in order to reduce the modified timing metric for $\theta \ne \theta_0$, it is sufficient to reduce $A_p(\boldsymbol{d})$ for every $1 \le p \le M - 2$. The right hand side of (10) is shown in a dashed line in Fig. 3. Note that $\widetilde{\Gamma}(\theta_0 + pL) = LA_p(\boldsymbol{d})$. The $A_p(\boldsymbol{d})$ values for the examples in Fig. 3 are (a) $(A_0, A_1, A_2, A_3, A_4) = (15, 10, 6, 3, 1)$, (b) $(15, 10, 6, 3, 1)$, (c) $(15, 4, 6, 1, 1)$, and (d) $(15, 4, 2, 1, 1)$. Note that the shapes of $\widetilde{\Gamma}(\theta)$ are completely different, although the $A_q(\boldsymbol{d})$ values for (a) and (b) are the same.

For the generalized Schmidl-Cox method, binary sequences with a low $A_p(\boldsymbol{d})$ ($p = 1, 2, \ldots, M - 1$) are desirable. We call binary sequences that minimize $\max_{p \ge 1} A_p(\boldsymbol{d})$ and $\sum_{p=1}^{M-1} A_p(\boldsymbol{X})$ *minimum maximum sidelobe (MMS) sequences* and *minimum total sidelobe (MTS) sequences*, respectively.

Next, consider the slope of $\widetilde{\Gamma}(\theta)$ at θ being very close to θ_0. Let $\theta = \theta_0 + \ell$ with $\ell \ll L$, then the first terms in the absolute operations in the summation in Eq. (9) are larger than the second terms. We have

Lemma 2: *For a noiseless case, the slope of the timing metric $\widetilde{\Gamma}(\theta)$ at $\theta = \theta_0 + \ell$ with $\ell \ll L$ only depends on $|\sum_{i=0}^{M-2} d_i d_{i+1}|$.*

Proof: Since $\sum_{q=1}^{M-1} B_{0,q}(\boldsymbol{d}) = \binom{M}{2}$ holds, we obtain

$$\widetilde{\Gamma}(\theta_0 + \ell) = (L-\ell) \binom{M}{2} + \ell \sum_{q=1}^{M-2} B_{1,q}(\boldsymbol{d})$$

$$= \binom{M}{2} L - \ell \left\{ \binom{M}{2} - \sum_{q=1}^{M-2} B_{1,q}(\boldsymbol{d}) \right\} \quad (\ell \ll L). \qquad (13)$$

This equation shows that the modified timing metric is steeper if

$$-A_1' = -\sum_{q=1}^{M-2} B_{1,q}(\boldsymbol{d})$$

is large. It is easily shown that $\left(\sum_{i=0}^{M-2} d_i d_{i+1}\right)^2 = 2A_1' + M - 1$. Hence,

$$-\sum_{q=1}^{M-2} B_{1,q}(\boldsymbol{d}) = \frac{M-1}{2} - \frac{1}{2}\left(\sum_{i=0}^{M-2} d_i d_{i+1}\right)^2, \tag{14}$$

which completes the proof.

Numerical results. The values $A_p(\boldsymbol{d})$ are calculated for every possible $\boldsymbol{d}$ for finding MTS and MMS sequences. In Table 1, for $4 \le M \le 15$, MTS sequences, i.e., sequences that attain minimum $\sum_p A_p$, are listed, where the code number implies $\sum_{j=1}^{M} d_j' 2^j$ with $d_j' = \frac{1-d_j}{2}$. For example, code number 6 for $M = 6$ implies

$$\boldsymbol{d} = (+1, +1, +1, -1, -1, +1).$$

For $M \ge 7$, the code number is expressed as an octal number. The number of minimum $\sum_p A_p$ sequences is not one. The earliest code number is shown in Table 1. By definition, we have $A_{M-1}(\boldsymbol{d}) = 1$ for any $\boldsymbol{d}$.

In Table 2, MMS sequences, i.e., sequences that attain the minimum $\max_p A_p$, are listed for $10 \le M \le 15$. The sequences for $4 \le M \le 9$ are the same as in Table 1. It is observed that $\max_p A_p$ in Table 2 is almost the same as M and that $\sum_p A_p$ in Table 1 is approximately $M^2/2$ for $M < 15$. This is not the case for a larger M. In Table 3, $\sum_p A_p$ and $\max_p A_p$ for $16 \le M \le 30$ are listed. It is shown that $\max A_p$ increases more than M and that $\sum A_p$ is much larger than $M^2/2$. Analysis of the behavior of $\sum_p A_p$ and $\max_p A_p$ remains to be solved.

Table 1. The list of minimum total sidelobe (MTS) sequences for $4 \le M \le 15$

M	Code number	$\sum_p A_p$	$\max A_p$	A_1	A_2	A_3	A_4	A_5	A_6	A_7	A_8	A_9	A_{10}	A_{11}	A_{12}	A_{13}
4	1	2	1	1	1											
5	1	4	2	2	1	1										
6	6	8	4	4	2	1	1									
7	$(11)_8$	13	5	5	4	2	1	1								
8	$(11)_8$	16	5	3	5	2	4	1	1							
9	$(46)_8$	24	6	6	5	5	4	2	1	1						
10	$(43)_8$	32	10	6	10	3	5	2	4	1	1					
11	$(144)_8$	43	11	11	8	6	5	5	4	2	1	1				
12	$(316)_8$	58	15	15	11	8	6	5	5	4	2	1	1			
13	$(1564)_8$	76	15	14	15	13	8	6	5	5	2	6	1	1		
14	$(1444)_8$	92	16	16	14	13	11	8	8	9	5	4	2	1	1	
15	$(5626)_8$	103	17	17	16	14	9	9	10	10	5	5	4	2	1	1

Table 2. The list of minimum maximum sidelobe (MMS) sequences for $10 \leq M \leq 15$

M	Code number	$\sum_p A_p$	$\max A_p$	A_1	A_2	A_3	A_4	A_5	A_6	A_7	A_8	A_9	A_{10}	A_{11}	A_{12}	A_{13}
10	$(243)_8$	36	8	8	6	5	7	4	2	3	1					
11	$(142)_8$	53	10	9	8	10	9	5	6	2	3	1				
12	$(726)_8$	64	11	11	11	10	10	3	5	10	2	1	1			
13	$(1114)_8$	78	13	12	13	13	8	10	9	5	4	2	1	1		
14	$(544)_8$	96	15	12	12	15	13	10	10	9	7	4	2	1	1	
15	$(1564)_8$	111	16	15	16	16	11	13	8	8	9	7	4	2	1	1

Table 3. A list of MTS (left) and MMS sequences (right) for $16 \leq M \leq 30$

	MTS sequences			MMS sequences		
M	Code number	$\sum_p A_p$	$\max A_p$	Code number	$\sum_p A_p$	$\max A_p$
16	$(20474)_8$	120	21	$(3110)_8$	132	18
17	$(30055)_8$	144	21	$(3321)_8$	148	18
18	$(103453)_8$	164	22	$(13347)_8$	172	20
19	$(30457)_8$	195	23	$(123107)_8$	215	23
20	$(353063)_8$	224	27	$(133476)_8$	230	25
21	$(1137166)_8$	254	32	$(64266)_8$	266	27
22	$(1137166)_8$	288	30	$(644721)_8$	312	29
23	$(3410324)_8$	329	36	$(4101511)_8$	339	32
24	$(4101511)_8$	370	37	$(322350)_8$	392	34
25	$(6113716)_8$	410	45	$(6167645)_8$	430	37
26	$(4663611)_8$	460	42	$(11145047)_8$	492	40
27	$(145660343)_8$	501	45	$(126401071)_8$	553	42
28	$(103153070)_8$	554	41	$(103153070)_8$	554	41
29	$(47323061)_8$	626	58	$(122676334)_8$	670	48
30	$(1262607143)_8$	680	58	$(1653275474)_8$	700	49

Table 4. The sidelobe values of the modified timing metric for the Barker sequences

M	Code number	$\sum_p A_p$	$\max A_p$	A_1	A_2	A_3	A_4	A_5	A_6	A_7	A_8	A_9	A_{10}	A_{11}
4	$(2)_8$	2	1	1	1									
5	$(2)_8$	8	4	4	3	1								
7	$(15)_8$	17	5	5	4	4	3	1						
11	$(355)_8$	97	21	21	18	18	13	15	4	4	3	1		
13	$(312)_8$	140	31	22	31	17	16	16	13	5	10	6	3	1

The absolute value of the aperiodic autocorrelation function of Barker sequences of nonzero delay is at most one, and hence Barker sequences are used for the synchronization process using cross-correlation between the received signal and a locally generated replica of the transmitted signal [15]. One may consider that Barker sequences are suitable for the generalized Schmidl-Cox method, as well. The values of A_p for Barker sequences are listed in Table 4, which indicates that a Barker sequence is optimal when $M = 4$ and nearly optimal when $M = 5$ and $M = 7$. However, $\sum_{p=1}^{M-1} A_p$ and $\max_{p \geq 1} A_p$ are significantly large when $M = 11$ and $M = 13$. Thus, Barker sequences with $M = 11$ and $M = 13$ cannot be recommended for the generalized Schmidl-Cox method.

4 Conclusion

In this paper, we consider the optimization of the performance of a timing synchronization method [9] that is a generalization of [2–4]. A reference block is divided into M repetitive parts, where each part is multiplied by a $\{+1, -1\}$-valued sequence. The original timing metric, $\Gamma(\theta)$, is slightly modified to make the analysis easy. An upper bound for the modified timing metric in the absence of noise is provided. Such an upper bound leads us to define two criteria for optimizing the sign patterns. Examples of optimal codes for $5 \leq M \leq 30$ are obtained by an exhaustive search. The slight modification of the timing metric is independent of the optimality of the sign patterns, as long as the decision rule $\hat{\theta}_0 = \arg\max_\theta \widetilde{\Gamma}(\theta)$ is used.

References

1. Morelli, M., Scott, I., Kuo, C.-C.J., Pun, M.-O.: Synchronization techniques for orthogonal frequency division multiple access (OFDMA): a tutorial review. Proc. of the IEEE **95**, 1394–1427 (2007)
2. Schmidl, T.M., Cox, D.C.: Robust frequency and timing synchronization for OFDM. IEEE Tran. Commun. **45**(12), 1613–1621 (1997)
3. Minn, H., Bhargava, V., Lataief, K.: A robust timing and frequency synchronization for OFDM systems. IEEE Tran. Wireless Commun. **2**(4), 822–839 (2003)
4. Shi, K., Serpedin, E.: Coarse frame and carrier synchronization of OFDM systems: a new metric and comparison. IEEE Tran. Wireless Commun. **3**(4), 1271–1284 (2004)
5. Tufvesson, F., Gezici, S., Molisch, A.: Ultra-wideband communications using hybrid matched filter correlation receivers. IEEE Tran. Commun. Lett. **14**(5), 378–380 (2010)
6. Yang, H., Jeong, K.-S., Yi, J.-H., You, Y.-H.: Integer frequency offset estimator by frequency domain spreading for UWB multiband OFDM. IEICE Trans. Fundam. **E93-A**(3), 648–650 (2010)
7. Liang, T., He, Q., Li, H., He, L.: A novel adaptive synchronization algorithm for intermediate frequency architecture CO-OFDM system. Optik **124**, 3406–3411 (2012)
8. Udupa, P., Sentieys, O., Scalart, P.: A novel hierarchical low complexity synchronization method for OFDM systems. In: IEEE 77th Vehicular Technology Conference (VTC Spring) 2013 (2013)
9. Ruan, M., Mark, M.C., Shi, Z.: Training symbol based coarse timing synchronization in OFDM systems. IEEE Trans. Wireless Commun. **8**(5), 2558–2569 (2009)

10. Erseghe, T., Vangelista, L.: Exact analytical expression of Schmidl-Cox signal detection performance in AWGN. IEEE Commun. Lett. **14**(5), 378–380 (2010)
11. Rick, R.R., Milstein, L.B.: Optimal decision strategies for acquisition of Spread-Spectrum signals in Frequency-Selective fading channel. IEEE Trans. Commun. **46**(5), 686–694 (1998)
12. Awoseyila, A.B., Kasparis, C., Evans, B.G.: Improved Preamble-Aided timing estimation for OFDM systems. IEEE Trans. Commun. Lett. **12**(11), 825–827 (2008)
13. Rathkanthiwar, A.P., Gandhi, A.S.: A new timing metric for timing error estimation in OFDM. In: IEEE International Conference on Wireless Communication, Vehicular Technology, Information Theory, and Aerospace and Electronic Systems Technology (Wireless VITAE), pp. 1–6 (2013)
14. Silva C, E.M., Harris, F.J., Dolecek, G.J.: On preamble design for timing and frequency synchronization of OFDM systems over Rayleigh fading channels. In: 18th IEEE International Conference Digital Signal Processing (DSP2013), July 2013, pp. 1–6 (2013)
15. Golomb, S.W., Gong, G.: Signal Design for Good Correlation. Cambridge University Press, Cambridge (2005)
16. Higuchi, T., Jitsumatsu, Y.: Performance analysis of a time synchronization method for multipath fading channels with doppler shift. In: 17th International Symposium on Wireless Personal Multimedia Communication (WPMC2014) (2014, to be presented)

A Novel Construction of Asymmetric Sequence Pairs Set with Zero-Correlation Zone

Longye Wang[1,2]([✉]), Xiaoli Zeng[2], and Hong Wen[1]

[1] National Key Laboratory of Science and Technology on Communications,
University of Electronic Science and Technology of China, Chengdu 611731, China
utibetwly@qq.com
[2] Engineering Research Center for Tibetan Information Technology,
Tibet University, Lhasa 850000, China

Abstract. An asymmetric zero-correlation zone (A-ZCZ) sequence pairs set is a zero-correlation zone (ZCZ) pairs set consisting of multiple ZCZ sequence pairs subsets, where two arbitrary sequence pairs which belong to different subsets should have a large zero cross-correlation zone (ZCCZ). Additionally, each subset is a typical ZCZ sequence pairs set. The new proposed A-ZCZ sequence pairs sets can be generated based on interleaved technique and perfect sequence pairs of length $P = Nmk$ with $N > 1$, $m \geq 1$, $k > 1$. The proposed A-ZCZ sequence pairs sets are applied to quasi-synchronous code-division multiple-access (QS-CDMA) systems and it can hypothetically achieve larger ZCCZ than typical ZCZ sequence pairs set, as well as eliminating both multiple access and multipath interference in the QS-CDMA system.

Keywords: ZCZ sequence pairs · QS-CDMA · Perfect sequence pairs · Interleaved sequences

1 Introduction

In quasi-synchronous code-division multiple-access (QS-CDMA) systems, synchronization condition is generally not very strict, and synchronization deviation of a few chips is allowed. It demands that the address code used for distinguishing different users has a realistic correlation characteristic in the synchronization deviation range. The concept of general orthogonal (GO) or ZCZ has been presented and a number of ZCZ sequence sets and ZCZ sequence pairs sets have been constructed [1,2]. The ZCZ sequence and sequence pairs set are used as address code which can eliminate both multiple access and multipath interference in the QS-CDMA system.

The concept of interleaved sequences was introduced by Gong [3], which can be used for the construction of ZCZ sequences and ZCZ sequence pairs. In recent years, the designs of ZCZ sequence set and ZCZ sequence pairs set have been researched based on interleaved technique and perfect sequences or sequence pairs [4–12]. Recently, the ZCZ sequence sets were composed of multiple sequence

© Springer International Publishing Switzerland 2014
K.-U. Schmidt and A. Winterhof (Eds.): SETA 2014, LNCS 8865, pp. 280–289, 2014.
DOI: 10.1007/978-3-319-12325-7_24

subsets, which also be said as A-ZCZ sequence sets, have been proposed and obtained [13–18]. Tang *et al.* presented several types of binary A-ZCZ sequence sets [13], and Hayashi *et al.* constructed several types of binary and ternary A-ZCZ sequence sets [14,15]. In addition, Zhang *et al.* got complementary A-ZCZ sequence sets [16]. Moreover, Torri *et al.* designed several types of polyphase A-ZCZ sequence sets [17,18].

However, the published construction of A-ZCZ sequence set is rare. Parameters in some construction methods based on interleaving perfect sequence must satisfy the rigorous requirements [17,18]. At the same time, the $p-$ary or $p-$phase perfect sequence is limited for a fixed length P of sequence. Therefore, only a few types of A-ZCZ sequence sets can be obtained by the previous method [17,18]. But the $p-$ary or $p-$phase perfect sequence pairs are richer than perfect sequences for a fixed length P.

In this paper, in order to enrich scope of spreading sequences, the concept of A-ZCZ sequence sets is extended to A-ZCZ sequence pair set. In addition, we propose a construction of A-ZCZ sequence pair set based on interleaved technique. The ZCCZ length between different sequence subsets of proposed A-ZCZ sequence pairs sets is larger than the mathematical upper bound of typical ZCZ sequence pairs sets [19].

The rest of this paper is organized as follows. Section 2 introduces the notation and the preliminaries required for the subsequent sections. The A-ZCZ sequence pairs sets from interleaved perfect sequence pairs are presented in Section 3. In Sect. 4, the construction method of A-ZCZ sequence set is illustrated through an example. Finally, Sect. 5 contains the concluding remarks.

2   Preliminaries

Let $x = (x_0, x_1, \cdots, x_{L-1})$ and $y = (y_0, y_1, \cdots, y_{L-1})$ be two complex-valued sequences of length L, they are called cyclically equivalent if there exists an integer k such that $x_j = y_{j+k}$ for all $j \geq 0$. Additionally, x and y are defined as a sequence pair (x, y) of length L. Then, the periodic autocorrelation function (ACF) of sequence pair (x, y) at a shift of τ is defined by $R_{(x,y)}(\tau) = \sum_{i=0}^{L-1} x_i y_{i+\tau}^*$, where the subscript $i + \tau = (i + \tau) \bmod L$ and the symbol y^* denotes the complex conjugate of the y. Further, if $x = y$, then $R_{(x,y)}(\tau)$ is simplified as $R_x(\tau)$.

Additionally, if $R_{(x,y)}(\tau) = 0$ for $0 < |\tau| < L$, then sequence pair (x, y) is called a perfect sequence pair.

Similarly, any two complex-valued sequence pairs (x, y) and (u, v) of length L, their periodic cross-correlation function (CCF) is defined by:

$$R_{(x,y)(u,v)}(\tau) = R_{(x,v)}(\tau) = \sum_{i=0}^{L-1} x_i v_{i+\tau}^* \tag{1}$$

Further, if $(x, y) = (u, v)$, then $R_{(x,y)(u,v)}(\tau)$ is simplified as $R_{(x,y)}(\tau)$.

Definition 1. *Suppose that A is a set of sequence pairs with M sequence pairs of length L, which can be written as $A = \{(x^{(i)}, y^{(i)})\}$, where $x^{(i)} = (x_0^{(i)}, x_1^{(i)}, \cdots, x_{L-1}^{(i)})$ and $y^{(i)} = (y_0^{(i)}, y_1^{(i)}, \cdots, y_{L-1}^{(i)})$, $i = 0, 1, 2, \cdots, M-1$. If all of the sequence pairs in A satisfy the following periodic correlation property,*

$$R_{(x^{(i_1)}, y^{(i_1)})(x^{(i_2)}, y^{(i_2)})}(\tau) = \begin{cases} 0, & \text{for } i_1 = i_2 \text{ and } 0 < |\tau| < z, \\ 0, & \text{for } i_1 \neq i_2 \text{ and } 0 \leq |\tau| < z. \end{cases} \tag{2}$$

then A is called a (typical) $Z(L, M, z)$ sequence pairs set and z is the ZCZ length of A.

It has been proved that the parameters L, M and z of a $Z(L, M, z)$ sequence set satisfy the following condition [19]:

$$Mz \leq L. \tag{3}$$

If the three parameters L, M and z of any $Z(L, M, z)$ sequence pairs set satisfy Eq. (3) with equality, then the $Z(L, M, z)$ is called an optimal ZCZ sequence pairs set.

Definition 2. *Suppose that $\mathcal{A} = \left\{ A^{(0)}, A^{(1)}, ..., A^{(n)}, ..., A^{(N-1)} \right\}$ is a set which consists of N sequence pairs subsets, and each subset $A^{(n)}$ is a ZCZ sequence pairs set $Z(P, Q, z)$. Here, $A^{(n)}$ can be represented as*

$$A^{(n)} = \left\{ A^{(n,0)}, A^{(n,1)}, ..., A^{(n,q)}, ..., A^{(n,Q-1)} \right\}, \tag{4a}$$

$$A^{(n,q)} = (X^{(n,q)}, Y^{(n,q)}), \tag{4b}$$

$$X^{(n,q)} = (x_0^{(n,q)}, x_1^{(n,q)}, ..., x_p^{(n,q)}, ..., x_{P-1}^{(n,q)}), \tag{4c}$$

$$Y^{(n,q)} = (y_0^{(n,q)}, y_1^{(n,q)}, ..., y_p^{(n,q)}, ..., y_{P-1}^{(n,q)}). \tag{4d}$$

If all of the sequence pairs in $A^{(n)}$ and $A^{(n')}$ ($n \neq n'$) satisfy the following cross-correlation property,

$$R_{A^{(n,q)}A^{(n',q')}}(\tau) = 0 \quad \text{for } n \neq n' \text{ and } 0 \leq |\tau| < z_A, \tag{5}$$

then $\mathcal{A}$ is called an A-ZCZ sequence pairs set $Z_A(P, [Q, N], [z, z_A])$ and z_A is the ZCCZ length of $\mathcal{A}$.

Definition 3. *Let a complex-valued sequence $a = (a_0, a_1, \cdots, a_{L-1})$ and a sequence $e = (e_0, e_1, \cdots, e_{T-1})$ over $\mathbb{Z}_L = \{0, 1, ..., L-1\}$. Then an L by T matrix U can be obtained as follows.*

$$U = (U_{i,j})_{L \times T} = \begin{pmatrix} a_{0+e_0} & a_{0+e_1} & \cdots & a_{0+e_{T-1}} \\ a_{1+e_0} & a_{1+e_1} & \cdots & a_{1+e_{T-1}} \\ \vdots & \vdots & \ddots & \vdots \\ a_{L-1+e_0} & a_{L-1+e_1} & \cdots & a_{L-1+e_{T-1}} \end{pmatrix}, \tag{6}$$

where the additions in subscripts are computed modulo L. By concatenating the successive rows of matrix U in Eq. (6), an interleaved sequence u can be obtained as

$$u_{iT+j} = U_{i,j}, \quad 0 \le i < L, \ 0 \le j < T. \tag{7}$$

For convenience, we rewrite the matrix U in Eq. (6) as

$$U = \left(L^{e_0}(a) \ L^{e_1}(a) \cdots L^{e_{T-1}}(a) \right), \tag{8}$$

and denote the interleaved sequence u by

$$u = I \left(L^{e_0}(a), L^{e_1}(a), \cdots, L^{e_{T-1}}(a) \right), \tag{9}$$

where $I(\cdot)$ and $L^e(\cdot)$ denote the interleaved operator and left cyclical shift operator, i.e., $L^i(a) = (a_i, a_{i+1}, \cdots, a_{L-1}, a_0, \cdots, a_{i-1})$ respectively, and the sequences a and e are called component and shift sequence of u respectively.

Similarly, let (x, y) be a sequence pairs and $e = (e_0, e_1, \cdots, e_{T-1})$ be a shift sequence, we define an interleaved sequence pairs (v, w) with component sequence pairs (x, y) and shift sequence e,

$$v = I(L^{e_0}(x), L^{e_1}(x), \cdots, L^{e_{T-1}}(x)) \ \text{and} \ w = I(L^{e_0}(y), L^{e_1}(y), \cdots, L^{e_{T-1}}(y)).$$

Furthermore, another interleaved sequence pairs (v', w') is defined by component sequence pairs (x, y) and shift sequence $f = (f_0, f_1, \cdots, f_{T-1})$. Then, the CCF at shift τ (where $\tau = T\tau_1 + \tau_2, 0 \le \tau_2 < T$) between (v, w) and (v', w') is given by [3,5],

$$R_{(v,w)(v',w')}(\tau) = \sum_{t=0}^{T-\tau_2-1} R_{(x,y)}(f_{t+\tau_2} - e_t + \tau_1)$$

$$+ \sum_{t=T-\tau_2}^{T-1} R_{(x,y)}(f_{t+\tau_2-T} - e_t + \tau_1 + 1). \tag{10}$$

Definition 4. *Let $A = (a_{i,j})_{L \times T}$ be an $L \times T$ matrix and H be a $T \times T$ matrix, here $h^{(i)} = (h_0^{(i)}, h_1^{(i)}, \cdots, h_{T-1}^{(i)})$ be the i-th row vector of H, then $H \odot A$ is defined as the set of T matrices $\left\{ h^{(0)} \odot A, h^{(1)} \odot A, ..., h^{(T-1)} \odot A \right\}$, where*

$$h^{(i)} \odot A = \begin{pmatrix} h_0^{(i)} a_{0,0} & h_1^{(i)} a_{0,1} & \cdots & h_{T-1}^{(i)} a_{0,T-1} \\ h_0^{(i)} a_{1,0} & h_1^{(i)} a_{1,1} & \cdots & h_{T-1}^{(i)} a_{1,T-1} \\ \vdots & \vdots & \ddots & \vdots \\ h_0^{(i)} a_{L-1,0} & h_1^{(i)} a_{L-1,1} & \cdots & h_{T-1}^{(i)} a_{L-1,T-1} \end{pmatrix}. \tag{11}$$

Furthermore, we shall refer to the operation $\odot$ as the orthogonality-preserving transformation of A if H is an orthogonal matrix.

3   Construction of A-ZCZ Sequence Set Based on Interleaved Perfect Sequence

In this section, we present a procedure for the construction of A-ZCZ sequence pairs set with subsets from interleaved perfect sequence pairs.

Given a perfect sequence pairs $(\boldsymbol{x}, \boldsymbol{y})$ of length $P = Nmk$ with $N > 1, m \geq 1, k > 1$, where N, m and k be integers. Let $\boldsymbol{H}^{(n)}$ be orthogonal matrices of order T with $0 \leq n < N$ and $T \leq k$, then a procedure of constructing A-ZCZ sequence pairs set is introduced as follows.

Construction 1: Construction of A-ZCZ sequence pairs set by interleaving perfect sequence pairs

(I) Generate an appropriate shift sequence $\boldsymbol{e} = (e_0, e_1, \cdots, e_{T-1})$ of length T.

(II) Construct sequences pairs set $\boldsymbol{B} = \{\boldsymbol{b}^{(0)}, \boldsymbol{b}^{(1)}, \cdots, \boldsymbol{b}^{(N-1)}\}$ from the perfect sequence pairs $(\boldsymbol{x}, \boldsymbol{y})$, where $\boldsymbol{b}^{(n)} = (\boldsymbol{v}^{(n)}, \boldsymbol{w}^{(n)}) = (L^{nmk}(\boldsymbol{x}), L^{nmk}(\boldsymbol{y}))$ and $0 \leq n < N$.

(III) Perform the interleaved operation on each sequence pairs $\boldsymbol{b}^{(n)}$ to get a sequence pairs set $(\boldsymbol{C}^{(n)}, \boldsymbol{D}^{(n)}) = (I(\boldsymbol{v}^{(n)}, \boldsymbol{e}), I(\boldsymbol{w}^{(n)}, \boldsymbol{e}))$, where

$$I(\boldsymbol{v}^{(n)}, \boldsymbol{e}) = I\left(L^{e_0}(\boldsymbol{v}^{(n)}) \, L^{e_1}(\boldsymbol{v}^{(n)}) \cdots L^{e_{T-1}}(\boldsymbol{v}^{(n)}) \right).$$

The expression of $I(\boldsymbol{w}^{(n)}, \boldsymbol{e})$ is similar with $I(\boldsymbol{v}^{(n)}, \boldsymbol{e})$. In addition, $\boldsymbol{e} = (e_0, e_1, \cdots, e_{T-1})$ is shift sequence which is defined as follows:
- If $\gcd(T, k) = 1$, and $e_t = st \pmod{k}$, where $s = T^{-1} \pmod{k}$.
- If $T | k$, and $e_t = st \pmod{k}$, where $s = \frac{k}{T} \pmod{k}$.

(IV) Perform the orthogonal transformation on each $(\boldsymbol{C}^{(n)}, \boldsymbol{D}^{(n)})$ to get a new set

$$\left(\boldsymbol{U}^{(n)}, \boldsymbol{V}^{(n)} \right) = \left(\boldsymbol{H}^{(n)} \odot \boldsymbol{C}^{(n)}, \boldsymbol{H}^{(n)} \odot \boldsymbol{D}^{(n)} \right)$$
$$= \left\{ (\boldsymbol{U}^{(n,0)}, \boldsymbol{V}^{(n,0)}), \cdots, (\boldsymbol{U}^{(n,T-1)}, \boldsymbol{V}^{(n,T-1)}) \right\},$$

where $\boldsymbol{H}^{(n)}$ are orthogonal matrices.

(V) Concatenate the successive rows of $\boldsymbol{U}^{(n,t)} \in \boldsymbol{U}^{(n)}$ and $\boldsymbol{V}^{(n,t)} \in \boldsymbol{V}^{(n)}$ respectively, where $(0 \leq n < N, 0 \leq t < T)$. Then we produce the interleaved sequence pairs $(\boldsymbol{u}^{(n,t)}, \boldsymbol{v}^{(n,t)})$ leading to ZCZ sequence pairs sets

$$\left(\mathcal{U}^{(n)}, \mathcal{V}^{(n)} \right) = \left\{ (\boldsymbol{u}^{(n,t)}, \boldsymbol{v}^{(n,t)}) \, | 0 \leq n < N, 0 \leq t < T \right\},$$

where

$$\left(\boldsymbol{u}^{(n,t)}, \boldsymbol{v}^{(n,t)} \right) = \left((u_0^{(n,t)}, \cdots, u_{TP-1}^{(n,t)}), (v_0^{(n,t)}, \cdots, v_{TP-1}^{(n,t)}) \right).$$

(VI) Join all sequence pairs subsets $(\mathcal{U}^{(n)}, \mathcal{V}^{(n)})(0 \leq n < N)$, and then we obtain union sequence pairs set $(\mathcal{U}, \mathcal{V}) = \bigcup_{n=0}^{N-1} (\mathcal{U}^{(n)}, \mathcal{V}^{(n)})$ leading to the desired A-ZCZ sequence pairs sets $(\mathcal{U}, \mathcal{V})$.

Lemma 1. *Let $(C^{(n)}, D^{(n)})$ be sequence pairs obtained from step (III) of **Construction 1**,*

(i) *If $\gcd(T, k) = 1$, and $e_t = st(\mathrm{mod}\, k)$, where $s = T^{-1}(\mathrm{mod}\, k)$, then each sequence pairs $(C^{(n)}, D^{(n)})$ is a $Z(TP, 1, k)$ sequence pairs.*

(ii) *If $T|k$, and $e_t = st(\mathrm{mod}\, k)$, where $s = \frac{k}{T}(\mathrm{mod}\, k)$, then each sequence pairs $(C^{(n)}, D^{(n)})$ is a $Z(TP, 1, k-1)$ sequence pairs.*

Proof. (i) According to Eq. (10), let $\tau = T\tau_1 + \tau_2$ and $0 \le \tau_2 < T - 1$, when $e_t = st\,(\mathrm{mod}\, k)$, the ACF of sequence pairs $(C^{(n)}, D^{(n)})$ for fixed n can be calculated as follows.

$$R_{(C^{(n)}, D^{(n)})}(\tau) = \sum_{t=0}^{T-1} R_{(x,y)}(s\tau_2\,(\mathrm{mod}\, k) + \tau_1) \qquad (12)$$

When $\tau_2 = 0$, then $R_{(C^{(n)}, D^{(n)})}(\tau) = TR_{(x,y)}(0) \ne 0$ for $\tau_1 = 0$ and $R_{(C^{(n)}, D^{(n)})}(\tau) = \sum_{t=0}^{T-1} R_{(x,y)}(\tau_1) = 0$ for $\tau_1 \ne 0 \bmod P$.

When $\tau_2 \ne 0$, let $\tau = T\tau_1 + \tau_2 = k$, then $\tau_2 = k + T\tau_1$, which lead to $(s\tau_2\,(\mathrm{mod}\, k) + \tau_1)(\mathrm{mod}\, P) = (s(k - T\tau_1)\,(\mathrm{mod}\, k) + \tau_1)(\mathrm{mod}\, P) = (-\tau_1 + \tau_1)(\mathrm{mod}\, P) = 0$ or $(s\tau_2\,(\mathrm{mod}\, k) + \tau_1)(\mathrm{mod}\, P) = (k - \tau_1 + \tau_1)(\mathrm{mod}\, P) = k$, thus Eq. (12) is not equivalent to 0 when $\tau = T\tau_1 + \tau_2 = k$. While, if $\tau = T\tau_1 + \tau_2 < k$, then $0 \ne (T\tau_1 + \tau_2)\,(\mathrm{mod}\, k) = (\tau_1 + s\tau_2)\,(\mathrm{mod}\, k) \ne 0$, such that $(\tau_1 + s\tau_2) \ne 0 \bmod P$, since $k|P$. It is said that Eq. (12) $= 0$ when $0 < \tau = T\tau_1 + \tau_2 < k$.

Therefore, if $\gcd(T, k) = 1$ and $0 < |\tau| < k$, then $R_{(C^{(n)}, D^{(n)})}(\tau) = 0$. Namely, the Lemma 1(i) is correct.

(ii) By using the same principle with the proof of Lemma 1(i), we can prove Lemma 1(ii) is correct. In other words, the Lemma 1 is correct.

Lemma 2. *When $0 \le i \ne j < N$, let $(C^{(i)}, D^{(i)})$ and $(C^{(j)}, D^{(j)})$ be two sequence pairs obtained from step (III) of **Construction 1**, then the CCF between the $(C^{(i)}, D^{(i)})$ and $(C^{(j)}, D^{(j)})$ satisfy the following property.*

(i) *If $\gcd(T, k) = 1$ and $|\tau| < [mk - t's]T + t'$, then $R_{(C^{(i)}, D^{(i)})(C^{(j)}, D^{(j)})}(\tau) = 0$, where t' satisfies condition $e_{t'} = \max\limits_{0 \le t < T} e_t$.*

(ii) *If $T|k$ and $|\tau| < [mk - (T-1)s]T + (T-1)$, then $R_{(C^{(i)}, D^{(i)})(C^{(j)}, D^{(j)})}(\tau) = 0$.*

Proof. (i) According to *step* (II) and *step* (III) of **Construction 1**, the pairs $(C^{(i)}, D^{(i)})$ and $(C^{(j)}, D^{(j)})$ can be simplified to the following form respectively.

$$(C^{(i)}, D^{(i)}) = (I(x, e'), I(y, e')) \quad \text{and} \quad (C^{(j)}, D^{(j)}) = (I(x, f'), I(y, f')).$$

Where

$$e' = (e_0 + imk, e_1 + imk, \cdots, e_{T-1} + imk)$$

and

$$f' = (e_0 + jmk, e_1 + jmk, \cdots, e_{T-1} + jmk).$$

Therefore, according to formula (10), let $\tau = T\tau_1 + \tau_2$ and $0 \leq \tau_2 < T - 1$, then the CCF of sequence pairs $(\boldsymbol{C}^{(i)}, \boldsymbol{D}^{(i)})$ and $(\boldsymbol{C}^{(j)}, \boldsymbol{D}^{(j)})$ for $0 \leq i \neq j < N$ can be calculated by

$$R_{(\boldsymbol{C}^{(i)}, \boldsymbol{D}^{(i)})(\boldsymbol{C}^{(j)}, \boldsymbol{D}^{(j)})}(\tau) = \sum_{t=0}^{T-1} R_{(\boldsymbol{x}, \boldsymbol{y})}((s\tau_2 \bmod k) + \tau_1 + (j - i)mk) \quad (13)$$

When $\tau_2 = 0$. $R_{(\boldsymbol{C}^{(i)}, \boldsymbol{D}^{(i)})(\boldsymbol{C}^{(j)}, \boldsymbol{D}^{(j)})}(\tau) = T R_{(\boldsymbol{x}, \boldsymbol{y})}((j - i)mk)$ for $\tau_1 = 0$. Hence, $mk \leq (j - i)mk \leq (N - 1)mk$ for $0 \leq i \neq j < N$, conclude $(j - i)mk \neq 0 \bmod P$ and $R_{(\boldsymbol{C}^{(i)}, \boldsymbol{D}^{(i)})(\boldsymbol{C}^{(j)}, \boldsymbol{D}^{(j)})}(0) = 0$. If $\tau_1 \neq 0$, then $\tau = T\tau_1$. When $\tau = T\tau_1 < (mk - t's)T + t' < (mk - t's)T + T$, then $\tau_1 < mk - t's + 1 < mk$. Thus, $mk \leq (j - i)mk + \tau_1 < mk + (j - i)mk \leq Nmk$. Namely, $(j - i)mk + \tau_1 \neq 0 \bmod P$ and $R_{(\boldsymbol{C}^{(i)}, \boldsymbol{D}^{(i)})(\boldsymbol{C}^{(j)}, \boldsymbol{D}^{(j)})}(\tau) = 0$ for $0 < \tau < (mk - t's)T + t'$.

When $\tau_2 \neq 0$. There exists integer v with $-m \leq v \leq m$, s.t., $\tau_1 = vk + \tau_1'$ for $0 \leq \tau_1' < k$. If $\tau = T\tau_1 + \tau_2 < (mk - t's)T + t'$, because $\tau_2 < (mk - t's)T + t' - T\tau_1$, $(s\tau_2 \bmod k) + \tau_1 + (j - i)mk < vk + (N - 1)mk = [v + (N - 1)m]k \leq Nmk$ for $v \leq m$, which lead to $s\tau_2 \bmod (2k + 1) + \tau_1 + (j - i)m(2k + 1) \neq 0 \bmod P$, namely Eq. (13) $= 0$.

From the above analysis, the conclusion is obtained as follows:

$$R_{(\boldsymbol{C}^{(i)}, \boldsymbol{D}^{(i)})(\boldsymbol{C}^{(j)}, \boldsymbol{D}^{(j)})}(\tau) = 0, \quad \text{if } |\tau| < (mk - t's)T + t'. \quad (14)$$

(ii) By using the same principle with the proof of *Lemma* 2(i), we can prove that *Lemma* 2(ii) is correct.

As a result, the *Lemma* 2 is correct.

Theorem 1. *Let sequence pairs set* $(\mathcal{U}, \mathcal{V}) = \bigcup_{n=0}^{N-1} (\mathcal{U}^{(n)}, \mathcal{V}^{(n)})$ *be obtained by* **Construction 1**, *then which is the A-ZCZ sequence pairs set that are composed of N sequence pairs subsets. And*

(i) If $\gcd(T, k) = 1$, *let* $\delta = [mk - t's]T + t'$, *then the parameters of* $(\mathcal{U}, \mathcal{V})$ *are represented* $(TP, [T, N], [k, \delta])$.

(ii) If $T | k$, *let* $\delta = [mk - (T - 1)s]T + (T - 1)$, *then the parameters of* $(\mathcal{U}, \mathcal{V})$ *are represented* $(TP, [T, N], [k - 1, \delta])$.

Proof. (i) Let $(\boldsymbol{U}^{(n)}, \boldsymbol{V}^{(n)})(0 \leq n < N)$ be sequence pairs obtained by *step* (IV) of the **Construction 1**. When $\tau = 0 (i.e., \tau_1 = \tau_2 = 0)$, it is easy to see that the orthogonality between the sequence pairs $(\boldsymbol{U}^{(n, t_1)}, \boldsymbol{V}^{(n, t_1)})$ and $(\boldsymbol{U}^{(n, t_2)}, \boldsymbol{V}^{(n, t_2)})$ is guaranteed by the orthogonality between $\boldsymbol{h}^{(n, t_1)}$ and $\boldsymbol{h}^{(n, t_2)}$, namely $R_{(\boldsymbol{U}^{(n, t_1)}, \boldsymbol{V}^{(n, t_1)})(\boldsymbol{U}^{(n, t_2)}, \boldsymbol{V}^{(n, t_2)})}(0) = 0$, whenever $0 \leq t_1 \neq t_2 < T$. Here $\boldsymbol{h}^{(n, t_1)}$ and $\boldsymbol{h}^{(n, t_2)}$ are the t_1-th and the t_2-th rows of the orthogonal matrix pair $\boldsymbol{H}^{(n)}$ respectively. When $0 < |\tau| < k$, the perfect sequence pairs $(\boldsymbol{x}, \boldsymbol{y})$ and the Lemma 1(1) guarantee that $R_{(\boldsymbol{U}^{(n, t_1)}, \boldsymbol{V}^{(n, t_1)})(\boldsymbol{U}^{(n, t_2)}, \boldsymbol{V}^{(n, t_2)})}(\tau) = 0$ with $0 \leq t_1 \neq t_2 < T$, as in the ACF case. When $|\tau| < \delta$, the Lemma 2 guarantees that $R_{(\boldsymbol{U}^{(n_1, t_1)}, \boldsymbol{V}^{(n_1, t_1)})(\boldsymbol{U}^{(n_2, t_2)}, \boldsymbol{V}^{(n_2, t_2)})}(\tau) = 0$, whenever $0 \leq t_1, t_2 < T$ and $0 \leq n_1 \neq n_2 < N$.

(ii) The proof of Theorem 1(ii) and Theorem 1(i) are the same, so we omit it.

Table 1. The A-ZCZ sequence pairs set $\left\{\left(\mathcal{U}^{(0)}, \mathcal{V}^{(0)}\right), \left(\mathcal{U}^{(1)}, \mathcal{V}^{(1)}\right)\right\}$

$\left(\mathcal{U}^{(0)}, \mathcal{V}^{(0)}\right)$	$\left(\boldsymbol{u}^{(0,0)}, \boldsymbol{v}^{(0,0)}\right)$	$(+ + + - + - - - - - - - - - + - +, + +$ $- - + - - + - - + - - + - -)$
	$\left(\boldsymbol{u}^{(0,1)}, \boldsymbol{v}^{(0,1)}\right)$	$(+ - + + + + - + - + - + - - - -, + -$ $- + + + - - - + + + - - - +)$
$\left(\mathcal{U}^{(1)}, \mathcal{V}^{(1)}\right)$	$\left(\boldsymbol{u}^{(1,0)}, \boldsymbol{v}^{(1,0)}\right)$	$(+ + + + + - + - - - - + - + + +, + +$ $- + + + - + + - - + + - + + -)$
	$\left(\boldsymbol{u}^{(1,1)}, \boldsymbol{v}^{(1,1)}\right)$	$(+ - + \ \ + + + + - + - - - - + -, +$ $- - - + + + - - + + - - - + +)$

Obviously, if all sequences of $(\mathcal{U}, \mathcal{V})$ are regarded as a sequence pairs set, then it is an optimal typical ZCZ sequence pairs set with parameters $Z(TP, NT, k)$ when $\gcd(T, k) = 1$ and $q = 1$. In addition, $(\mathcal{U}, \mathcal{V})$ is a quasi-optimal typical ZCZ sequence pairs set with parameters $Z(TP, NT, k-1)$ if $T|k$ and $q = 1$. If $N = 1$ and $q = 1$, in particular, the proposed method corresponding to part of method in [10, 11].

4   Example

In this section, we give an example.

Given a binary sequences $(\boldsymbol{x}, \boldsymbol{y}) = (+ + + - - - - -, + - + - - + - -)$ of length 8, where the symbol "+" and "−" represents "+1" and "−1" respectively. When $N = 2$, $m = 1$ and $k = 4$, then $P = 8 = 2 \times 1 \times 4$. Let $\boldsymbol{H}^{(0)}$ and $\boldsymbol{H}^{(1)}$ be Hadamard matrices of order $T = 2$, where

$$\boldsymbol{H}^{(0)} = \begin{pmatrix} 1 & 1 \\ 1 & -1 \end{pmatrix} \quad \text{and} \quad \boldsymbol{H}^{(1)} = \begin{pmatrix} -1 & -1 \\ -1 & 1 \end{pmatrix}.$$

Then a binary A-ZCZ sequence pairs set $(\mathcal{U}, \mathcal{V}) = \left\{(\mathcal{U}^{(0)}, \mathcal{V}^{(0)}), (\mathcal{U}^{(1)}, \mathcal{V}^{(1)})\right\}$ can be obtained from **Construction 1**. All sequence pairs of this A-ZCZ sequence set $(\mathcal{U}, \mathcal{V})$ are given in Table 1.

For example, the absolute value of the ACF of $(\boldsymbol{u}^{(n,i)}, \boldsymbol{v}^{(n,i)})$ can be obtained by calculating

$$\left| R_{\left(\boldsymbol{u}^{(0,0)}, \boldsymbol{v}^{(0,0)}\right)}(\tau) \right| = (8, 0, 0, X, 0, 0, 0, 0, 0, 0, 0, 0, 0, X, 0, 0)$$

On the other hand, the absolute value of the CCF between $(\boldsymbol{u}^{(n,i)}, \boldsymbol{v}^{(n,i)})$ and $(\boldsymbol{u}^{(n,j)}, \boldsymbol{v}^{(n,j)})$ can be obtained by calculating

$$\left| R_{\left(\boldsymbol{u}^{(0,0)}, \boldsymbol{v}^{(0,0)}\right)\left(\boldsymbol{u}^{(0,1)}, \boldsymbol{v}^{(0,1)}\right)}(\tau) \right| = (0, 0, 0, X, 0, 0, 0, 0, 0, 0, 0, 0, 0, X, 0, 0)$$

Additionally, the absolute value of the CCF between $(\boldsymbol{u}^{(n_1,i)}, \boldsymbol{v}^{(n_1,i)})$ and $(\boldsymbol{u}^{(n_2,j)}, \boldsymbol{v}^{(n_2,j)})$ can be obtained by calculating

$$\left| R_{(\boldsymbol{u}^{(0,0)}, \boldsymbol{v}^{(0,0)})(\boldsymbol{u}^{(1,0)}, \boldsymbol{v}^{(1,0)})}(\tau) \right| = (0,0,0,0,0, X, X, X, X, X, X, X, 0,0,0,0)$$

Therefore, the sequence set $(\mathcal{U}, \mathcal{V}) = \{(\mathcal{U}^{(0)}, \mathcal{V}^{(0)}), (\mathcal{U}^{(1)}, \mathcal{V}^{(1)})\}$ is an A-ZCZ sequence pairs set $Z_A(16, [2,2], [3,5])$.

5   Conclusion

In this paper, the authors presented a new construction of A-ZCZ sequence pairs set, which is based on interleaved any perfect sequence pairs of length $P = Nmk$ with $N > 1, m \geq 1, k > 1$, according to an orthogonal matrix of order T. In addition, if sequence pairs subsets of an A-ZCZ sequence set are assigned to adjacent cells, the asymmetric property can be useful in reducing or avoiding inter-cell interference because of the larger ZCCZ length between different sequence pairs subsets. The proposed A-ZCZ sequence pairs set is expected to be useful for designing spreading sequences for QS-CDMA systems.

Acknowledgments. The authors gratefully acknowledgments very valuable comments given by anonymous reviewers. This work was supported by China Nation Natural Science Foundation of China (NSFC) under Grant No.61261021, 61032003, 61271172, Research Fund for the Doctoral Program of Higher Education of China(RFDP) under Grant No.20120185110030, 20130185130002, SRF for ROCS, SEM and Sichuan International Corporation Project under Grant No. 2013HH0005.

References

1. Fan, P.Z., Hao, L.: Generalized orthogonal sequences and their applications in synchronous CDMA systems. IEICE Trans. Fundam. **E83–A**(11), 1–16 (2000)
2. Deng, X.M., Fan, P.Z.: Spreading sequence sets with zero correlation zone. Electron. Lett. **36**(11), 982–983 (2000)
3. Gong, G.: New designs for signal sets with low cross correlation, balance property, and large linear span: GF(p) case. IEEE Trans. Inf. Theory **48**(11), 2847–2867 (2002)
4. Hayashi, T.: Zero-correlation zone sequence set constructed from a perfect sequence. IEICE Trans. Fundam. **E90–A**(5), 1–5 (2007)
5. Tang, X.H., Mow, W.H.: A new systematic construction of zero correlation zone sequences based on interleaved perfect sequences. IEEE Trans. Inf. Theory **54**(12), 5729–5734 (2008)
6. Matsufuji, S.: Two types polyphase sequence sets for approximately synchronized CDMA systems. IEICE Trans. Fundam. **E86–A**(1), 229–234 (2003)
7. Torii, H., Nakamura, M., Suehiro, N.: A new class of polyphase sequence sets with optimal zero-correlation zones. IEICE Trans. Fundam. **E88–A**(7), 1987–1994 (2005)

8. Zhou, Z.C., Pan, Z., Tang, X.H.: New families of optimal zero correlation zone sequences based on interleaved technique and perfect sequences. IEICE Trans. Fundam. **E91–A**(12), 3691–3697 (2008)
9. Wang, L.Y., Zeng, X.L.: A new class of sequences with zero correlation zone based on interleaving perfect sequence. In: 5th International Workshop on Signal Design and Its Applications in Communications, pp. 29–31. IEEE Press, New York (2011)
10. Gao, J.P., Li, Q., Dai, J.F.: A new construction technique of ZCZ sequence pairs set. In: 2nd IET International Communication Conference on Wireless, Mobile and Sensor Networks, pp. 970–973. IET, Stevenage (2007)
11. Shi, R.H., Zhao, X.Q., Li, L.Z.: Research on construction method of ZCZ sequence pairs set. J. Converg. Inf. Technol. **6**(1), 15–23 (2011)
12. Tang, X.H., Mow, W.H.: Design of spreading codes for quasi-synchronous CDMA with intercell interference. IEEE J. Sel. Areas Commun. **24**(1), 84–93 (2006)
13. Tang, X.H., Fan, P.Z., Lindner, J.: Multiple binary ZCZ sequence sets with good cross-correlation property based on complementary sequence sets. IEEE Trans. Inf. Theory **56**(8), 4038–4045 (2010)
14. Hayashi, T., Maeda, T., Matsufuji, S., Okawa, S.: A ternary zero-correlation zone sequence set having wide inter-subset zero-correlation zone. IEICE Trans. Fundam. **E94–A**(11), 2230–2235 (2011)
15. Hayashi, T., Maeda, T., Okawa, S.: A generalized construction of zero-correlation zone sequence set with sequence subsets. IEICE Trans. Fundam. **E94–A**(7), 1597–1602 (2011)
16. Zhang, Z.Y., Zeng, F.X., Xuan, G.X.: A class of complementary sequences with multi-width zero cross-correlation zone. IEICE Trans. Fundam. **E93–A**(8), 1508–1517 (2010)
17. Torii, H., Nakamura, M.: A study of asymmetric ZCZ sequence sets. In: 11th WSEAS International Conference on Multimedia Systems and Signal Processing, pp. 79–86, WSEAS, Athens (2011)
18. Torii, H., Matsumoto, T., Nakamura, M.: A new method for constructing asymmetric ZCZ sequence sets. IEICE Trans. Fundam. **E95–A**(9), 1577–1586 (2012)
19. Li, G., Xu, C.Q., Liu, K., Liang, Q.M.: Some novel results on 1-dimension and 2-dimension sequence pairs with zero correlation zone. In: Proceedings of Global Mobile Congress, pp. 551–552. Delson Group Inc., Cupertino (2007)

Frequency-Hopping Sequences

On Low-Hit-Zone Frequency-Hopping Sequence Sets with Optimal Partial Hamming Correlation

Hongyu Han[✉], Daiyuan Peng, and Xing Liu

Key Laboratory of Information Coding and Transmission,
Southwest Jiaotong University, Chengdu 610031, Sichuan, People's Republic of China
hyhan@my.swjtu.edu.cn, dypeng@swjtu.edu.cn, liuxing4@126.com

Abstract. In quasi-synchronous frequency-hopping code division multiple-access systems, frequency-hopping sequences (FHSs) with low-hit-zone (LHZ) are commonly employed to minimize multiple-access interferences. Usually, the length of correlation window is shorter than the period of the chosen FHSs due to the limited synchronization time or hardware complexity. Therefore, the study of the partial Hamming correlation properties of an FHS set with LHZ is particularly important. In this paper, we prove the nonexistence of an LHZ-FHS set with strictly optimal maximum partial Hamming correlation in some conditions. A sufficient condition for an LHZ-FHS set with strictly optimal average partial Hamming correlation is also given. In addition, a concatenated construction method is presented. The LHZ-FHS sets with optimal maximum partial Hamming correlation and the LHZ-FHS sets with strictly optimal average partial Hamming correlation whose sequence length can be infinite are constructed by the new construction, respectively.

Keywords: Frequency-hopping sequences · Low-hit-zone · Partial hamming correlation · Quasi-synchronous frequency-hopping communication

1 Introduction

Frequency-hopping code-division multiple-access (FH-CDMA) techniques have been widely employed in modern communication systems such as ultra-wideband (UWB), military, and radar applications [7]. In such systems, it is required to design frequency-hopping sequences (FHSs) with good Hamming correlation in order to reduce the MA interference caused by hits of frequencies. There are two kinds of measurement on the Hamming correlation of an FHS set: one is the maximum Hamming correlation which represents its worst-case performance and the other is the average Hamming correlation among FHSs which measures its average performance.

This work was supported by the National Science Foundation of China (Grant No. 61271244).

K.-U. Schmidt and A. Winterhof (Eds.): SETA 2014, LNCS 8865, pp. 293–304, 2014.
DOI: 10.1007/978-3-319-12325-7_25

Traditionally the periodic Hamming correlation of FHSs has received most attention. There are several known constructions [3–5,8,9,22] for FHS sets having optimal maximum periodic Hamming correlation with respect to the Peng-Fan bound [18]. While the average periodic Hamming correlation indicates the average interference performance of the FH-CDMA systems, the design of optimal FHSs with respect to the average periodic Hamming correlation property is meaningful as well. There exist several constructions of FHS sets with optimal average periodic Hamming correlation [1,11,20] with respect to the Peng-Niu-Tang bound [20].

Compared with the traditional periodic Hamming correlation, the partial Hamming correlation of FHSs (where correlation is computed over only subsequences of FHSs) is much less studied. However, FHSs with good partial Hamming correlation properties are important for certain application scenarios where an appropriate window length shorter than the period of the chosen sequences is employed to minimize the synchronization time or to reduce the hardware complexity of the FH-CDMA receivers [6]. In 2004, Eun et al. [6] generalized the Lempel-Greenberger bound [10] to the case of partial correlation. In 2012, Zhou et al. [23] obtained a bound on the maximum partial Hamming correlation of an FHS set. Subsequently, both individual FHSs with (strictly) optimal maximum partial autocorrelation and FHS sets with (strictly) optimal maximum partial correlation were constructed [17,23]. In 2013, Ren et al. [21] discussed the average partial Hamming correlation of FHSs.

Different from the conventional FHSs design, the FHSs design with low-hit-zone (LHZ) aims at making Hamming correlation values equal to a very low value within a correlation zone [19]. The significance of an LHZ sequence set is that the number of hits between difference sequences will always be very small as long as the relative delay does not exceed certain zone, thus reducing the mutual interference. In recent years, several optimal LHZ-FHS sets meeting the Peng-Fan-Lee bound [19] were constructed [2,13,16].

The bounds on the maximum partial Hamming correlation and the average partial Hamming correlation of an LHZ-FHS set were established in [14,15], respectively. Recently, Liu et al. [12] gave a construction of the LHZ-FHS sets with strictly optimal maximum partial Hamming correlation. In this paper, we will pay attention to the LHZ-FHS sets with optimal maximum/average partial Hamming correlation.

The outline of this paper is as follows. In Sect. 2, some preliminaries on FHSs are presented. We prove the nonexistence of an LHZ-FHS set with strictly optimal maximum partial Hamming correlation in some conditions, and give a sufficient condition for an LHZ-FHS set with strictly optimal average partial Hamming correlation. In Sect. 3, a concatenated construction method is presented. The LHZ-FHS sets with optimal maximum partial Hamming correlation and the LHZ-FHS sets with strictly optimal average partial Hamming correlation whose sequence length can be infinite are constructed by the new construction, respectively. Finally, we give some concluding remarks in Sect. 4.

2   Preliminaries

Let $\mathcal{F} = \{f_1, f_2, \cdots, f_q\}$ be a set of q available frequencies, also called the *alphabet*. A sequence $X = \{x_j\}_{j=0}^{N-1}$ is called an FHS of length N over $\mathcal{F}$ if $x_j \in \mathcal{F}$ for all $0 \leq j \leq N-1$. For any two FHSs $X = \{x_j\}_{j=0}^{N-1}$, $Y = \{y_j\}_{j=0}^{N-1}$ of length N over $\mathcal{F}$, their *periodic Hamming correlation* $H_{X,Y}(\tau)$ at time delay τ is defined by

$$H_{X,Y}(\tau) = \sum_{j=0}^{N-1} h(x_j, y_{j+\tau}), \quad 0 \leq \tau < N$$

where $h(x_j, y_{j+\tau}) = 1$ if $x_j = y_{j+\tau}$, and 0 otherwise, and all operations among the position indices are performed modulo N. When $X = Y$, $H_{X,X}(\tau)$ is called *periodic Hamming autocorrelation* of X, and denoted by $H_X(\tau)$ for short. When $X \neq Y$, $H_{X,Y}(\tau)$ is called the *periodic Hamming cross-correlation* of X and Y.

Let $\mathcal{S}$ be an (N, M, q) FHS set, that is, an FHS set consisting of M FHSs of length N over an alphabet $\mathcal{F}$ of size q. The *maximum periodic Hamming correlation* $H(\mathcal{S})$ of the sequence set $\mathcal{S}$ is defined by

$$H(\mathcal{S}) = \max \left\{ \max_{X \in \mathcal{S}, 1 \leq \tau < N} \{H_X(\tau)\}, \max_{X,Y \in \mathcal{S}, X \neq Y, 0 \leq \tau < N} \{H_{X,Y}(\tau)\} \right\}.$$

Throughout this paper, let (N, M, q, α) denote an FHS set of M FHSs of length N over an alphabet of size q, with the maximum periodic Hamming correlation $\alpha = H(\mathcal{S})$.

The following lower bound on the maximum periodic Hamming correlation of an FHS set was derived by Peng and Fan [18].

Lemma 1 [18]. *Let $\mathcal{S}$ be an (N, M, q, α) FHS set. Then we have*

$$\alpha \geq \left\lceil \frac{(MN - q)N}{(MN - 1)q} \right\rceil. \tag{1}$$

Definition 1. *It is said that an FHS set $\mathcal{S}$ is optimal with respect to the bound (1), if the equality in (1) is achieved. In this case, $\mathcal{S}$ is called an FHS set with optimal maximum periodic Hamming correlation.*

The FHS set $\mathcal{S}$ is called an LHZ-FHS set with respect to the LHZ L_H, if there is a nonnegative integer β such that

$$H_{X,Y}(\tau) \leq \beta,$$

for any $X, Y \in \mathcal{S}$, $0 < \tau \leq L_H$ when $X = Y$, and $0 \leq \tau \leq L_H$ when $X \neq Y$.

Throughout this paper, we use (N, M, q, L_H, β) to denote an LHZ-FHS set of M FHSs of length N over an alphabet of size q, with the maximum periodic Hamming correlation β within the LHZ L_H.

Peng et al. [19] established the following lower bound on the maximum periodic Hamming correlation of an LHZ-FHS set.

Lemma 2 [19]. *Let S be an (N, M, q, L_H, β) LHZ-FHS set. Then we have*

$$\beta \geq \left\lceil \frac{(ML_H + M - q)N}{(ML_H + M - 1)q} \right\rceil. \tag{2}$$

Definition 2. *It is said that an FHS set S is optimal with respect to the bound (2), if the equality in (2) is achieved. In this case, S is called an LHZ-FHS set with optimal maximum periodic Hamming correlation.*

The *partial Hamming correlation* of two FHSs $X, Y \in S$, for the correlation window length L starting at k is defined by

$$H_{X,Y}(k|L;\tau) = \sum_{j=k}^{k+L-1} h(x_j, y_{j+\tau}), \quad 0 \leq \tau, k < N, 1 \leq L \leq N \tag{3}$$

where all operations among the position indices are performed modulo N. When $X = Y$, $H_{X,X}(k|L;\tau)$ is called the *partial Hamming autocorrelation* of X, and denoted by $H_X(k|L;\tau)$ for short. When $X \neq Y$, $H_{X,Y}(k|L;\tau)$ is called the *partial Hamming cross-correlation* of X and Y. In particular, when $L = N$, the partial Hamming correlation in (3) becomes the periodic Hamming correlation $H_{X,Y}(\tau)$.

For the FHS set S, the *maximum partial Hamming correlation* $H(S; L)$ of S for the correlation window length L is defined by

$$H(S; L) = \max \left\{ \max_{\substack{X \in S, 1 \leq \tau < N, \\ 0 \leq k < N}} \{H_X(k|L;\tau)\}, \max_{\substack{X, Y \in S, X \neq Y, \\ 0 \leq \tau, k < N}} \{H_{X,Y}(k|L;\tau)\} \right\}.$$

Throughout this paper, let (N, M, q, L, δ) denote an FHS set of M FHSs of length N over an alphabet of size q, with the maximum partial Hamming correlation $\delta = H(S; L)$ with respect to the correlation window length L.

A lower bound on the maximum partial Hamming correlation of an FHS set was established by Zhou et al. [23] as follows.

Lemma 3 [23]. *Let S be an (N, M, q, L, δ) FHS set. Then we have*

$$\delta \geq \left\lceil \frac{(MN - q)L}{(MN - 1)q} \right\rceil. \tag{4}$$

Definition 3. *It is said that an FHS set S is optimal with respect to the bound (4) and the given correlation window length L, if the equality in (4) is achieved for the given correlation window length L. In this case, the FHS set S is called an FHS set with optimal maximum partial Hamming correlation. If the equality in (4) is achieved for all correlation window length $1 \leq L \leq N$, then the FHS set S is said to be strictly optimal with respect to the bound (4). In this case, the FHS set S is called an FHS set with strictly optimal maximum partial Hamming correlation.*

The FHS set S is called an LHZ-FHS set with respect to the LHZ L_{PH} and correlation window length L, if there is a nonnegative integer ε such that

$$H_{X,Y}(k|L;\tau) \leq \varepsilon,$$

for any $X, Y \in S$, $0 \leq k < N$, $0 < \tau \leq L_{PH}$ when $X = Y$, and $0 \leq \tau \leq L_{PH}$ when $X \neq Y$.

Throughout this paper, we use $(N, M, q, L_{PH}, L, \varepsilon)$ to denote an LHZ-FHS set of M FHSs of length N over an alphabet of size q, with the maximum partial Hamming correlation ε with respect to the correlation window length L within the LHZ L_{PH}.

Niu et al. [15] obtained the following lower bound on the maximum partial Hamming correlation of an LHZ-FHS set.

Lemma 4 [15]. *Let S be an $(N, M, q, L_{PH}, L, \varepsilon)$ LHZ-FHS set. Then we have*

$$\varepsilon \geq \left\lceil \frac{(ML_{PH} + M - q)L}{(ML_{PH} + M - 1)q} \right\rceil. \tag{5}$$

Definition 4. *It is said that an FHS set S is optimal with respect to the bound (5) and the given correlation window length L, if the equality in (5) is achieved for the given correlation window length L. In this case, the FHS set S is called an LHZ-FHS set with optimal maximum partial Hamming correlation. If the equality in (5) is achieved for all correlation window length $1 \leq L \leq N$, then the FHS set S is said to be strictly optimal with respect to the bound (5). In this case, the FHS set S is called an LHZ-FHS set with strictly optimal maximum partial Hamming correlation.*

Theorem 1. *There does not exist an LHZ-FHS set of family size M, length N, and LHZ L_{PH} over an alphabet of size q which is strictly optimal with respect to the bound (5), provided that $M(L_{PH} + 1) > q^2$ and $q \geq 2$.*

Proof: For the correlation window length $L = 2$, it is not difficult to get that the right-hand side term in inequality (5)

$$\left\lceil \frac{2(ML_{PH} + M - q)}{(ML_{PH} + M - 1)q} \right\rceil \leq 1 \tag{6}$$

because of $q \geq 2$.

Suppose that there exists an LHZ-FHS set $\mathcal{U} = \{U^i = (U_0^i, U_1^i, \cdots, U_{N-1}^i) | 0 \leq i < M\}$ of family size M, length N, and LHZ L_{PH} over an alphabet of size q with strictly optimal maximum partial Hamming correlation, where $M(L_{PH}+1) > q^2$ and $q \geq 2$. From the Definition 4 and inequality (6), we get that

$$\varepsilon = \left\lceil \frac{2(ML_{PH} + M - q)}{(ML_{PH} + M - 1)q} \right\rceil \leq 1.$$

It indicates that among the set

$$\mathcal{T} = \left\{ (U_j^i, U_{j+1}^i) \mid 0 \leq i < M, \ k \leq j \leq k + L_{PH} \right\}$$

where $0 \leq k < N$ and all operations among the position indices are performed modulo N, they are different from each other. The size of the sequence set T is $M(L_{PH} + 1)$. There are at most q^2 different vectors in the 2-dimensional vector space over an alphabet of size q. Thus, we have $M(L_{PH} + 1) \leq q^2$ which contradicts the condition $M(L_{PH} + 1) > q^2$. □

Let S be an $(N, M, q, L_{PH}, L, \varepsilon)$ LHZ-FHS set. For the correlation window length L, the overall number of partial Hamming autocorrelation and partial Hamming cross-correlation of S within the LHZ L_{PH} are defined as follows, respectively:

$$S_a(L) = \sum_{X \in S} \sum_{\tau=1}^{L_{PH}} \sum_{k=0}^{N-1} H_X(k|L; \tau), \tag{7}$$

$$S_c(L) = \frac{1}{2} \sum_{X,Y \in S, X \neq Y} \sum_{\tau=0}^{L_{PH}} \sum_{k=0}^{N-1} H_{X,Y}(k|L; \tau). \tag{8}$$

For the correlation window length L, the *average partial Hamming autocorrelation* $P_a(L)$ $(0 \leq P_a(L) \leq \varepsilon)$ and the *average partial Hamming cross-correlation* $P_c(L)$ $(0 \leq P_c(L) \leq \varepsilon)$ of S within the LHZ L_{PH} are defined as follows, respectively:

$$P_a(L) = \frac{S_a(L)}{MNL_{PH}}, \tag{9}$$

$$P_c(L) = \frac{2S_c(L)}{MN(M-1)(L_{PH}+1)}. \tag{10}$$

Throughout this paper, we use $(N, M, q, L_{PH}, L, P_a(L), P_c(L))$ to denote an LHZ-FHS set of M FHSs of length N over an alphabet of size q, with the average partial Hamming autocorrelation $P_a(L)$ and the average partial Hamming cross-correlation $P_c(L)$ of S with respect to the correlation window length L within the LHZ L_{PH}.

Niu et al. [14] stated the following bound on the average partial Hamming autocorrelation and average partial Hamming cross-correlation of an LHZ-FHS set.

Lemma 5 [14]. *Let S be an $(N, M, q, L_{PH}, L, P_a(L), P_c(L))$ LHZ-FHS set. Then we have*

$$qL_{PH}P_a(L) + q(M-1)(L_{PH}+1)P_c(L) \geq (L_{PH}+1)LM - Lq. \tag{11}$$

Definition 5. *It is said that an FHS set S is optimal with respect to the bound (11) and the given correlation window length L, if the equality in (11) is achieved for the given correlation window length L. In this case, the FHS set S is called an LHZ-FHS set with optimal average partial Hamming correlation. If the equality in (11) is achieved for all correlation window length $1 \leq L \leq N$, then the FHS set S is said to be strictly optimal with respect to the bound (11). In this case, the FHS set S is called an LHZ-FHS set with strictly optimal average partial Hamming correlation.*

Let $\mathcal{F} = \{f_1, f_2, \cdots, f_q\}$, $\mathcal{S} = \{S^i = (S_0^i, S_1^i, \cdots, S_{N-1}^i) | \ 0 \leq i \leq M - 1\}$, and $W^k = (S_k^0, S_k^1, \cdots, S_k^{M-1})$, $0 \leq k \leq N - 1$. For any frequency $f \in \mathcal{F}$, let

$$n(W^k, f) = \sum_{i=0}^{M-1} h(S_k^i, f)$$

be the number of occurrences of frequency f in W^k.

Theorem 2. *Let $\mathcal{S}$ be an $(N, M, q, L_{PH}, L, P_a(L), P_c(L))$ LHZ-FHS set. If $n(W^k, f_i) = \frac{M}{q}$ for any $i = 1, 2, \cdots, q$, $k = 0, 1, \cdots, N - 1$, then $\mathcal{S}$ is an LHZ-FHS set with strictly optimal average partial Hamming correlation.*

Proof: For arbitrary correlation window length $1 \leq L \leq N$, from Eqs. (7), (8), (9), and (10), we get

$$\sum_{S^i, S^j \in \mathcal{S}} \sum_{\tau=0}^{L_{PH}} H_{S^i, S^j}(\tau) = MN + \sum_{S^i \in \mathcal{S}} \sum_{\tau=1}^{L_{PH}} H_{S^i}(\tau) + \sum_{S^i, S^j \in \mathcal{S}, i \neq j} \sum_{\tau=0}^{L_{PH}} H_{S^i, S^j}(\tau)$$

$$= MN + \sum_{S^i \in \mathcal{S}} \sum_{\tau=1}^{L_{PH}} H_{S^i}(0|N; \tau) + \sum_{S^i, S^j \in \mathcal{S}, i \neq j} \sum_{\tau=0}^{L_{PH}} H_{S^i, S^j}(0|N; \tau)$$

$$= MN + \frac{1}{L} \sum_{S^i \in \mathcal{S}} \sum_{\tau=1}^{L_{PH}} \sum_{k=0}^{N-1} H_{S^i}(k|L; \tau) + \frac{1}{L} \sum_{S^i, S^j \in \mathcal{S}, i \neq j} \sum_{\tau=0}^{L_{PH}} \sum_{k=0}^{N-1} H_{S^i, S^j}(k|L; \tau)$$

$$= MN + \frac{1}{L} S_a(L) + \frac{2}{L} S_c(L)$$

$$= MN + \frac{1}{L} M N L_{PH} P_a(L) + \frac{1}{L} MN(M - 1)(L_{PH} + 1)P_c(L). \tag{12}$$

On the other hand, we have

$$\sum_{S^i, S^j \in \mathcal{S}} \sum_{\tau=0}^{L_{PH}} H_{S^i, S^j}(\tau) = \sum_{S^i, S^j \in \mathcal{S}} \sum_{\tau=0}^{L_{PH}} \sum_{k=0}^{N-1} h(S_k^i, S_{k+\tau}^j) = \sum_{i,j=0}^{M-1} \sum_{k=0}^{N-1} \sum_{\tau=0}^{L_{PH}} h(S_k^i, S_{k+\tau}^j)$$

$$= \sum_{k=0}^{N-1} \sum_{\tau=0}^{L_{PH}} \sum_{i=0}^{M-1} n(W^{k+\tau}, S_k^i) = \sum_{k=0}^{N-1} \sum_{\tau=0}^{L_{PH}} \sum_{i=1}^{q} n(W^k, f_i) \times n(W^{k+\tau}, f_i).$$

Because $n(W^k, f_i) = \frac{M}{q}$ for any $i = 1, 2, \cdots, q$, $k = 0, 1, \cdots, N - 1$. Then we get that

$$\sum_{S^i, S^j \in \mathcal{S}} \sum_{\tau=0}^{L_{PH}} H_{S^i, S^j}(\tau) = \sum_{k=0}^{N-1} \sum_{\tau=0}^{L_{PH}} \sum_{i=1}^{q} n(W^k, f_i) \times n(W^{k+\tau}, f_i)$$

$$= \sum_{k=0}^{N-1} \sum_{\tau=0}^{L_{PH}} \sum_{i=1}^{q} \left(\frac{M}{q}\right)^2 = \frac{M^2 N(L_{PH} + 1)}{q}. \tag{13}$$

From Eqs. (12) and (13), it is easy to get

$$qL_{PH}P_a(L) + q(M-1)(L_{PH}+1)P_c(L) = (L_{PH}+1)LM - Lq.$$

According to the Definition 5, $\mathcal{S}$ is strictly optimal with respect to the bound (11).

$\square$

3   Constructions of LHZ-FHS Sets with Optimal Partial Hamming Correlation

In this section, a concatenated construction method is presented. The LHZ-FHS sets with optimal maximum partial Hamming correlation and the LHZ-FHS sets with strictly optimal average partial Hamming correlation whose sequence length can be infinite are constructed by the new construction, respectively.

Construction 1 *Step 1. Choose an (N, M, q) FHS set*

$$\mathcal{A} = \{A^i = (A_0^i, A_1^i, \cdots, A_{N-1}^i)| \ 0 \leq i < M\}$$

Step 2. For a positive integer l, define an FHS set

$$\mathcal{B} = \{B^i = (B_0^i, B_1^i, \cdots, B_{lN-1}^i)| \ 0 \leq i < M\}$$

where $B_j^i = A_{(j)_N}^i$, $0 \leq j \leq lN-1$, $0 \leq i < M$. Here $(j)_N = j \ mod \ N$.

Theorem 3. *Assume that $\mathcal{A}$ is an (N, M, q, α) FHS set with optimal maximum periodic Hamming correlation. Then the FHS set $\mathcal{B}$ constructed by Construction 1 is an $(lN, M, q, N-1, N, \alpha)$ LHZ-FHS set with optimal maximum partial Hamming correlation.*

Proof: Let $B^{i_1}, B^{i_2} \in \mathcal{B}, 0 \leq i_1, i_2 < M$. The partial Hamming correlation $H_{B^{i_1}B^{i_2}}(k|N; \tau)$ between B^{i_1} and B^{i_2} for the correlation window length N starting at k within the LHZ $N-1$ is given by

$$H_{B^{i_1}B^{i_2}}(k|N; \tau) = \sum_{j=k}^{k+N-1} h(B_j^{i_1}, B_{j+\tau}^{i_2}) = \sum_{j=0}^{N-1} h(A_j^{i_1}, A_{j+\tau}^{i_2}) = H_{A^{i_1}A^{i_2}}(\tau)$$

where $0 \leq k \leq lN-1$, $0 < \tau \leq N-1$ when $i_1 = i_2$, and $0 \leq \tau \leq N-1$ when $i_1 \neq i_2$. Because $\mathcal{A}$ is an (N, M, q, α) FHS set. Then we have

$$H_{B^{i_1}B^{i_2}}(k|N; \tau) = H_{A^{i_1}A^{i_2}}(\tau) \leq \alpha.$$

Since the FHS set $\mathcal{A}$ is optimal with respect to the bound (1), then we have

$$\alpha = \left\lceil \frac{(MN-q)N}{(MN-1)q} \right\rceil = \left\lceil \frac{(M(N-1)+M-q)N}{(M(N-1)+M-1)q} \right\rceil.$$

From the Definition 4, the $(lN, M, q, N-1, N, \alpha)$ LHZ-FHS set $\mathcal{B}$ constructed by Construction 1 is optimal with respect to the bound (5). $\square$

Theorem 4. *Assume that $\mathcal{A}$ is an (N, M, q, L_H, β) LHZ-FHS set with optimal maximum periodic Hamming correlation. Then the FHS set $\mathcal{B}$ constructed by Construction 1 is an $(lN, M, q, L_H, N, \beta)$ LHZ-FHS set with optimal maximum partial Hamming correlation.*

Proof: Let $B^{i_1}, B^{i_2} \in \mathcal{B}, 0 \leq i_1, i_2 < M$. The partial Hamming correlation $H_{B^{i_1} B^{i_2}}(k|N; \tau)$ between B^{i_1} and B^{i_2} for the correlation window length N starting at k within the LHZ L_H is given by

$$H_{B^{i_1} B^{i_2}}(k|N; \tau) = \sum_{j=k}^{k+N-1} h(B_j^{i_1}, B_{j+\tau}^{i_2}) = \sum_{j=0}^{N-1} h(A_j^{i_1}, A_{j+\tau}^{i_2}) = H_{A^{i_1} A^{i_2}}(\tau)$$

where $0 \leq k \leq lN - 1$, $0 < \tau \leq L_H$ when $i_1 = i_2$, and $0 \leq \tau \leq L_H$ when $i_1 \neq i_2$. Because $\mathcal{A}$ is an (N, M, q, L_H, β) LHZ-FHS set. Then we have

$$H_{B^{i_1} B^{i_2}}(k|N; \tau) = H_{A^{i_1} A^{i_2}}(\tau) \leq \beta.$$

Since the FHS set $\mathcal{A}$ is optimal with respect to the bound (2), then we have

$$\beta = \left\lceil \frac{(ML_H + M - q)N}{(ML_H + M - 1)q} \right\rceil.$$

From the Definition 4, the $(lN, M, q, L_H, N, \beta)$ LHZ-FHS set $\mathcal{B}$ constructed by Construction 1 is optimal with respect to the bound (5). □

Theorem 5. *Assume that $\mathcal{A}$ is an (N, M, q, L, δ) FHS set with optimal maximum partial Hamming correlation. Then the FHS set $\mathcal{B}$ constructed by Construction 1 is an $(lN, M, q, N - 1, L, \delta)$ LHZ-FHS set with optimal maximum partial Hamming correlation.*

Proof: Let $B^{i_1}, B^{i_2} \in \mathcal{B}, 0 \leq i_1, i_2 < M$. The partial Hamming correlation $H_{B^{i_1} B^{i_2}}(k|L; \tau)$ between B^{i_1} and B^{i_2} for the correlation window length L starting at k within the LHZ $N - 1$ is given by

$$H_{B^{i_1} B^{i_2}}(k|L; \tau) = \sum_{j=k}^{k+L-1} h(B_j^{i_1}, B_{j+\tau}^{i_2})$$

$$= \sum_{j=(k)_N}^{(k)_N+L-1} h(A_j^{i_1}, A_{j+\tau}^{i_2})$$

$$= H_{A^{i_1} A^{i_2}}((k)_N|L; \tau),$$

where $0 \leq k \leq lN - 1$, $0 < \tau \leq N - 1$ when $i_1 = i_2$, and $0 \leq \tau \leq N - 1$ when $i_1 \neq i_2$. Because $\mathcal{A}$ is an (N, M, q, L, δ) FHS set. Then we have

$$H_{B^{i_1} B^{i_2}}(k|L; \tau) = H_{A^{i_1} A^{i_2}}((k)_N|L; \tau) \leq \delta.$$

Since the FHS set $\mathcal{A}$ is optimal with respect to the bound (4), then we have

$$\delta = \left\lceil \frac{(MN-q)L}{(MN-1)q} \right\rceil = \left\lceil \frac{(M(N-1)+M-q)L}{(M(N-1)+M-1)q} \right\rceil.$$

From the Definition 4, the $(lN, M, q, N-1, L, \delta)$ LHZ-FHS set $\mathcal{B}$ constructed by Construction 1 is optimal with respect to the bound (5). $\qquad\square$

Theorem 6. *Assume that $\mathcal{A}$ is an $(N, M, q, L_{PH}, L, \varepsilon)$ LHZ-FHS set with optimal maximum partial Hamming correlation. Then the FHS set $\mathcal{B}$ constructed by Construction 1 is an $(lN, M, q, L_{PH}, L, \varepsilon)$ LHZ-FHS set with optimal maximum partial Hamming correlation.*

Proof: Let $B^{i_1}, B^{i_2} \in \mathcal{B}, 0 \leq i_1, i_2 < M$. The partial Hamming correlation $H_{B^{i_1}B^{i_2}}(k|L; \tau)$ between B^{i_1} and B^{i_2} for the correlation window length L starting at k within the LHZ L_{PH} is given by

$$H_{B^{i_1}B^{i_2}}(k|L;\tau) = \sum_{j=k}^{k+L-1} h(B_j^{i_1}, B_{j+\tau}^{i_2})$$

$$= \sum_{j=(k)_N}^{(k)_N+L-1} h(A_j^{i_1}, A_{j+\tau}^{i_2})$$

$$= H_{A^{i_1}A^{i_2}}((k)_N|L;\tau),$$

where $0 \leq k \leq lN-1$, $0 < \tau \leq L_{PH}$ when $i_1 = i_2$, and $0 \leq \tau \leq L_{PH}$ when $i_1 \neq i_2$. Because $\mathcal{A}$ is an $(N, M, q, L_{PH}, L, \varepsilon)$ LHZ-FHS set. Then we have

$$H_{B^{i_1}B^{i_2}}(k|L;\tau) = H_{A^{i_1}A^{i_2}}((k)_N|L;\tau) \leq \varepsilon.$$

Since the FHS set $\mathcal{A}$ is optimal with respect to the bound (5), then we have

$$\varepsilon = \left\lceil \frac{(ML_{PH}+M-q)L}{(ML_{PH}+M-1)q} \right\rceil.$$

From the Definition 4, the $(lN, M, q, L_{PH}, L, \varepsilon)$ LHZ-FHS set $\mathcal{B}$ constructed by Construction 1 is optimal with respect to the bound (5). $\qquad\square$

Theorem 7. *Assume that $\mathcal{A}$ is an LHZ-FHS set of family size M, length N, and LHZ L_{PH} over an alphabet of size q with strictly optimal average partial Hamming correlation. Then the FHS set $\mathcal{B}$ constructed by Construction 1 is an LHZ-FHS set of family size M, length lN, and LHZ L_{PH} over an alphabet of size q with strictly optimal average partial Hamming correlation.*

Proof: For the FHS set $\mathcal{B}$, it is easy to get $n(W^k, f_i) = \frac{M}{q}$ for any $i = 1, 2, \cdots, q$, $k = 0, 1, \cdots, lN-1$. Then $\mathcal{B}$ is an LHZ-FHS set with strictly optimal average partial Hamming correlation according to the Theorem 2. $\qquad\square$

Remark 1. *It should be noted that both the bound (5) and bound (11) are independent of the sequence length. Thus, the sequence length of the FHS set $\mathcal{B}$ in Theorems 3–7 can be infinite.*

4 Conclusion Remarks

In this paper, we prove the nonexistence of an LHZ-FHS set of family size M, length N, and LHZ L_{PH} over an alphabet of size q with strictly optimal maximum partial Hamming correlation, provided that $M(L_{PH} + 1) > q^2$ and $q \geq 2$. A sufficient condition for an LHZ-FHS set with strictly optimal average partial Hamming correlation is also given. In addition, a simple and useful concatenated construction method is presented. The LHZ-FHS sets with optimal maximum partial Hamming correlation and the LHZ-FHS sets with strictly optimal average partial Hamming correlation whose sequence length can be infinite are constructed by the new construction, respectively. Some new LHZ-FHS sets with optimal maximum partial Hamming correlation are listed in Table 1. It is expected that the proposed LHZ-FHS sets will be helpful in quasi-synchronous FH-CDMA systems to eliminate MA interference.

Table 1. Some new LHZ-FHS sets with optimal maximum partial Hamming correlation (p: a prime, q: a prime power)

Parameters $(N, M, q, L_{PH}, L, \varepsilon)$	Constraints	Based on optimal FHS sets	Remarks
$\left(l(\frac{q^r-1}{s}), s, q, \frac{q^r-1}{s}-1, \frac{q^r-1}{s}, \frac{q^{r-1}-1}{s}\right)$	$s = 2$, q and r are odd	[3]	Theorem 3
	$s = q - 1$, $\gcd(r, s) = 1$	[4]	Theorem 3
	$\gcd(r, s) = 1$, $s\|(q-1)$	[8]	Theorem 3
$\left(l(q^2+1), q^2-1, q, q^2, q^2+1, q+1\right)$	q is even	[5]	Theorem 3
$\left(l(q-1), r, r+1, q-2, q-1, \frac{q-1}{r}\right)$	$r\|(q-1), r+1 > \frac{q-1}{r}$	[9]	Theorem 3
$\left(l(q^r-1), q^s, q^s, q^r-2, q^r-1, q^{r-s}\right)$	$1 \leq s \leq r$	[22]	Theorem 3
$\left(lp^2(q-1), pq, pq, \min\{p^2-1, q-2\},\right.$ $\left. p^2(q-1), p\right)$	$\gcd(p, q-1)=1, 2p\leq q-1$	[2]	Theorem 4
$\left(ls(q^n-1), m, q, Z, s(q^n-1), s(q^{n-1}-1)\right)$	$q^n-1=m(Z+1)$, $\gcd(s, q^n-1)=1, s<m$	[13]	Theorem 4
$\left(\frac{lsr(q^n-1)}{k}, m\lfloor\frac{k}{r}\rfloor, q, Z, \frac{sr(q^n-1)}{k},\right.$ $\left.\frac{sr(q^{n-1}-1)}{k}\right)$	$1\leq r\leq k, \frac{q^n-1}{k}=m(Z+1),$ $\gcd(k, n)=1,$ $s\equiv 1 \bmod (Z+1),$ $s < \frac{kq(q^n-2)}{q-1}$	[16]	Theorem 4
$\left(l\left(\frac{q^r-1}{m}\right), m, q^k, \frac{q^r-1}{m}-1, aT+b,\right.$ $\left.\left\lceil(aT+b)\frac{q^{r-k}-1}{q^r-1}\right\rceil\right)$	$T=\frac{q^r-1}{q^k-1}, 0\leq a\leq\frac{q^k-1}{m},$ $1\leq b\leq T, b=1$ or $\left\lceil\frac{q^{r-k}-1}{q^k-1}\right\rceil=\left[b\frac{q^{r-k}-1}{q^r-1}\right]$	[17]	Theorem 5

References

1. Chung, J.H., Yang, K.: New frequency-hopping sequence sets with optimal average and good maximum Hamming correlations. IET Commun. **6**(13), 2048–2053 (2012)
2. Chung, J.H., Yang, K.: New classes of optimal low-hit-zone frequency-hopping sequence sets by Cartesian product. IEEE Trans. Inf. Theory **59**(1), 726–732 (2013)
3. Ding, C., Moisio, M.J., Yuan, J.: Algebraic constructions of optimal frequency-hopping sequences. IEEE Trans. Inf. Theory **53**(7), 2606–2610 (2007)

4. Ding, C., Yin, J.: Sets of optimal frequency-hopping sequences. IEEE Trans. Inf. Theory **54**(8), 3741–3745 (2008)
5. Ding, C., Fuji-Hara, R., Fujiwara, Y., Jimbo, M., Mishima, M.: Sets of frequency hopping sequences: Bounds and optimal constructions. IEEE Trans. Inf. Theory **55**(7), 3297–3304 (2009)
6. Eun, Y.C., Jin, S.Y., Hong, Y.P., Song, H.Y.: Frequency hopping sequences with optimal partial autocorrelation properties. IEEE Trans. Inf. Theory **50**(10), 2438–2442 (2004)
7. Fan, P.Z., Darnell, M.: Sequence Design for Communications Applications. RSP-John Wiley & Sons, London (1996)
8. Ge, G., Miao, Y., Yao, Z.: Optimal frequency hopping sequences: Auto- and cross-correlation properties. IEEE Trans. Inf. Theory **55**(2), 867–879 (2009)
9. Han, Y.K., Yang, K.: On the sidel'nikov sequences as frequency-hopping sequences. IEEE Trans. Inf. Theory **55**(9), 4279–4285 (2009)
10. Lempel, A., Greenberger, H.: Families of sequences with optimal Hamming correlation properties. IEEE Trans. Inf. Theory **20**(1), 90–94 (1974)
11. Liu, F., Peng, D.Y., Zhou, Z.C., Tang, X.H.: A new frequency-hopping sequence set based upon generalized cyclotomy. Des. Codes Crypt. **69**(2), 247–259 (2013)
12. Liu, X., Peng, D.Y., Han, H.Y.: Low-hit-zone frequency hopping sequence sets with optimal partial Hamming correlation properties. Designs, Codes and Cryptography (2013)
13. Ma, W.P., Sun, S.H.: New designs of frequency hopping sequences with low hit zone. Des. Codes Crypt. **60**(2), 145–153 (2011)
14. Niu, X.H., Peng, D.Y., Liu, F.: Lower bounds on the average partial hamming correlations of frequency hopping sequences with low hit zone. In: Proceedings of the 6th International Conference Sequences and Their Applications, pp. 67–75 (2010)
15. Niu, X.H., Peng, D.Y., Liu, F., Liu, X.: Lower bounds on the maximum partial correlations of frequency hopping sequence set with low hit zone. IEICE Trans. Fundam. Electron. Commun. Comput. Sci. **93–A**(11), 2227–2231 (2010)
16. Niu, X.H., Peng, D.Y., Zhou, Z.C.: New classes of optimal low hit zone frequency hopping sequences with new parameters by interleaving technique. IEICE Trans. Fundam. Electron. Commun. Comput. Sci. **95–A**(11), 1835–1842 (2012)
17. Niu, X.H., Peng, D.Y., Zhou, Z.C.: Frequency/time hopping sequence sets with optimal partial Hamming correlation properties. Sci. China Ser. F. Inf. Sci. **55**(10), 2207–2215 (2012)
18. Peng, D.Y., Fan, P.Z.: Lower bounds on the Hamming auto- and cross correlations of frequency-hopping sequences. IEEE Trans. Inf. Theory **50**(9), 2149–2154 (2004)
19. Peng, D.Y., Fan, P.Z., Lee, M.H.: Lower bounds on the periodic Hamming correlations of frequency hopping sequences with low hit zone. Sci. China Ser. F. Inf. Sci. **49**(2), 1–11 (2006)
20. Peng, D.Y., Niu, X.H., Tang, X.H.: Average Hamming correlation for the cubic polynomial hopping sequences. IET Commun. **4**(15), 1775–1786 (2010)
21. Ren, W.L., Fu, F.W., Zhou, Z.C.: On the average partial Hamming correlation of frequency hopping sequences. IEICE Trans. Fundam. Electron. Commun. Comput. Sci. **96–A**(5), 1010–1013 (2013)
22. Zhou, Z.C., Tang, X.H., Peng, D.Y., Udaya, P.: New constructions for optimal sets of frequency-hopping sequences. IEEE Trans. Inf. Theory **57**(6), 3831–3840 (2011)
23. Zhou, Z.C., Tang, X.H., Niu, X.H., Udaya, P.: New classes of frequency-hopping sequences with optimal partial correlation. IEEE Trans. Inf. Theory **58**(1), 453–458 (2012)

Improved Singleton Bound on Frequency Hopping Sequences

Xing Liu$^{(\boxtimes)}$, Daiyuan Peng, and Hongyu Han

Provincial Key Lab of Information Coding and Transmission, Institute of Mobile Communications, Southwest Jiaotong University, Chengdu, China
liuxing4@126.com, dypeng@swjtu.edu.cn, hongyuhanswjtu@163.com

Abstract. In this paper, a new bound on the frequency hopping (FH) sequences with respect to the size of the frequency slot set, the sequence length, the family size, and the maximum periodic Hamming correlation is established. The new bound is tighter than the Singleton bounds on the FH sequences derived by Ding et al. (2009) and Yang et al. (2011) and the bound derived by Liu and Peng (2013).

Keywords: Frequency hopping sequences · Frequency hopping spread spectrum · Hamming correlation · The Peng-Fan bound · The Singleton bound

1 Introduction

In wireless communication systems, frequency hopping (FH) spread spectrum and direct sequence spread spectrum are two main spread coding technologies. FH multiple access spread spectrum systems, with anti-jamming, secure, and multiple access properties, have found many applications in Bluetooth, military radio communications, mobile communications, and modern radar and sonar echolocation systems [1–3]. In such systems each user is represented by a sequence of hopping frequencies. Simultaneous transmission by any two users over the same frequency band results in collisions of signals, and hence, it is very desirable that such collisions over the same frequency band are minimized. The degree of such collisions is clearly related to the Hamming correlation properties of the FH sequence [2–4]. In order to evaluate the goodness of FH sequence design, the periodic Hamming correlation function is used as an important measure and the periodic Hamming correlation is considered in almost all papers [5–12].

FH sequence design normally involves five parameters: the size of the frequency slot set, the sequence length, the family size, the maximum periodic Hamming autocorrelation and the maximum periodic Hamming crosscorrelation. Generally speaking, the five parameters are bounded by certain theoretical

This work was supported by National Science Foundation of China (Grant No. 61271244).

© Springer International Publishing Switzerland 2014
K.-U. Schmidt and A. Winterhof (Eds.): SETA 2014, LNCS 8865, pp. 305–314, 2014.
DOI: 10.1007/978-3-319-12325-7_26

limits. In order to evaluate the performance of the FH sequence, it is important to find the theoretical limits which set a bounded relation among these parameters.

Let $F = \{f_1, f_2, \cdots, f_q\}$ be a frequency slot set with size q, S a set of M FH sequences of length N. For any two FH sequences $x = \{x_0, x_1, \cdots, x_{N-1}\}$, $y = \{y_0, y_1, \cdots, y_{N-1}\} \in S$, any positive integer $\tau, 0 \leq \tau < N$, the periodic Hamming correlation function $H_{xy}(\tau)$ of x and y at time delay τ is defined as follow:

$$H_{xy}(\tau) = \sum_{k=0}^{N-1} h(x_k, y_{k+\tau}), (\tau = 0, 1, \cdots, N-1), \tag{1}$$

where $h(a, b) = 1$ if $a = b$, and 0 otherwise. The subscript addition $k + \tau$ is performed modulo N.

For any given FH sequence set S, the maximum Hamming autocorrelation $H_a(S)$, the maximum Hamming crosscorrelation $H_c(S)$ and the maximum Hamming correlation are defined as follows, respectively:

$$H_a(S) = \max\{H_{xx}(\tau)|x \in S, \tau=1, 2, \cdots, N-1\},$$
$$H_c(S) = \max\{H_{xy}(\tau)|x, y \in S, x \neq y, \tau=0, 1, \cdots, N-1\},$$
$$H_m(S) = \max\{H_a(S), H_c(S)\}.$$

For simplicity and convenience, let $H_a = H_a(S)$, $H_c = H_c(S)$, $H_m = H_m(S)$.

Generally speaking, we need varieties of sequence designs with various size and frequency set. However, there exist different bounds involving various parameters of the FH sequence design. For different FH sequences, the different bounds can be used to evaluate the performance of the FH sequences.

As early as 1974, Lempel and Greenberger [4] established the following bound on the maximum Hamming autocorrelation of an FH sequence.

Lemma 1. *(The Lempel-Greenberger bound) For any FH sequence of length N over a frequency slot set F of size q, we have*

$$H_a \geq \frac{(N-r)(N+r-q)}{(N-1)q} \tag{2}$$

where r is the least nonnegative residue of N modulo q.

In 2004, Peng and Fan [13] obtained lower bounds on the maximum Hamming autocorrelation and the maximum Hamming crosscorrelation of an FH sequence set. The bounds are given by the following lemma.

Lemma 2. *(The Peng-Fan bounds) For a set of M FH sequences of length N over a given frequency slot set F with size q, we have*

$$(N-1)qH_a + (M-1)NqH_c \geq (MN-q)N \tag{3}$$

and

$$(N-1)MH_a + (M-1)MNH_c \geq 2IMN - (I+1)Iq \tag{4}$$

where $I = \left\lfloor \frac{MN}{q} \right\rfloor$.

The Peng-Fan bound (4) includes the Lempel-Greenberger bound as a special case.

In 2009, Ding et al. [5] derived the following upper bound on the number of the FH sequences from the Singleton bound on error correcting code [14].

Lemma 3. *(The Singleton bound on the family size of FH sequence set) For a set of M FH sequences of length N over a given frequency slot set F with size q, we have*

$$M \leq \left\lfloor \frac{q^{H_m+1}}{N} \right\rfloor. \tag{5}$$

In 2011, Yang et al. [8] derived the following lower bound on the maximum Hamming correlation of the FH sequences from the Singleton bound on error correcting code [14].

Lemma 4. *(The Singleton bound on the maximum Hamming correlation of FH sequences) For a set of M FH sequences of length N over a given frequency slot set F with size q, we have*

$$H_m \geq \lceil \log_q MN \rceil - 1. \tag{6}$$

In [8], the authors pointed out that if the bound (6) is met, then the bound (5) may not be met.

In 2013, Liu et al. [15] improved the Singleton bound on FH sequences. The bound is given by the following lemma.

Lemma 5. *For a set of M FH sequences of length N over a given frequency slot set F with size q, we have*

$$\begin{cases} q^{H_m+1} \geq MN, & \text{if } \gcd(H_m+1, N) = 1 \\ q^{H_m+1} - q^{\frac{H_m+1}{\gcd(H_m+1,N)}} \geq MN, & \text{otherwise} . \end{cases} \tag{7}$$

By the above bounds, an FH sequence set is said to be optimal if one of the bounds is met with equality.

The rest of this paper is organized as follows. In Sect. 2, a new bound on the FH sequences is derived. In Sect. 3, the new bound is compared with the previous bounds. Finally, some concluding remarks are given in Sect. 4.

2   New Bound on FH Sequences

We are now ready to state the main theorem on the new bound on an FH sequence set.

Theorem 1. *For a set of M FH sequences of length N over a given frequency slot set F with size q, we have*

$$\sum_{d|\gcd(H_m+1,N)} \mu(d)q^{\frac{H_m+1}{d}} \geq MN \tag{8}$$

where $\mu(d)$ is the Möbius function defined as follows:

$$\mu(d) = \begin{cases} 1, & d = 1 \\ (-1)^r, & d \text{ is a product of } r \text{ different primes} \\ 0, & \text{otherwise.} \end{cases}$$

Proof. Without loss of the fairness, let $F = \{0, 1, \cdots, q-1\}$ be a frequency slot set with size q, $S = \{S_0, S_1, \cdots, S_{M-1}\}$ a set of M FH sequences of length N, where $S_k = (s_k(0), s_k(1), \cdots, s_k(N-1))$, $k = 0, 1, \cdots, M-1$. For simplicity and convenience, let $\lambda = \gcd(H_m(S) + 1, N)$, $\eta = \frac{H_m(S)+1}{\lambda}$, $\zeta = \frac{N}{\lambda}$.

We now construct an FH sequence set Q where $Q = \{Q_{k,j} | k = 0, 1, \cdots, M-1, j = 0, 1, \cdots, \zeta - 1\}$ and $Q_{k,j} = (Q_{k,j}(0), Q_{k,j}(1), \cdots, Q_{k,j}(\lambda - 1))$. Let

$$Q_{k,j}(i) = s_k(i\zeta + j) + q \times s_k(i\zeta + j + 1) + \cdots + q^{\eta-1} \times s_k(i\zeta + j + \eta - 1),$$
$$i = 0, 1, \cdots, \lambda - 1. \tag{9}$$

Note that all operations among the brackets are performed modulo N. It is clear that Q is a set of FH sequences of family size ζM and length λ over a frequency slot set with size q^η.

For any two FH sequences $Q_{k_1,j_1}, Q_{k_2,j_2} \in Q, k_1 \neq k_2$ or $j_1 \neq j_2$, we have

$$H(Q_{k_1,j_1}, Q_{k_1,j_1}; \tau) = \sum_{i=0}^{\lambda-1} h(Q_{k_1,j_1}(i), Q_{k_1,j_1}(i+\tau)) \quad (\tau = 1, 2, \cdots, \lambda - 1) \tag{10}$$

and

$$H(Q_{k_1,j_1}, Q_{k_2,j_2}; \tau) = \sum_{i=0}^{\lambda-1} h(Q_{k_1,j_1}(i), Q_{k_2,j_2}(i+\tau)) \quad (\tau = 0, 1, \cdots, \lambda - 1). \tag{11}$$

If $H(Q_{k_1,j_1}, Q_{k_1,j_1}; \tau) = \lambda$ or $H(Q_{k_1,j_1}, Q_{k_2,j_2}; \tau) = \lambda$, then based on the definition of $Q_{k,j}$ in (9) we have

$$s_{k_1}(i\zeta + j_1 + m) = s_{k_1}(i\zeta + j_1 + m + \zeta\tau)$$
$$(i = 0, 1, \cdots, \lambda - 1, m = 0, 1, \cdots, \eta - 1) \tag{12}$$

or

$$s_{k_1}(i\zeta + j_1 + m) = s_{k_2}(i\zeta + j_2 + m + \zeta\tau)$$
$$(i = 0, 1, \cdots, \lambda - 1, m = 0, 1, \cdots, \eta - 1). \tag{13}$$

Thus, we have

$$H(S_{k_1}, S_{k_1}; \zeta\tau)$$

$$= \sum_{i=0}^{N-1} h(s_{k_1}(i+j_1), s_{k_1}(i+j_1+\zeta\tau))$$

$$\geq \sum_{i=0}^{\eta-1} h(s_{k_1}(i+j_1), s_{k_1}(i+j_1+\zeta\tau)) + \sum_{i=\zeta}^{\zeta+\eta-1} h(s_{k_1}(i+j_1), s_{k_1}(i+j_1+\zeta\tau))$$

$$+ \sum_{i=2\zeta}^{2\zeta+\eta-1} h(s_{k_1}(i+j_1), s_{k_1}(i+j_1+\zeta\tau))$$

$$+ \cdots + \sum_{i=(\lambda-1)\zeta}^{(\lambda-1)\zeta+\eta-1} h(s_{k_1}(i+j_1), s_{k_1}(i+j_1+\zeta\tau))$$

$$= \sum_{i=0}^{\eta-1}\sum_{m=0}^{\lambda-1} h(s_{k_1}(m\zeta+i+j_1), s_{k_1}(m\zeta+i+j_1+\zeta\tau))$$

$$\overset{(12)}{=} \sum_{i=0}^{\eta-1}\sum_{m=0}^{\lambda-1} 1 = \lambda\eta = H_m(S) + 1 \tag{14}$$

and

$$H(S_{k_1}, S_{k_2}; j_2 - j_1 + \zeta\tau)$$

$$= \sum_{i=0}^{N-1} h(s_{k_1}(i+j_1), s_{k_2}(i+j_2+\zeta\tau))$$

$$\geq \sum_{i=0}^{\eta-1} h(s_{k_1}(i+j_1), s_{k_2}(i+j_2+\zeta\tau)) + \sum_{i=\zeta}^{\zeta+\eta-1} h(s_{k_1}(i+j_1), s_{k_2}(i+j_2+\zeta\tau))$$

$$+ \sum_{i=2\zeta}^{2\zeta+\eta-1} h(s_{k_1}(i+j_1), s_{k_2}(i+j_2+\zeta\tau))$$

$$+ \cdots + \sum_{i=(\lambda-1)\zeta}^{(\lambda-1)\zeta+\eta-1} h(s_{k_1}(i+j_1), s_{k_2}(i+j_2+\zeta\tau))$$

$$= \sum_{i=0}^{\eta-1}\sum_{m=0}^{\lambda-1} h(s_{k_1}(m\zeta+i+j_1), s_{k_2}(m\zeta+i+j_2+\zeta\tau))$$

$$\overset{(13)}{=} \sum_{i=0}^{\eta-1}\sum_{m=0}^{\lambda-1} 1 = \lambda\eta = H_m(S) + 1. \tag{15}$$

Since $H_m(S)$ is the maximum periodic Hamming correlation of S, the inequalities (14) and (15) contradict the definition of S. Therefore, we have

$$H_m(Q) \leq \lambda - 1. \tag{16}$$

Define

$$L^\tau(Q_{k,j}) = (Q_{k,j}(\tau), Q_{k,j}(\tau+1), \cdots, Q_{k,j}(\tau+\lambda-1)) \quad (\tau = 0, 1, \cdots, \lambda-1)$$

where all operations among the brackets are performed modulo λ. The inequality (16) indicates that among the sequence set

$$R = \{L^\tau(Q_{k,j}) | k = 0, 1, \cdots, M-1, j = 0, 1, \cdots, \zeta-1, \tau = 0, 1, \cdots, \lambda-1\}$$

they are different from each other. The size of the sequence set is $\lambda\zeta M$.

In the λ-dimensional vector space over a finite alphabet q^η, there are at most $q^{\lambda\eta}$ different vectors defined by V. But the sequence $(f_0, f_1, \cdots, f_{d-1}, f_0, f_1, \cdots, f_{d-1}, \cdots, f_0, f_1, \cdots, f_{d-1}) \in V$ do not belong to R, where $1 \le d < \lambda$, $d|\lambda$ and $f_0, f_1, \cdots, f_{d-1} = 0, 1, \cdots, q^\eta - 1$. If $L^\tau(Q_{k,j}) = (f_0, f_1, \cdots, f_{d-1}, f_0, f_1, \cdots, f_{d-1}, \cdots, f_0, f_1, \cdots, f_{d-1})$, then $L^\tau(Q_{k,j}) = L^{\tau+id}(Q_{k,j})$ where $i = 1, 2, \cdots, \frac{\lambda}{d} - 1$ and the addition in the superscript is performed modulo λ, which contradicts the definition of R. d is called the cycle length of the sequence. Two sequences are said to be equivalent if one is a cyclic left shift of another. By means of this equivalent relation, all the sequences in V can be classified into equivalent classes. In each equivalent class, all the sequences are pairwise equivalent and have same minimum cycle length d. Obviously, the number of elements in equivalent class is equal to cycle length d. Let S_d be the number of equivalent classes which have cycle length d, where $1 \le d \le \lambda$ and $d|\lambda$. Then the family size of V can be represented by

$$|V| = \sum_{d|\lambda} dS_d = q^{\lambda\eta} \tag{17}$$

Thus the maximum number of elements in R can be calculated by λS_λ. By the Möbius Inversion Formulas [16,17], (17) becomes

$$\lambda S_\lambda = \sum_{d|\lambda} \mu(\frac{\lambda}{d}) q^{d\eta} = \sum_{d|\lambda} \mu(d) q^{\frac{\lambda\eta}{d}}$$

where $\mu(d)$ is the Möbius function. Then we have

$$\sum_{d|\lambda} \mu(d) q^{\frac{\lambda\eta}{d}} \ge \lambda\zeta M$$

which indicates that

$$\sum_{d| \gcd(H_m(S)+1, N)} \mu(d) q^{\frac{H_m(S)+1}{d}} \ge MN.$$

$\square$

It is easily seen that the new bound (8) is tighter than the Singleton bounds on the FH sequences (5) and (6), since

$$\sum_{d| \gcd(H_m+1, N)} \mu(d) q^{\frac{H_m+1}{d}} \le q^{H_m+1}.$$

The new bound (8) is tighter than the bound (7), since

$$\sum_{d|\gcd(H_m+1,N)} \mu(d)q^{\frac{H_m+1}{d}} \le q^{H_m+1} - q^{\frac{H_m+1}{\gcd(H_m+1,N)}}.$$

3   Comparison of New Bound and Previous Bounds

In order to compare the new bound with previous bounds, let q be a prime power, and let k be an integer with $1 \le k \le q - 1$. Define

$$\mathrm{GF}(q)[x]_k = \left\{ \sum_{i=1}^{k} g_i x^i : g_i \in \mathrm{GF}(q), i = 1, 2, \cdots, k \right\}. \tag{18}$$

Define $n = q - 1$ and

$$C_{RS} = \left\{ (g(1), g(\alpha), \cdots, g(\alpha^{n-1})) : g(x) \subset \mathrm{GF}(q)[x]_k \right\}$$

where α is a generator of $\mathrm{GF}(q)^*$. Let $F = \{f_0, f_1, \cdots, f_{q-1}\}$ be an abelian group of size q. Define

$$F^n = \{(c_0, c_1, \cdots, c_{n-1}) : c_i \in F \text{ for all } i\}.$$

The Hamming distance between two vectors in F^n is the total number of coordinate positions in which they differ. An (n, M, q, d') code is an M subset of the space F^n with the minimum Hamming distance d'. A code is called equidistant if the distance between every pair of distinct codewords is the same. An $[n, k, q, d']$ code is a linear subspace of F^n with dimension k such that the minimum Hamming distance between all pairs of distinct codewords is d'. It is well known that the code C_{RS} has parameters $[n, k, q, d' = n - k + 1]$ and is cyclic.

Define the following cyclic permutation ρ of an element $x = (x_0, x_1, \cdots, x_{n-1})$ as $\rho x = (x_1, \cdots, x_{n-1}, x_0)$. For any x, y in C_{RS}, if $x = \rho^m y$ for some integer m, the x and y are said to be ρ-equivalent. ρ-equivalent gives a partition of C_{RS} into disjoint subsets, called cyclic equivalence classes. The number of codewords in an equivalent class is the cycle length of the equivalent class. Thus picking up one element from each equivalence class gives a subcode of C_{RS}, say $\overline{C}_{RS}$, which has the property that the cyclic shifts of two distinct codewords of $\overline{C}_{RS}$ do not overlap in more than $n - d'$ places.

Next, define a subset S_{RS} of $\overline{C}_{RS}$ as follows.

Definition 1. *S_{RS} consists of those codewords $x \in \overline{C}_{RS}$ such that $\rho^j x \neq x$ for $j = 1, 2, \cdots, n - 1$.*

Thus, each codeword of S_{RS} has the cycle length n. Then S_{RS} is called the full-cycle equivalent class. Let S_{RS} be an FH sequence set, we know the sequence length is $n = q - 1$ and the maximum periodic Hamming correlation of S_{RS} is $k - 1$. From Lemma 19 in [8], the size of S_{RS} is

$$|S_{RS}| = \frac{1}{n} \sum_{d|n} \mu(d)q^{\lfloor \frac{k}{d} \rfloor},$$

where $\mu(d)$ is the Möbius function. Let $n = q - 1 = p_1^{e_1} \times p_2^{e_2} \times \cdots \times p_r^{e_r}$, where p_i's are distinct primes and e_j's are positive integers, $j = 1, 2, \cdots, r$. We know $p_1, p_2, \cdots, p_r$ are all the prime factors of n. If $k = \lambda p_1 p_2 \cdots p_r$, where λ is a positive integer, i.e., k includes all the prime factors of n, then we have

$$\gcd(k, n) = p_1^{u_1} \times p_2^{u_2} \times \cdots \times p_r^{u_r}$$

where $1 \leq u_j \leq e_j$, $j = 1, 2, \cdots, r$. It can be seen that $\gcd(k, n)$ includes all the prime factors of n and does not include other prime factors. Based on the definition of the Möbius function, we have

$$\frac{1}{n} \sum_{d|n} \mu(d) q^{\lfloor \frac{k}{d} \rfloor} = \frac{1}{n} \sum_{d| \gcd(k,n)} \mu(d) q^{\lfloor \frac{k}{d} \rfloor}.$$

Hence, when $k = \lambda p_1 p_2 \cdots p_r$, where λ is a positive integer, the size of S_{RS} is

$$|S_{RS}| = \frac{1}{n} \sum_{d| \gcd(k,n)} \mu(d) q^{\lfloor \frac{k}{d} \rfloor}.$$

According to the new bound (8), when $k = \lambda p_1 p_2 \cdots p_r$, S_{RS} is an optimal FH sequence set with respect to the family size.

Without loss of the fairness, we choose the Singleton bound (5), the bound (7) and the new bound (8) for comparison, as shown in Table 1.

Example 1. *Let $q = 17$, $n = q - 1 = 16$, $k = 4$. In this case, one can obtain a set S_{RS} of 5202 FH sequences, as shown below (only 20 sequences are shown for simplicity):*

$S_{RS} =$
$\{(10,15,14,4,6,9,5,16,7,2,3,13,11,8,12,1), (6,14,1,5,2,8,9,0,3,1,7,14,7,7,16,2),$
$(2,13,5,6,15,7,13,1,16,0,11,15,3,6,3,3), (15,12,9,7,11,6,0,2,12,16,15,16,16,5,7,4),$
$(11,11,13,8,7,5,4,3,8,15,2,0,12,4,11,5), (7,10,0,9,3,4,8,4,4,14,6,1,8,3,15,6),$
$(3,9,4,10,16,3,12,5,0,13,10,2,4,2,2,7), (16,8,8,11,12,2,16,6,13,12,14,3,0,1,6,8),$
$(12,7,12,12,8,1,3,7,9,11,1,4,13,0,10,9), (8,6,16,13,4,0,7,8,5,10,5,5,9,16,14,10),$
$(4,5,3,14,0,16,11,9,1,9,9,6,5,15,1,11), (0,4,7,15,13,15,15,10,14,8,13,7,1,14,5,12),$
$(13,3,11,16,9,14,2,11,10,7,0,8,14,13,9,13), (9,2,15,0,5,13,6,12,6,6,4,9,10,12,13,14),$
$(5,1,2,1,1,12,10,13,2,5,8,10,6,11,0,15), (1,0,6,2,14,11,14,14,15,4,12,11,2,10,4,16),$
$(14,16,10,3,10,10,1,15,11,3,16,12,15,9,8,0), (3,9,10,13,5,15,11,16,14,8,7,4,12,2,6,1),$
$(13,10,11,7,1,5,12,6,12,1,9,13,3,4,7,5), (9,9,15,8,14,4,16,7,8,0,13,14,16,3,11,6),$
$\cdots\}$

By the Singleton bound (5), we have

$$M' \leq 5220.0625$$

Table 1. Parameters of the FH sequence sets S_{RS}

Sequence length n	k	Family size	Hamming correlation	q	Singleton bound (5)	Bound (7)	New bound (8)
4	2	5	1	5	not optimal	optimal	optimal
8	2	9	1	9	not optimal	optimal	optimal
8	4	810	3	9	not optimal	not optimal	optimal
8	6	66339	5	9	not optimal	optimal	optimal
12	6	402038	5	13	not optimal	not optimal	optimal
16	2	17	1	17	not optimal	optimal	optimal
16	4	5202	3	17	not optimal	not optimal	optimal
16	6	1508291	5	17	not optimal	optimal	optimal
16	8	435979620	7	17	not optimal	not optimal	optimal
18	6	2613260	5	19	not optimal	not optimal	optimal
24	6	10171850	5	25	not optimal	not optimal	optimal
36	6	71268734	5	37	not optimal	not optimal	optimal
48	6	288357650	5	49	not optimal	not optimal	optimal
72	6	2101858778	5	73	not optimal	not optimal	optimal
96	6	8676782114	5	97	not optimal	not optimal	optimal

By the bound (7), we have

$$M' \leq 5219$$

By the new bound (8), we have

$$M' \leq 5202$$

Thus S_{RS} is an optimal FH sequence set with respect to the family size according to the new bound (8). But according to the Singleton bound (5) and the bound (7), S_{RS} is not an optimal FH sequence set with respect to the family size.

4 Conclusions

In this paper, a new bound on the FH sequences with respect to the size of the frequency slot set, the sequence length, the family size, and the maximum periodic Hamming correlation is established. The new bound is tighter than the Singleton bounds on the FH sequences derived by Ding et al. and Yang et al. and the bound derived by Liu and Peng. It is expected that the new bound will be useful in designing and evaluating new FH sequence designs.

References

1. Specification of the Bluetooth Systems-Core. The Bluetooth Special Interest Group (SIG). http://www.bluetooth.com
2. Fan, P.Z., Darnell, M.: Sequence Design for Communications Applications. Research Studies Press (RSP), John Wiley and Sons, London, UK (1996)
3. Golomb, S.W., Gong, G.: Signal Design for Good Correlation: For Wireless Communication, Cryptography and Radar. Cambridge University Press, Cambridge (2005)
4. Lempel, A., Greenberger, H.: Families of sequences with optimal Hamming correlation properties. IEEE Trans. Inf. Theor. **20**(1), 90–94 (1974)
5. Ding, C., Fuji-Hara, R., Fujiwara, Y., Jimbo, M., Mishima, M.: Sets of frequency hopping sequences: Bounds and optimal constructions. IEEE Trans. Inf. Theor. **55**(7), 3297–3304 (2009)
6. Ding, C., Yang, Y., Tang, X.H.: Optimal sets of frequency hopping sequences from linear cyclic codes. IEEE Trans. Inf. Theor. **55**(7), 3605–3612 (2010)
7. Chu, W., Colbourn, C.J.: Optimal frequency-hopping sequences via cyclotomy. IEEE Trans. Inf. Theor. **51**(3), 1139–1141 (2005)
8. Yang, Y., Tang, X.H., Udaya, P., Peng, D.Y.: New bound on frequency hopping sequence sets and its optimal constructions. IEEE Trans. Inf. Theor. **57**(11), 7605–7613 (2011)
9. Ge, G., Miao, Y., Yao, Z.: Optimal frequency hopping sequences: Auto- and cross-correlation properties. IEEE Trans. Inf. Theor. **55**(2), 867–879 (2009)
10. Udaya, P., Siddiqi, M.U.: Optimal large linear span frequency hopping patterns derived from polynomial residue class rings. IEEE Trans. Inf. Theor. **44**(4), 1492–1503 (1998)
11. Chung, J.H., Yang, K.: Optimal frequency-hopping sequences with new parameters. IEEE Trans. Inf. Theor. **56**(4), 1685–1693 (2010)
12. Peng, D.Y., Fan, P.Z., Lee, M.H.: Lower bounds on the periodic Hamming correlations of frequency hopping sequences with low hit zone. Sci. China Ser. F. Inf. Sci. **49**(2), 208–218 (2006)
13. Peng, D.Y., Fan, P.Z.: Lower bounds on the Hamming auto- and cross-correlations of frequency- hopping sequences. IEEE Trans. Inf. Theor. **50**(9), 2149–2154 (2004)
14. Lin, S., Xing, C.: Coding Theory: A First Course. Cambridge University Press, Cambridge (2004)
15. Liu, X., Peng, D.Y.: Theoretical bound on frequency hopping sequence set. Electron. Lett. **49**(10), 654–656 (2013)
16. Golomb, S.W.: A mathematical theory of discrete classification. In: Proceedings of 4th London Symposium Information Theory, pp. 404–425. Butterworths, London (1961)
17. Golomb, S.W., Gordon, B., Welch, L.R.: Comma-free codes. Can. J. Math. **10**, 202–209 (1958)

Author Index

Printed in the United States
By Bookmasters